Michael Růžička

Electrorheological Fluids: Modeling and Mathematical Theory

Springer

Author

Michael Růžička
Institute of Applied Mathematics
University of Freiburg
Eckerstr. 1
79104 Freiburg, Germany

E-mail: rose@mathematik.uni-freiburg.de

Cataloging-in-Publication Data applied for

Die Deutsche Bibliothek - CIP-Einheitsaufnahme

Růžička, Michael:
Electrorheological fluids : modeling and mathRematical theory / Michael Růžička. - Berlin ;
Heidelberg ; New York ; Barcelona ; Hong Kong ; London ; Milan ; Paris ; Singapore ; Tokyo
: Springer, 2000
  (Lecture notes in mathematics ; 1748)
  ISBN 3-540-41385-5

Mathematics Subject Classification (2000): 76Axx, 35Q35, 76D03, 35Dxx,
35K55, 35J60, 3502

ISSN 0075-8434
ISBN 3-540-41385-5 Springer-Verlag Berlin Heidelberg New York

Springer-Verlag Berlin Heidelberg New York
a member of BertelsmannSpringer Science+Business Media GmbH

© Springer-Verlag Berlin Heidelberg 2000
Printed in Germany

Typesetting: Camera-ready TEX output by the author
SPIN: 10724266   41/3142-543210 - Printed on acid-free paper

MEINEN ELTERN

Introduction

Electrorheological fluids are special viscous liquids, that are characterized by their ability to undergo significant changes in their mechanical properties when an electric field is applied. This property can be exploited in technological applications, e.g. actuators, clutches, shock absorbers, and rehabilitation equipment to name a few. Winslow [131] is credited for the first observation of the behaviour of electrorheological fluids in 1949. Since then great strides have been made to overcome the impediments of early electrorheological fluids, as the abrasive nature and the instability of the suspension and the enormous voltage requirements that are necessary for a significant change in the material properties. Nowadays existing electrorheological fluids, which make the above mentioned devices possible, are the result of intensive efforts to manufacture materials without these impediments.

In this book we present a theoretical investigation of electrorheological fluids. Firstly, we develop a model for these liquids within the framework of Rational Mechanics, which takes into account the complex interactions between the electro-magnetic fields and the moving liquid. Secondly, we carry out a mathematical analysis of the resulting system of partial differential equations which possesses so-called non-standard growth conditions. We discuss the functional setting, namely generalized Lebesgue and generalized Sobolev spaces, and show existence and uniqueness of weak and strong solutions respectively. In this introduction we want to give a general overview of the issues which are discussed in the following chapters in detail.

Electrorheological fluids can be modeled in many ways. One possibility consists in the investigation of the underlying microstructure to obtain a macroscopic description of the material (cf. Klingenberg, van Swol, Zukoski [54], Halsey, Toor [43], Bonnecaze, Brady [16], Parthasarathy, Klingenberg [97]). Another approach uses the framework of continuum mechanics and treats the electrorheological fluid as a homogenized single continuum (cf. Atkin, Shi, Bullogh [6], Rajagopal, Wineman [107], Wineman, Rajagopal [130]) or models it using the theory of mixture (cf. Rajagopal, Yalamanchili, Wineman [108]). A completely different perspective is provided by modeling based on direct numerical simulations taking into account the dynamics and interaction of particles (cf. Whittle [133], Bailey, Gillies, Heyes, Sutcliffe [7]). In all these models the electric field is treated as a constant parameter. However, there is strong evidence in the literature (cf. Katsikopoulos, Zukoski [52], Abu-Jdayil, Brunn [1], [2], [3], Wunderlich, Brunn [134]) that the behaviour of electrorheological fluids is strongly influenced by non-homogeneous electric fields. In many experiments one tries to modify the geometry in such a way that the so-called ER-effect (i.e. the reduction in volumetric flow rate at constant pressure drop) increases at the same field strength. In order to capture such effects we develop a model that takes into account the interaction of the electro-magnetic fields and the moving liquid and thus, the electric field is treated as a variable, that has to be determined.

The interaction of the electro-magnetic fields and the moving material is described on the basis of the "dipole current-loop" model (cf. Grot [41], Pao [96]). In the first part of Chapter 1 we carry out the modeling, which builds on ideas from Rajagopal, Růžička [104], [105]. More precisely, we start with the general balance laws for mass, linear momentum, angular momentum, energy, the second law of thermodynamics in the form of the Clausius-Duhem inequality and Maxwell's equations in their Minkowskian formulation. We then choose dependent and independent variables, reflecting the nature of the processes we are interested in, and impose the requirement of Galilean invariance of both the constitutive relations and the balance laws. In the next step we simplify the system by incorporating the physical properties of an electrorheological fluid, namely that it can be considered as a non-conducting dielectric and that the mechanical response does not change if a magnetic field is applied. In the last step we carry out a dimensional analysis and a subsequent approximation, restricting the validity of the resulting system to certain but typical situations. The main assumptions are that the magnetic field is of secondary importance and that the oscillations of the electro-magnetic fields are not larger than the reaction time of the fluid. The final result of the whole procedure is the system (1.2.32)–(1.2.36) and (1.2.43)–(1.2.46), which governs the motion of an electrorheological fluid. For example, if the fluid is assumed to be incompressible it turns out that the relevant equations are

$$\operatorname{div} \left(\mathbf{E} + \mathbf{P}\right) = 0 \,, \\ \operatorname{curl} \mathbf{E} = \mathbf{0} \,, \tag{0.1}$$

$$\rho_0 \frac{\partial \mathbf{v}}{\partial t} - \operatorname{div} \mathbf{S} + \rho_0 [\nabla \mathbf{v}] \mathbf{v} + \nabla \phi = \rho_0 \mathbf{f} + [\nabla \mathbf{E}] \mathbf{P} \,, \\ \operatorname{div} \mathbf{v} = 0 \,, \tag{0.2}$$

where $\mathbf{E}$ is the electric field, $\mathbf{P}$ the polarization, ρ_0 the density, $\mathbf{v}$ the velocity, $\mathbf{S}$ the extra stress, ϕ the pressure, and $\mathbf{f}$ the mechanical force. In the remaining part of Chapter 1 we discuss various special forms of the constitutive relation for the extra stress $\mathbf{S}$. In Section 1.3 we consider models in which the extra stress $\mathbf{S}$ depends linearly on the symmetric velocity gradient $\mathbf{D}$, when the fluid is either compressible or incompressible or mechanically incompressible but electrically compressible. In particular, the additional restrictions on the material response function imposed by the Clausius-Duhem inequality are examined. Experimental evidence for concrete electrorheological fluids however suggests that a linear dependence is in general too simple. In the last section we therefore propose a model capable of explaining many of the observed phenomena. In this model the extra stress has the form

$$\mathbf{S} = \alpha_{21}\big((1+|\mathbf{D}|^2)^{\frac{p-1}{2}} - 1\big)\mathbf{E} \otimes \mathbf{E} + (\alpha_{31} + \alpha_{33}|\mathbf{E}|^2)(1+|\mathbf{D}|^2)^{\frac{p-2}{2}}\mathbf{D}$$
$$+ \alpha_{51}(1+|\mathbf{D}|^2)^{\frac{p-2}{2}}(\mathbf{DE} \otimes \mathbf{E} + \mathbf{E} \otimes \mathbf{DE}), \qquad (0.3)$$

where α_{ij} are material constants and where the material function p depends on the strength of the electric field $|\mathbf{E}|^2$ and satisfies

$$1 < p_\infty \le p(|\mathbf{E}|^2) \le p_0 < \infty. \qquad (0.4)$$

To illustrate the features of (0.3) we solve the boundary value problem for a flow driven by a pressure drop between infinite parallel planes. Finally, we discuss the consequences of the Clausius-Duhem inequality for the model (0.3), which from the mathematical point of view is related to the coercivity of the elliptic operator, and conditions which ensure the uniform monotonicity of the operator induced by $-\operatorname{div}\mathbf{S}$ (cf. Lemma 1.4.46, Lemma 1.4.64).

Let us now have a closer look at the system (0.1)–(0.4) under the assumption that the polarization depends linearly on the electric field, i.e. $\mathbf{P} = \chi^E\mathbf{E}$. Fortunately, the system is separated into the quasi-static Maxwell's equations (0.1) and the equation of motion and the conservation of mass (0.2), where $\mathbf{E}$ can be viewed as a parameter. Maxwell's equations are widely studied in the literature and well understood. For details, we refer the reader to the overview article Milani, Picard [83] within the context of Hilbert space methods and to Schwarz [115], where questions concerning the regularity of solutions to (0.1) are discussed. In Section 2.3 we have collected and indicated the proofs of the results for the system (0.1) that we need in the sequel. The situation for the system (0.2)–(0.4) is quite different and hence the remaining part of this book is devoted to the study of it. Since the material function p, which essentially determines $\mathbf{S}$, depends on the magnitude of the electric field $|\mathbf{E}|^2$, we have to deal with an elliptic or parabolic system of partial differential equations with so-called non-standard growth conditions, i.e. the elliptic operator $\mathbf{S}$ satisfies

$$\mathbf{S}(\mathbf{D},\mathbf{E}) \cdot \mathbf{D} \ge c_0(1+|\mathbf{E}|^2)(1+|\mathbf{D}|^2)^{\frac{p_\infty-2}{2}}|\mathbf{D}|^2,$$
$$|\mathbf{S}(\mathbf{D},\mathbf{E})| \le c_1(1+|\mathbf{D}|^2)^{\frac{p_0-1}{2}}|\mathbf{E}|^2. \qquad (0.5)$$

The problem of regularity of solutions to equations, variational integrals and systems with non-standard growth conditions has been extensively studied in the last ten years, starting with counterexamples by Giaquinta [37] and Marcellini [76]. At the beginning of Chapter 2 we give a short overview of some of the results which have been achieved in the last years. Since the solution $\mathbf{E}$ of Maxwell's equations is in general not constant it is clear from the form of the extra stress $\mathbf{S}$ that the canonical functional setting are the spaces $L^{p(x)}(\Omega)$ and $W^{1,p(x)}(\Omega)$, so-called generalized Lebesgue and generalized

Sobolev spaces, respectively. These spaces have been studied e.g. by Hudzik [44] and Kováčik, Rákosník [55]. Since they are not so well known and because we also need weighted versions of them, we will discuss their properties in some detail in Section 2.2. We would like to point out that generalized Lebesgue and generalized Sobolev spaces have a lot in common with the classical Lebesgue and classical Sobolev spaces, however there are also many open fundamental questions, which are well understood in the theory of classical Lebesgue and classical Sobolev spaces. For example it is not known, even for very "nice" functions $p(x)$, whether smooth functions are dense in the space $W^{1,p(x)}(\Omega)$ or whether the maximal operator is continuous from $L^{p(x)}(\Omega)$ into $L^{p(x)}(\Omega)$.

In Chapter 3 we study steady flows of incompressible shear dependent electrorheological fluids. Their motion is governed by the system

$$
\begin{aligned}
\operatorname{div} \mathbf{E} &= 0 \\
\operatorname{curl} \mathbf{E} &= 0 \qquad \text{in } \Omega\,, \\
\mathbf{E} \cdot \mathbf{n} &= \mathbf{E}_0 \cdot \mathbf{n} \qquad \text{on } \partial\Omega\,,
\end{aligned}
\tag{0.6}
$$

$$
\begin{aligned}
-\operatorname{div} \mathbf{S}(\mathbf{D}, \mathbf{E}) + [\nabla \mathbf{v}]\mathbf{v} + \nabla\phi &= \mathbf{f} + \chi^E[\nabla \mathbf{E}]\mathbf{E}\,, \\
\operatorname{div} \mathbf{v} &= 0 \qquad \text{in } \Omega\,, \\
\mathbf{v} &= 0 \qquad \text{on } \partial\Omega\,,
\end{aligned}
\tag{0.7}
$$

where Ω is a bounded domain and the data $\mathbf{f}$ and $\mathbf{E}_0$ are given. The extra stress $\mathbf{S}$ is given by (0.3), (0.4) and the coefficients α_{ij} are such that the operator is coercive and uniformly monotone. In Section 3.2 we show the existence of weak solutions to the system (0.6), (0.7) whenever the lower bound p_∞ of the material function p is larger than 9/5 (cf. Theorem 3.2.4). The solution is unique if the data are small enough (cf. Proposition 3.2.38). The existence result is accomplished by adapting the theory of monotone operators and a compactness argument to our situation. This is possible since the energy estimate ensures that, within the context of generalized weighted Sobolev spaces, the operator induced by $-\operatorname{div} \mathbf{S}(\mathbf{D}, \mathbf{E})$ maps the natural energy space[1] $E_{p(x),\nu}$ into its dual.

In the remaining part of Chapter 3 we present a different approach to the steady problem, which ensures the existence of strong solutions to the system (0.6), (0.7), (0.3), (0.4), but does not use the theory of monotone operators. The importance of this method will become clear when unsteady problems are treated. The main problem of course consists in the identification of the limiting element for the sequence $\mathbf{S}(\mathbf{D}(\mathbf{v}^n), \mathbf{E})$, where $\mathbf{v}^n$ is some approximate solution to (0.7). This problem can be solved by using Vitali's theorem if we derive apriori estimates that ensure the almost everywhere convergence of $\mathbf{D}(\mathbf{v}^n)$. To this end we can build on ideas initiated by Nečas [93] and developed by Málek, Nečas, Růžička [71], [73], Bellout, Bloom, Nečas [9], Málek, Nečas, Rokyta, Růžička [70] in their study of unsteady flows of generalized

---

[1] see (2.2.47) for the definition of $E_{p(x),\nu}$

Newtonian fluids[2]. The desired estimate will be obtained using essentially $-\Delta \mathbf{v}^n \xi_0^2$ as a test function in the weak formulation. Since this test function is not divergence free also terms containing the pressure ϕ have to be estimated. The regularity properties of ϕ have to be determined from (0.7), viewed as an operator equation in some negative Sobolev space. It turns out that due to the growth conditions of the extra stress $\mathbf{S}$ it is not possible to handle the pressure terms. Therefore, we approximate $\mathbf{S}(\mathbf{D}, \mathbf{E})$ by some sequence $\mathbf{S}^A(\mathbf{D}, \mathbf{E})$, for $A \to \infty$, which is constructed explicitly. The approximation $\mathbf{S}^A(\mathbf{D}, \mathbf{E})$ is chosen such that on the one hand the growth properties allow us to estimate (in dependence on A) the pressure terms and that on the other hand we do not lose too much in the coercivity and monotonicity properties of $\mathbf{S}^A(\mathbf{D}, \mathbf{E})$ compared with the coercivity and monotonicity properties of $\mathbf{S}(\mathbf{D}, \mathbf{E})$. What this vague statement means precisely is formulated and proven in Section 3.3.1. The final approximation of the system (0.7), which we will denote by $(0.7)_{\varepsilon,A}$, is obtained by a mollification of the convective term. It is easy to see that the system $(0.7)_{\varepsilon,A}$ possesses weak solutions (cf. Proposition 3.4.2). Using the difference quotient method one can show that locally second order derivatives belong to an appropriate Lebesgue space. This property in turn enables us to use $-\Delta \mathbf{v}^{\varepsilon,A} \xi^{2\alpha}$, $\alpha > 1$, as a test function and to derive local estimates for the second order derivatives $\nabla^2 \mathbf{v}^{\varepsilon,A}$, the gradient of the extra stress $\nabla \mathbf{S}^A(\mathbf{D}(\mathbf{v}^{\varepsilon,A}), \mathbf{E})$ and the pressure $\phi^{\varepsilon,A}$, which are independent of A (cf. Proposition 3.5.7). We would like to point out that these local estimates are possible because we have at our disposal some global information from the energy estimate, which is sufficient to compensate the non-local nature of the pressure.

These estimates enable the limiting process $A \to \infty$ and we arrive at an approximation of the system (0.7), where only the convective term is mollified. In a final step, it remains to estimate the convective term, which stands on the right-hand side of our estimates (cf. Proposition 3.5.42), by the information we have on the left-hand sides[3]. It turns out that this procedure works under certain restrictions on p_0 in terms of p_∞ that can be found in the formulation of Theorem 3.3.7, our main result concerning strong solutions of the steady system (0.6), (0.7), (0.3), (0.4). Let us point out that the lower bound for p_∞, which is governed by the convective term, is the same as in Theorem 3.2.4, where monotonicity methods have been used. However, in contrast to Theorem 3.2.4 upper bounds for p_0 appear, which are in the case $p_\infty \geq 2$ due to the chosen approximation, and in the case $p_\infty < 2$ due to the non-standard growth conditions of the elliptic operator $\mathbf{S}$. Finally, we would like to mention that Theorem 3.2.4 and Theorem 3.3.7 are to my knowledge the only existence result for elliptic systems with non-standard growth conditions with a nonlinear right-hand side $\mathbf{b}(\mathbf{u}, \nabla \mathbf{u})$. In the case of elliptic equations with non-standard growth conditions there

---

[2]Generalized Newtonian fluids can by considered as a special case of electrorheological fluids if we set $\mathbf{E} = 0$ and thus $p = $ const. The aim in these papers was to establish the existence of weak solutions to the equations of motion when the theory of monotone operators fails because p_∞ is too small (cf. Ladyzhenskaya [58], Lions [68]).

[3]In this and the previous step various technical assertions are needed, which are proved in the Appendix.

are existence results by Marcellini [78] and Boccardo, Gallouet, Marcellini [12] under the assumption that the right-hand side $b(x)$ belongs to the space L^∞_{loc} and L^1, respectively. Of course, in both the scalar case and the vectorial case, the existence of minimizers of variational integrals is clear (using lower semicontinuity arguments) under certain conditions on the integrand.

In the last chapter we turn our attention to the study of unsteady flows of shear dependent electrorheological fluids. We treat the system (0.1), (0.2) together with appropriate boundary and initial conditions for two different models for the extra stress tensor $\mathbf{S}$. The main part of the chapter is devoted to the case when we set in (0.3) $\alpha_{21} = \alpha_{51} = 0$ and $\alpha_{31} = \alpha_{33}$, i.e. $\mathbf{S}$ has the form

$$\mathbf{S}(\mathbf{D}(\mathbf{v}), \mathbf{E}) = \alpha_{31}(1 + |\mathbf{E}|^2)(1 + |\mathbf{D}(\mathbf{v})|^2)^{\frac{p-2}{2}} \mathbf{D}(\mathbf{v}), \tag{0.8}$$

where $\alpha_{31} > 0$ and $p = p(|\mathbf{E}|^2)$ satisfies (0.4) with $p_\infty \geq 2$. In the second model that we consider we assume that the extra stress tensor $\mathbf{S}$ depends linearly on $\mathbf{D}$, i.e. in (0.3) we set $\alpha_{21} = 0$ and $p \equiv 2$. In both cases we again assume that the polarization is given by $\mathbf{P} = \chi^E \mathbf{E}$. The second model behaves very similar to the unsteady Navier-Stokes system and we prove global in time existence of weak solutions for large data (cf. Theorem 4.1.20). The proof of this result is not carried out since only minor changes in the standard proofs of the corresponding result in the Navier-Stokes theory are necessary. On the other hand, the system (0.1), (0.2) with $\mathbf{S}$ given by (0.8) behaves quite differently. First of all and in contrast to the steady system (0.6), (0.7) we cannot use monotonicity methods to prove the existence of solutions to the system (0.1), (0.2), (0.8), (0.4) due to the non-standard growth of the operator. More precisely, the canonical functional setting would be the space $L^{p(t,x)}(Q_T)$, since the energy method provides the estimate

$$\int\limits_0^T \int\limits_\Omega (1 + |\mathbf{D}(\mathbf{v})|^2)^{\frac{p(t,x)-2}{2}} |\mathbf{D}(\mathbf{v})|^2 \, dx \, dt \leq c, \tag{0.9}$$

which implies that $\mathbf{D}(\mathbf{v}) \in L^{p(t,x)}(Q_T)$. For the treatment of parabolic problems it is essential to treat the time and the space variables differently and to employ the equivalence of the spaces

$$L^q(Q_T) \qquad \text{and} \qquad L^q(I, L^q(\Omega)). \tag{0.10}$$

In the case $p = \text{const.}$ the estimate (0.9) therefore yields $\mathbf{v} \in L^q(I, W^{1,q}(\Omega))$ and the technique of monotone operators together with compactness arguments works. However, in our case it is not clear how an equivalence similar to (0.10) can be achieved for the space $L^{p(t,x)}(Q_T)$, which would allow us to adapt the arguments of the classical setting. Fortunately, the second approach presented in Chapter 3, using Vitali's theorem and almost everywhere convergence can also be used in the unsteady case. In fact, we employ ideas from Málek, Nečas, Růžička [73], which have been developed in the context of generalized Newtonian fluids, to treat situations where monotonicity methods fail. The procedure in Chapter 4 is very similar to the one used in Chapter 3 and we concentrate on the new aspects. One essential difference is, that now not only the regularity properties of the pressure ϕ but also the regularity properties of the time derivative of the velocity $\frac{\partial v}{\partial t}$ have to be computed from the system (0.2). This in fact prevents us from working with local estimates only, because we would run into a circle argument. Therefore we need to derive global estimates.

The main result of this chapter is the existence of global in time weak and strong solutions to the system (0.1), (0.2), (0.8) for large data under certain restrictions on p_∞ and p_0, which are formulated in Theorem 4.1.14. We also prove that strong solutions are unique. We would like to point out that weak solutions exist for $p_\infty \geq 2$, which clearly shows the advantage of the chosen approach, since, even if the monotonicity method would work, the best lower bound for p_∞ which could be achieved is $11/5$. For the proof of Theorem 4.1.14 we again approximate the elliptic operator $\mathbf{S}$ by an appropriate sequence $\mathbf{S}^A$ and we also mollify the convective term. We denote the resulting system by $(0.2)_{\varepsilon,A}$. In Section 4.1 we collect the relevant properties of the approximative elliptic operator $\mathbf{S}^A$ and define weak and strong solutions. In the next section we show the existence of strong solutions to the system (0.1), $(0.2)_{\varepsilon,A}$, (0.8), (0.4), which belong to the space $L^2(I, W^{2,2}(\Omega))$ (cf. Proposition 4.2.1). For that we apply the difference quotient method in the interior and in the tangential directions near the boundary. To estimate the normal derivatives we use the pressure eliminating operator curl to the system $(0.2)_{\varepsilon,A}$ and the internal constraint div $\mathbf{v} = 0$. In the case when p_∞ is near 2 we need additional estimates, which are proved in Lemma 4.3.8. The next step is to derive estimates that are independent of A. In fact, this step is crucial for the whole procedure, and here various difficulties occur due to the non-standard growth of the system and the fact that the operators $\mathbf{S}^A$ are only partially potential like. As a consequence of these problems an upper bound for p_0 appears (cf. Lemma 4.3.8, Lemma 4.3.41). However, the estimates for the velocity $\mathbf{v}^{\varepsilon,A}$, the pressure $\phi^{\varepsilon,A}$ and the elliptic operator $\mathbf{S}^A(\mathbf{D}(\mathbf{v}^{\varepsilon,A}), \mathbf{E})$ are sufficient for the limiting process $A \to \infty$ (cf. Proposition 4.3.57). At the end of Section 4.3 we derive "weighted" estimates (cf. Lemma 4.3.82, Lemma 4.3.88), which are crucial for the treatment of p_∞ near 2. In the last section it remains to find conditions on p_∞ and p_0, that allow to handle the convective term, which appears on the right-hand side of the above mentioned estimates. To this end we combine all possible information we have in a rather delicate way. Basically, we derive the following inequality

$$\frac{1}{1-\mu}\left(1 + \|\mathbf{D}(\mathbf{v}^\varepsilon(t))\|_{p(|\mathbf{E}(t)|^2)}\right)^{1-\mu} \tag{0.11}$$

$$+ \int_0^t \|\nabla\mathbf{S}(\mathbf{D}(\mathbf{v}^\varepsilon(\tau)), \mathbf{E}(\tau))\|_r^u \left(1 + \|\mathbf{D}(\mathbf{v}^\varepsilon(\tau))\|_{p(|\mathbf{E}(\tau)|^2)}\right)^{-\mu} d\tau \leq c(\mathbf{f}, \mathbf{v}_0, \mathbf{E}_0),$$

where $u = \frac{2}{p_0-1}$, $r = \frac{6}{p_0+1}$ and $0 < \mu$. In this inequality of course μ depends on p_∞ and p_0. The size of μ decides whether we obtain strong or weak solutions. In fact, for $\mu \leq 1$

we get strong solutions, while $\mu > 1$ yields weak solutions. In all cases the estimate (0.11) ensures the almost everywhere convergence of ∇v^ε. With this information at hand the limiting process $\varepsilon \to 0$ is no problem and the proof of Theorem 4.1.14 can be finished. I would like to mention the difference to the corresponding Theorem 4.1.14 in Růžička [113], where the system (0.1), (0.2) with an elliptic operator $\mathbf{S}$ of the form (0.8), however with $\mathbf{D}(v)$ replaced by ∇v, is treated. Due to this dependence this system does not describe the motion of an electrorheological fluid, but is only an unsteady system with non-standard growth conditions motivated by the study of electrorheological fluids. Theorem 4.1.14 in [113] proves the existence of solutions for a larger range for p compared with Theorem 4.1.14 here. The reason is a completely different behaviour of the normal derivatives of the velocity gradient when $\mathbf{S}$ depends on $\mathbf{D}(v)$ or on ∇v. Let me finally mention that Theorem 4.1.14 together with my results in [113], [114] are the first results for parabolic systems with non-standard growth conditions.

At this point I would like to mention, that the present book is based on my habilitation thesis (University of Bonn, 1998). The first three chapters, up to minor corrections of misprints and small changes for a better understandability of the text, are the same as in this thesis [113]. However Chapter 4 is completely new.

Finally it is my pleasure to thank J. Nečas and J. Frehse who showed me the beautiful world of mathematics and accompanied me on my way not only with professional guidance but also, and probably more importantly, with their personal advice. I also want to express my sincere thanks to K.R. Rajagopal who strongly influenced my views of mechanical engineering, which is much more than a huge source of fascinating questions for a mathematician. I am thankful to G.P. Galdi for initiating my stays at the Universities of Pittsburgh and Ferrara and for his hospitality. During these visits, which were supported by fellowships from DFG and CNR, I started my investigations on electrorheological fluids. Of course I also wish to thank my friends and colleagues J. Málek, A. Novotný, M. Padula, A. Passerini, L. Pick, A. Romano, A. Sequeira, A. Srinivasa, M. Steinhauer and G. Thäter, who influenced me and my work in different situations and ways.

Last but not least I want to express my gratitude to I. Schmelzer for her excellent typing of my sometimes chaotic manuscript. I am very grateful to W. Eckart, H. Kastrup, L. Pick, M. Specovius-Neugebauer and G. Thäter for reading large parts of a previous version of the manuscript. The time they invested has certainly improved this final version.

Contents

1 Modeling of Electrorheological Fluids

1.1 General Balance Laws

Electrorheological fluids are special viscous fluids, which are characterized by their ability to undergo significant changes in their mechanical properties due to the application of an electric field. When they are modeled within the frame-work of continuum mechanics, the result is a special case of the theory which describes the interaction of electro-magnetic fields with moving deformable bodies. Beside classical textbooks on electrodynamics as e.g. Landau, Lifschitz [66], Jackson [49], Sommerfeld [117], Feynman, Leighton, Sands [28], which also touch this subject, there are many monographs and research articles as e.g. Penfield, Haus [98], Panofsky, Phillips [95], de Groot, Suttorp [40], Truesdell, Toupin [123], Hutter, van de Ven [48], Eringen, Maugin [27], Grot [41], Pao [96], Maugin, Eringen [82], Toupin [122] and Dixon, Eringen [21], [22], which discuss this subject in detail. Nevertheless the theory is not at all unified and we refer the reader to Pao [96] for a beautiful review and comparison of different approaches.

Here we present an approach to the modeling of electrorheological fluids which uses ideas, which have been developed in Rajagopal, Růžička [104], [105]. More precisely, we start with the general balance laws for mass, linear momentum, angular momentum, energy, the second law of thermodynamics in the form of the Clausius-Duhem inequality and Maxwell's equations. The terms representing the interaction of the electro-magnetic fields and the moving deformable body are described on the basis of the "dipole current-loop" model (cf. Pao [96], Grot [41])[1].

Let $\Omega_0 \subset \mathbb{R}^3$ denote the reference configuration of an abstract body $\mathcal{B}$. By the motion of the body $\mathcal{B}$ we mean a one-to-one mapping χ that assigns to each particle $\mathbf{X} \in \Omega_0$ a position $\mathbf{x}$ in the three dimensional Euclidean space, at the instant of time t, i.e.

$$\mathbf{x} = \chi(t, \mathbf{X}). \tag{1.1}$$

The velocity $\mathbf{v}(t, \mathbf{x})$ and the acceleration $\mathbf{a}(t, \mathbf{x})$ are defined through

$$\mathbf{v} = \frac{\partial \chi}{\partial t}, \quad \text{and} \quad \mathbf{a} = \frac{\partial^2 \chi}{\partial t^2}. \tag{1.2}$$

---

[1] It is shown in Pao [96] that the final equations based on the "dipole current-loop" model and on the statistical approach, which is very popular (cf. Eringen, Maugin [27]), are identical.

The velocity gradient $\mathbf{L}(t, \mathbf{x})$ is a tensor given by

$$\mathbf{L} = \nabla \mathbf{v} = \left(\frac{\partial v_i}{\partial x_j} \right)_{i,j=1,2,3},$$

where ∇ is the partial derivative with respect to $\mathbf{x}$. We denote the total time derivative by $\frac{d}{dt}$. The symmetric part of the velocity gradient $\mathbf{L}$ is denoted by $\mathbf{D}$, i.e. $\mathbf{D} = \frac{1}{2}(\mathbf{L} + \mathbf{L}^T)$, where $\mathbf{L}^T$ is the transpose of the tensor $\mathbf{L}$; and the skew part of $\mathbf{L}$ is denoted by $\mathbf{W}$. In this section we assume that all field variables are sufficiently smooth in order to make all operations that are carried out meaningful.

We shall start by recording the local forms of the thermo-mechanical balance laws. The conservation of mass and the balance of linear momentum, respectively, are given by

$$\frac{\partial \rho}{\partial t} + \operatorname{div}(\rho \mathbf{v}) = 0, \tag{1.3}$$

$$\rho \frac{d\mathbf{v}}{dt} - \operatorname{div} \mathbf{T} = \rho \mathbf{f} + \mathbf{f}_e, \tag{1.4}$$

where ρ is the density, $\mathbf{T}$ is the Cauchy stress tensor, $\mathbf{f}$ the external mechanical body force, and $\mathbf{f}_e$ the electro-magnetic body force given by

$$\mathbf{f}_e = q_e \mathbf{E} + \frac{1}{c} \mathbf{J} \times \mathbf{B} + \frac{1}{c} \left(\frac{d\mathbf{P}}{dt} + (\operatorname{div} \mathbf{v})\mathbf{P} \right) \times \mathbf{B} + \frac{1}{c} \mathbf{v} \times ([\nabla \mathbf{B}]\mathbf{P})$$
$$+ [\nabla \mathbf{B}]^T \mathcal{M} + [\nabla \mathbf{E}]\mathbf{P}. \tag{1.5}$$

The form of the electro-magnetic body force (1.5) as well as all other terms representing the interaction of the electro-magnetic fields with the moving deformable body is based on the "dipole current-loop" model (cf. Pao [96], Grot [41]).

Here and in the sequel we shall use for vectors $\mathbf{u}, \mathbf{w}$ and tensors $\mathbf{S}, \mathbf{T}$ the notation

$$\operatorname{div} \mathbf{u} = \frac{\partial u_i}{\partial x_i}, \qquad \mathbf{S} \cdot \mathbf{T} = S_{ij} T_{ij},$$

$$[\nabla \mathbf{u}]\mathbf{w} = \left(w_j \frac{\partial u_i}{\partial x_j} \right)_{i=1,2,3}, \qquad \operatorname{div} \mathbf{S} = \left(\frac{\partial S_{ij}}{\partial x_j} \right)_{i=1,2,3},$$

in which the summation convention over repeated indices is employed. We will apply this convention throughout the book. Moreover the operator $\times$ denotes the usual vector product of two vectors.

In (1.5) we have used the notation that q_e is the electric charge density, $\mathbf{E}$ the electric field, $\mathbf{J}$ the total current, $\mathbf{P}$ the electric polarization, $\mathbf{B}$ the magnetic induction, and $\mathcal{M}$ is defined by[2]

$$\mathcal{M} = \mathbf{M} + \frac{1}{c} \mathbf{v} \times \mathbf{P}, \tag{1.6}$$

---

[2]Note, that $\mathcal{M}$ can be interpreted as the magnetization in the co-moving frame. An analogue interpretation can be given for $\mathcal{E}$ and $\mathcal{J}$, which will be defined later.

where $\mathbf{M}$ is the magnetic polarization and $c \approx 3 \cdot 10^{10}$ cm sec^{-1} denotes the speed of light. The balance of angular momentum takes the form

$$\mathbf{x} \times \rho \frac{d\mathbf{v}}{dt} - \operatorname{div}(\mathbf{x} \times \mathbf{T}) = \mathbf{x} \times \rho \mathbf{f} + \mathbf{l}_e \,, \tag{1.7}$$

in which $\mathbf{l}_e$ denotes the electro-magnetic torque density given by

$$\mathbf{l}_e = \mathbf{x} \times \mathbf{f}_e + \mathbf{P} \times \boldsymbol{\mathcal{E}} + \boldsymbol{\mathcal{M}} \times \mathbf{B}\,, \tag{1.8}$$

where $\boldsymbol{\mathcal{E}}$ is the electromotive intensity defined through

$$\boldsymbol{\mathcal{E}} = \mathbf{E} + \frac{1}{c}\mathbf{v} \times \mathbf{B}\,. \tag{1.9}$$

The balance of energy can be written as

$$\rho \frac{d}{dt}\left(e + \frac{1}{2}|\mathbf{v}|^2\right) + \operatorname{div}\mathbf{q} = \operatorname{div}(\mathbf{Tv}) + \rho\mathbf{f}\cdot\mathbf{v} + \rho r + w_e\,. \tag{1.10}$$

Here we denoted by e the specific internal energy, by $\mathbf{q}$ the heat flux vector, by r the heat source density, and the energy production density w_e is given by

$$w_e = \mathbf{f}_e \cdot \mathbf{v} + \rho \boldsymbol{\mathcal{E}} \cdot \frac{d}{dt}\left(\frac{\mathbf{P}}{\rho}\right) - \boldsymbol{\mathcal{M}} \cdot \frac{d\mathbf{B}}{dt} + \boldsymbol{\mathcal{J}} \cdot \boldsymbol{\mathcal{E}}\,, \tag{1.11}$$

where the conduction current $\boldsymbol{\mathcal{J}}$ is defined through

$$\boldsymbol{\mathcal{J}} = \mathbf{J} - q_e\mathbf{v}\,. \tag{1.12}$$

We interpret the second law of thermodynamics in the form of the Clausius-Duhem inequality[3]:

$$\rho \frac{d\eta}{dt} + \operatorname{div}\left(\frac{\mathbf{q}}{\theta}\right) - \rho\frac{r}{\theta} \geq 0\,, \tag{1.13}$$

where θ is the absolute temperature and η the specific entropy.

Using (1.3) and (1.4) we easily get the reduced forms of (1.7) and (1.10), which read

$$\epsilon\left(\mathbf{T} + \mathbf{P} \otimes \boldsymbol{\mathcal{E}} + \boldsymbol{\mathcal{M}} \otimes \mathbf{B}\right) = 0\,, \tag{1.14}$$

in which ϵ is the complete skew symmetric Levi-Civita tensor and $\otimes$ denotes the usual tensor product of two vectors, and

$$\rho \dot{e} + \operatorname{div}\mathbf{q} = \mathbf{T}\cdot\mathbf{L} + \boldsymbol{\mathcal{E}}\cdot\dot{\mathbf{P}} - \boldsymbol{\mathcal{M}}\cdot\dot{\mathbf{B}}$$
$$+ \boldsymbol{\mathcal{J}}\cdot\boldsymbol{\mathcal{E}} + (\boldsymbol{\mathcal{E}}\cdot\mathbf{P})\operatorname{div}\mathbf{v} + \rho r\,. \tag{1.15}$$

---

[3]We also refer the reader to Liu, Müller [69] and Hutter, van de Ven [48], Hutter [46], [47] for different formulations of the second law of thermodynamics within the context of the motion of a deformable body under the influence of electro-magnetic fields.

Here and in the following a superposed dot denotes the total time derivative. Introducing the specific Helmholtz potential ψ through

$$\psi = e - \eta\,\theta - \frac{1}{\rho}\,\mathcal{E}\cdot\mathbf{P}\,, \tag{1.16}$$

and substituting (1.16) into (1.13) we obtain the dissipation inequality

$$-\rho\big(\dot{\psi} + \eta\,\dot{\theta}\big) + \mathbf{T}\cdot\mathbf{L} - \frac{\mathbf{q}\cdot\nabla\theta}{\theta} \\ -\,\dot{\mathcal{E}}\cdot\mathbf{P} - \mathcal{M}\cdot\dot{\mathbf{B}} + \mathcal{J}\cdot\mathcal{E} \geq 0\,. \tag{1.17}$$

Finally, we list Maxwell's equations[4] for the moving liquid[5]. Gauss' law reads

$$\operatorname{div}\mathbf{D}_e = q_e\,, \tag{1.18}$$

where $\mathbf{D}_e$ is the electric displacement given by

$$\mathbf{D}_e = \mathbf{P} + \mathbf{E}\,. \tag{1.19}$$

Faraday's law is given by

$$\operatorname{curl}\mathbf{E} + \frac{1}{c}\frac{\partial\mathbf{B}}{\partial t} = \mathbf{0}\,, \tag{1.20}$$

and the conservation of magnetic flux takes the form

$$\operatorname{div}\mathbf{B} = 0\,. \tag{1.21}$$

Ampere's law reads

$$\operatorname{curl}\mathbf{H} = \frac{1}{c}\frac{\partial\mathbf{D}_e}{\partial t} + \frac{1}{c}\left(\mathcal{J} + q_e\mathbf{v}\right), \tag{1.22}$$

where the magnetic field $\mathbf{H}$ is given by

$$\mathbf{H} = \mathbf{B} - \mathbf{M}\,. \tag{1.23}$$

The conservation of electric charge, which is a consequence of (1.18) and (1.22), reads

$$\frac{\partial q_e}{\partial t} + \operatorname{div}\left(\mathcal{J} + q_e\mathbf{v}\right) = 0\,. \tag{1.24}$$

---

[4]Throughout this chapter we use Heavyside Lorentz units (cf. Maugin, Eringen [82], Jackson [49]).

[5]The approach to the theory of electrodynamics for moving media, which we follow here, goes back to Minkowski [85]. In his formulation of the field equations, which is identical to equations (1.18), (1.20)–(1.22), only the variables $\mathbf{E}$, $\mathbf{B}$, $\mathbf{D}_e$, $\mathbf{H}$, $\mathcal{J}$ and q_e are used. The system is completed by constitutive relations (cf. (1.39)) and transformation formulae relating the fields in the co-moving and the laboratory frame (cf. (1.33)). Note, that there exist many different formulations for the electrodynamics of moving media, see e.g. Pao [96] or Hutter, van de Ven [48] for a comparison.

The system (1.3), (1.4), (1.14), (1.15), (1.17) and (1.18), (1.20)–(1.22), (1.24), which describes the motion of the body, has far more unknowns than equations. It is rendered determinate by providing appropriate constitutive relations, reflecting the material properties. We will assume that

$$\rho, \mathbf{v}, \mathbf{D}, \mathbf{E}, \mathbf{B}, \theta, \nabla\theta \tag{1.25}$$

are the independent variables and thus we provide constitutive relations for

$$e, \psi, \mathbf{T}, \mathbf{q}, \mathcal{M}, \mathbf{P}, \mathcal{J} \tag{1.26}$$

of the form

$$f = \hat{f}(\rho, \mathbf{v}, \mathbf{D}, \mathbf{E}, \mathbf{B}, \theta, \nabla\theta), \tag{1.27}$$

where f stands for any of the quantities in (1.26). The choice of dependent and independent mechanical variables is quite standard and unified, the situation is different for the electro-magnetic variables. We have made the above choice because it seems to be adequate for our purposes (cf. Grot [41], Eringen, Maugin [27]) and it reflects the connection of the variables to the material in the most direct way (cf. Thiersten [121]).

Both the material and the balance equations are subject to invariance requirements. It is well known that the thermo-mechanical balance laws (1.3), (1.4), (1.14), (1.15) and (1.17), without the terms due to the interaction of the electro-magnetic fields and the material, are form invariant under Galilean transformations of the frame given by

$$\begin{aligned} \mathbf{x}^* &= \mathbf{Q}\mathbf{x} - \mathbf{v}_0 t + \mathbf{b}_0, \\ t^* &= t, \end{aligned} \tag{1.28}$$

where $\mathbf{v}_0, \mathbf{b}_0$ are given vectors and $\mathbf{Q}$ is a time independent orthogonal tensor, i.e. $\mathbf{Q}\mathbf{Q}^\mathsf{T} = \mathbf{I}$, if one requires that the fields representing material properties are covariant, i.e. they transform in the following way

$$\begin{aligned} \mathbf{T}^* &= \mathbf{Q}\mathbf{T}\mathbf{Q}^\mathsf{T}, \quad \mathbf{q}^* = \mathbf{Q}\mathbf{q}, \\ \psi^* &= \psi, \quad e^* = e, \end{aligned} \tag{1.29}$$

and that the other quantities transform in the usual way

$$\theta^* = \theta, \quad \rho^* = \rho, \quad r^* = r, \quad \mathbf{f}^* = \mathbf{Q}\mathbf{f}. \tag{1.30}$$

In all transformation formulae the quantities on the left-hand side and on the right-hand side have to be evaluated in the same material point, e.g. $\rho^*(\mathbf{x}^*) = \rho(\mathbf{x})$, where $\mathbf{x}^*$ and $\mathbf{x}$ are related by $(1.28)_1$. The transformation rules (1.29) are a special case of the principle of material frame indifference, which requires that (1.29) has to hold under changes of frame given by

$$\begin{aligned} \mathbf{x}^* &= \mathbf{Q}(t)\mathbf{x} + \mathbf{c}(t), \\ t^* &= t, \end{aligned} \tag{1.31}$$

where $\mathbf{Q}(t)$ is a time dependent orthogonal tensor and $\mathbf{c}(t)$ is a given time dependent vector. Note that one usually assumes that the constitutive relations depend on $\mathbf{L}$ instead of $\mathbf{D}$, and then one deduces from the principle of material frame indifference that this dependence on $\mathbf{L}$ has to reduce to a dependence on $\mathbf{D}$ only. In fact, this is for us the only relevant consequence of the stronger requirement of material frame indifference which cannot be obtained from the requirement that the material properties are invariant under Galilean transformations (1.28) only. Moreover, there seems to be no firm agreement how the electro-magnetic quantities should transform under transformations of the form (1.31) (cf. Grot [41] pp. 157, 183), while the transformation properties for Galilean transformations are on firm grounds. Therefore we have decided to include from the beginning $\mathbf{D}$ into the list of independent variables and to work only with the requirement of Galilean invariance.

It is also well-known that Maxwell's equations (1.18), (1.20)–(1.22) and (1.24) are form invariant under Lorentz transformations, e.g. if we consider a motion of the frame with velocity $\mathbf{v}_0 = v\mathbf{e}_1$ in direction $\mathbf{e}_1$ the Lorentz transformation is given by

$$
x_1^* = \frac{x_1 - vt}{\sqrt{1 - v^2/c^2}}, \quad x_2^* = x_2, \quad x_3^* = x_3,
$$
$$
t^* = \frac{1}{\sqrt{1 - v^2/c^2}}\left(t - v\frac{x_1}{c^2}\right). \tag{1.32}
$$

Since we are interested in non-relativistic effects we shall assume that $|\mathbf{v}_0| \ll c$. Neglecting terms of order $\left|\frac{v}{c}\right|^2$ in (1.32) we recover $(1.28)_1$ (with $\mathbf{Q} = \mathbf{I}, \mathbf{b}_0 = 0$), but $(1.32)_2$ reduces to[6]

$$
t^* = t - \frac{v}{c}\frac{x_1}{c},
$$

where the second term can be neglected for x_1 belonging to compact sets, but not uniformly for all x_1. Thus, the Galilean transformation (1.28) is not a uniform approximation of the Lorentz transformation. Nevertheless, for the study of effects in a finite body (1.28) can be viewed as a good approximation of (1.32). Similar difficulties manifest themselves if one tries to deduce approximate transformation formulae for $\mathbf{B}, \mathbf{E}, \mathbf{D}_e, \mathbf{H}, \mathbf{J}, q_e, \mathbf{P}$ and $\mathbf{M}$ from the transformation formulae for these quantities that arise from the Lorentz transformation[7]. One can avoid these difficulties if one requires the Maxwell's equations (1.18), (1.20)–(1.22) and (1.24) to be form invariant under the Galilean transformation (1.28). This requirement renders the following

---

[6]This observation can be found in the literature, cf. Lévy-Leblond [66], Le Bellac, Lévy-Leblond [8] where this issue is discussed in detail. It is clearly pointed out there that beside the assumption $|\mathbf{v}_0| \ll c$ one needs additional requirements to reduce the Lorentz transformation to the Galilean transformation. For example it is assumed in Lévy-Leblond [66] that the space-time interval is of "large time-like" type, i.e. $\Delta x \ll c\Delta t$. We have brought up again attention to this point, because unfortunately one often finds in textbooks statements like "Galilean transformation is a good approximation of Lorentz transformation if $|\mathbf{v}_0| \ll c$ ", which might be misleading.

[7]Especially the transformation formulas for $q_e, \mathbf{P}$ and $\mathbf{M}$ coming from the requirement of form invariance of Maxwell's equations under Galilean transformations are not approximations of the formulae coming from the Lorentz transformation.

transformation formulae if $\mathbf{Q} = \mathbf{I}$, $\mathbf{b}_0 = \mathbf{0}$

$$\mathbf{B}^* = \mathbf{B}, \quad \mathbf{E}^* = \mathbf{E} + \frac{1}{c}\mathbf{v}_0 \times \mathbf{B},$$

$$\mathbf{D}_e^* = \mathbf{D}_e, \quad \mathbf{H}^* = \mathbf{H} - \frac{1}{c}\mathbf{v}_0 \times \mathbf{D}_e, \qquad (1.33)$$

$$q_e^* = q_e.$$

It is worth noticing that the transformation formulae for $\mathbf{P}$ and $\mathbf{M}$, which would follow from (1.33), (1.19) and (1.23) are not reasonable. But in the formulation of Maxwell's equations these fields do not occur.

Based on the above discussion we shall make the following invariance requirements: We assume that the quantities (1.26), describing the material properties, are co-variant under Galilean transformations (1.28). Moreover we require that all balance laws (1.3), (1.4)[8], (1.14), (1.15), (1.17) and (1.18), (1.20)–(1.22), (1.24) are form invariant under Galilean transformations (1.28). These two requirements imply the following transformation laws[9]

$$\mathbf{T}^* = \mathbf{Q}\mathbf{T}\mathbf{Q}^T, \quad \mathbf{q}^* = \mathbf{Q}\mathbf{q},$$
$$\psi^* = \psi, \quad e^* = e, \qquad (1.34)$$

$$\mathbf{P}^* = \mathbf{Q}\mathbf{P}, \quad \mathcal{M}^* = \mathbf{Q}\mathcal{M}, \quad \mathcal{J}^* = \mathbf{Q}\mathcal{J}, \qquad (1.35)$$

$$\mathbf{B}^* = \mathbf{Q}\mathbf{B}, \quad \mathbf{E}^* = \mathbf{Q}\left(\mathbf{E} + \frac{1}{c}(\mathbf{Q}^T\mathbf{v}_0) \times \mathbf{B}\right),$$

$$\mathbf{D}_e^* = \mathbf{Q}\mathbf{D}_e, \quad \mathbf{H}^* = \mathbf{Q}\left(\mathbf{H} - \frac{1}{c}(\mathbf{Q}^T\mathbf{v}_0) \times \mathbf{D}_e\right), \qquad (1.36)$$

$$\mathcal{E}^* = \mathbf{Q}\mathcal{E}, \quad q_e^* = q_e, \quad \mathbf{f}_e^* = \mathbf{Q}\mathbf{f}_e,$$

$$\rho^* = \rho, \quad \theta^* = \theta, \quad \nabla^*\theta^* = \mathbf{Q}\nabla\theta,$$
$$\mathbf{f}^* = \mathbf{Q}\mathbf{f}, \quad \mathbf{v}^* = \mathbf{Q}\mathbf{v} - \mathbf{v}_0, \quad \mathbf{D}^* = \mathbf{Q}\mathbf{D}\mathbf{Q}^T, \quad r^* = r. \qquad (1.37)$$

The Galilean invariance of the quantities (1.26) implies that the constitutive relations (1.27) are isotropic functions of its arguments, which transform according to (1.34)–(1.37). The system (1.3), (1.4), (1.14), (1.15), (1.17) and (1.18), (1.20)–(1.22), (1.24) together with the constitutive equations for the quantities in (1.25) describes the behaviour of many materials in a variety of situations. This is much too general, since in our investigation we are only interested in a description of electrorheological fluids.

---

[8]Note, that one should use the identity $q_e\mathbf{E} + \frac{1}{c}\mathbf{J} \times \mathbf{B} = q_e\mathcal{E} + \frac{1}{c}\mathcal{J} \times \mathbf{B}$ to rewrite the electromagnetic body force $\mathbf{f}_e$ in a more convenient form.

[9]Note, that this could only lead to inconsistencies if both $\mathbf{P}$ and $\mathbf{D}_e$ or $\mathcal{M}$ and $\mathbf{H}$ would appear simultaneously in one equation, which is never the case.

Therefore we will make some simplifying assumptions, which make the system easier, but leave it general enough for our purposes. Since our main interest in this investigation is not the thermal behaviour[10] of the electrorheological fluid we shall start by simplifying the form of the thermal response of the material. We assume that the heat flux vector $\mathbf{q}$ is given by Fourier's law of heat conduction

$$\mathbf{q} = -k\nabla\theta, \tag{1.38}$$

where k is the thermal conductivity, which is assumed to be constant. In all other constitutive relations we will drop the dependence on $\nabla\theta$. This and the invariance requirements imply that (1.27) has to be replaced by (use the transformation formulae for $\mathbf{Q} = \mathbf{I}$, $\mathbf{v}_0 = \mathbf{v}$; see also Grot [41])

$$f = \hat{f}(\rho, \mathbf{D}, \boldsymbol{\mathcal{E}}, \mathbf{B}, \theta). \tag{1.39}$$

In addition to restrictions placed on the constitutive response functions by the invariance requirements we have additional strictures due to the requirement of the second law of thermodynamics. We shall now determine the restrictions imposed by requiring that all admissible processes of the body, i.e. processes compatible with the balance laws and constitutive response functions, meet the Clausius-Duhem inequality (1.17). It immediately follows from (1.17) and (1.3) that

$$-\rho\left(\frac{\partial\psi}{\partial\theta} + \eta\right)\dot{\theta} - \rho\frac{\partial\psi}{\partial\mathbf{D}}\cdot\dot{\mathbf{D}} - \left(\boldsymbol{\mathcal{M}} + \rho\frac{\partial\psi}{\partial\mathbf{B}}\right)\cdot\dot{\mathbf{B}} + \left(\mathbf{T} + \rho^2\frac{\partial\psi}{\partial\rho}\mathbf{I}\right)\cdot\mathbf{D}$$
$$+ \mathbf{T}\cdot\mathbf{W} + k\frac{|\nabla\theta|^2}{\theta} + \left(\rho\frac{\partial\psi}{\partial\boldsymbol{\mathcal{E}}} - \mathbf{P}\right)\cdot\dot{\boldsymbol{\mathcal{E}}} + \boldsymbol{\mathcal{J}}\cdot\boldsymbol{\mathcal{E}} \geq 0. \tag{1.40}$$

Using the linearity of (1.40) with respect to the dotted quantities and $\mathbf{W}$ and their independence on the arguments appearing in the constitutive relations (1.39) one easily deduces (cf. Coleman, Noll [19], Truesdell, Noll [124], Grot [41])

$$\eta = -\frac{\partial\psi}{\partial\theta}, \quad \frac{\partial\psi}{\partial\mathbf{D}} = \mathbf{0},$$
$$\mathbf{P} = -\rho\frac{\partial\psi}{\partial\boldsymbol{\mathcal{E}}}, \quad \boldsymbol{\mathcal{M}} = -\rho\frac{\partial\psi}{\partial\mathbf{B}}, \tag{1.41}$$
$$\mathbf{T}^T = \mathbf{T}$$

and the reduced dissipation inequality

$$\left(\mathbf{T} + \rho^2\frac{\partial\psi}{\partial\rho}\mathbf{I}\right)\cdot\mathbf{D} + k\frac{|\nabla\theta|^2}{\theta} + \boldsymbol{\mathcal{J}}\cdot\boldsymbol{\mathcal{E}} \geq 0. \tag{1.42}$$

---

[10]However, it is clear that for practical applications thermal effects play an important role. This point will be investigated after a better understanding of the mechanical effects.

Summarizing, we are now dealing with the following system of balance laws

$$\dot{\rho} + \rho \, \mathrm{div} \, \mathbf{v} = 0 \,, \tag{1.43}$$

$$\rho \dot{\mathbf{v}} - \mathrm{div} \, \mathbf{T} = \rho \mathbf{f} + \mathbf{f}_e \tag{1.44}$$

$$\rho \dot{e} - k \Delta \theta = \mathbf{T} \cdot \mathbf{D} + \dot{\mathbf{P}} \cdot \boldsymbol{\mathcal{E}} + (\mathbf{P} \cdot \boldsymbol{\mathcal{E}}) \, \mathrm{div} \, \mathbf{v} \tag{1.45}$$
$$- \boldsymbol{\mathcal{M}} \cdot \dot{\mathbf{B}} + \boldsymbol{\mathcal{J}} \cdot \boldsymbol{\mathcal{E}} + \rho r$$

$$\epsilon (\mathbf{P} \otimes \boldsymbol{\mathcal{E}} + \boldsymbol{\mathcal{M}} \otimes \mathbf{B}) = 0 \,, \tag{1.46}$$

$$(\mathbf{T} + \phi \mathbf{I}) \cdot \mathbf{D} + k \frac{|\nabla \theta|^2}{\theta} + \boldsymbol{\mathcal{J}} \cdot \boldsymbol{\mathcal{E}} \geq 0 \,, \tag{1.47}$$

$$\mathrm{div} \, \mathbf{D}_e = q_e \,, \tag{1.48}$$

$$\mathrm{curl} \, \mathbf{E} + \frac{1}{c} \frac{\partial \mathbf{B}}{\partial t} = 0 \,, \tag{1.49}$$

$$\mathrm{div} \, \mathbf{B} = 0 \,, \tag{1.50}$$

$$\mathrm{curl} \, \mathbf{H} = \frac{1}{c} \frac{\partial \mathbf{D}_e}{\partial t} + \frac{1}{c} (\boldsymbol{\mathcal{J}} + q_e \mathbf{v}) \,, \tag{1.51}$$

$$\frac{\partial q_e}{\partial t} + \mathrm{div} \, (\boldsymbol{\mathcal{J}} + q_e \mathbf{v}) = 0 \,, \tag{1.52}$$

where $\mathbf{f}_e$ is given by

$$\mathbf{f}_e = q_e \, \boldsymbol{\mathcal{E}} + \frac{1}{c} \boldsymbol{\mathcal{J}} \times \mathbf{B} + \frac{1}{c} (\dot{\mathbf{P}} + (\mathrm{div} \, \mathbf{v}) \mathbf{P}) \times \mathbf{B} + \frac{1}{c} \mathbf{v} \times ([\nabla \mathbf{B}] \mathbf{P})$$
$$+ [\nabla \mathbf{B}]^T \boldsymbol{\mathcal{M}} + [\nabla \mathbf{E}] \mathbf{P} \,, \tag{1.53}$$

and where the thermodynamic pressure ϕ is defined through

$$\phi \equiv \rho^2 \frac{\partial \psi}{\partial \rho} \,. \tag{1.54}$$

1.2 Electrorheological Fluids

Now we shall make further simplifications of the system (1.43)–(1.52) based on our understanding of the behaviour of electrorheological fluids. We will make assumptions on the form of some constitutive relations which reflect the observed behaviour of electrorheological fluids and then carry out a dimensional analysis and a subsequent approximation which restricts the validity of the resulting system to special but typical situations. We would like to point out that some of our assumptions in this section are rather based on general observations than on careful experimental evidence, which is missing in the literature. Nevertheless, we believe that these assumptions are reasonable, however they have to be carefully tested when appropriate experimental data are available.

Firstly, we assume that the stress tensor does not depend on the magnetic induction $\mathbf{B}$, i.e.

$$\mathbf{T} = \hat{\mathbf{T}}(\rho, \mathbf{D}, \boldsymbol{\mathcal{E}}, \theta) \,. \tag{2.1}$$

A more detailed discussion on the role of the magnetic induction in electrorheological fluids will be provided when the non-dimensionalization is carried out. Secondly, we shall assume that the fluid is non-conducting, i.e. (cf. Grot [41] Section 2.2)

$$\mathcal{J} \equiv 0, \tag{2.2}$$

and thirdly, as we are dealing with a dielectric (cf. Grot [41] Section 2.2)

$$\mathcal{M} \equiv 0. \tag{2.3}$$

These assumptions already simplify the system (1.43)–(1.52). Note, that from $(1.41)_2$ and (2.3) it follows that

$$\psi = \hat{\psi}(\rho, \theta, \mathcal{E}) \tag{2.4}$$

and therefore also (cf. $(1.41)_2$)

$$\mathbf{P} = \hat{\mathbf{P}}(\rho, \theta, \mathcal{E}). \tag{2.5}$$

Our invariance requirements imply that $\hat{\psi}$ is an isotropic function of its arguments and thus

$$\mathbf{P} = -2\rho \frac{\partial \hat{\psi}}{\partial |\mathcal{E}|^2} \mathcal{E}, \tag{2.6}$$

which in turn implies together with (2.3) that (1.46) is trivially satisfied and we will drop it from the list of equations. Further we get from (1.6) and (2.3) that

$$\mathbf{M} = -\frac{1}{c} \mathbf{v} \times \mathbf{P}. \tag{2.7}$$

Next, we discuss the issue concerning the importance of the magnetic induction. With regard to this point, there is not a convincing body of systematic experimental evidence that can throw light on this matter, though the popular wisdom seems to be of the opinion that it is of secondary importance (see Filisko [29], Wineman [129]). In fact, to our knowledge no experimental paper even reports on the role of the magnetic induction. Most of the experiments measure at best global quantities like the flow rate and not even the mechanical quantities such as the velocity field locally, as the flow regions of interest are so small. However, in order to determine and retain terms in (1.43)–(1.52) that are dominant and discard others that are insignificant, it is necessary to carry out a dimensional analysis of this system.

We shall introduce the following non-dimensional quantities:

$$\bar{\mathbf{E}} = \frac{\mathbf{E}}{E_0}, \quad \bar{\mathbf{B}} = \frac{\mathbf{B}}{B_0}, \quad \bar{q}_e = \frac{q_e}{q_0}, \quad \bar{\mathbf{v}} = \frac{\mathbf{v}}{V_0}, \quad \bar{\mathbf{x}} = \frac{\mathbf{x}}{L_0},$$
$$\bar{t} = \frac{t}{T_0}, \quad \bar{\mathbf{P}} = \frac{\mathbf{P}}{E_0}, \quad \bar{\rho} = \frac{\rho}{\rho_0}, \quad \bar{\mathbf{f}} = \frac{\mathbf{f}}{f_0}, \quad \bar{\theta} = \frac{\theta}{\theta_0}, \tag{2.8}$$

where the quantities with the suffix zero are appropriate representative quantities. While T_0 could be associated with the period of oscillation of the electric field when an ac field is involved, a characteristic time for the problem would have to be introduced if a dc field is used. One could use e.g. the reaction time of the material to form structures. Here, we shall not go into the details of that issue.

In typical problems, we envisage that

$$E_0 \sim 10^{-2} - 10^2 \text{ statvolts cm}^{-1},$$
$$L_0 \sim 10^{-1} - 1 \text{ cm}, \qquad (2.9)$$
$$T_0 \sim 10^{-4} - 1 \text{ sec},$$

and thus we conclude that

$$\frac{L_0}{c T_0} \sim 10^{-11} - 10^{-6}. \qquad (2.10)$$

In the analysis that we shall carry out, we chose to define a Reynolds number Re through

$$Re = \frac{\rho_0 L_0 V_0}{\mu_0}, \qquad (2.11)$$

where ρ_0 and μ_0 are the density and viscosity of the fluid in the absence of an electric field. We shall be interested in problems wherein the Reynolds number[11] Re lies between 1 and 10^2. We shall be concerned with materials for which a typical viscosity μ_0 is of order of 10^{-2} g cm^{-1} s^{-1} and a typical density is of order of 1 g cm^{-3}, reflecting the situation for a large class of fluids. This and (2.11) in turn imply, that the typical velocity field ranges over 10^{-2} to 10 cm s^{-1}, implying

$$\frac{V_0}{c} \sim 10^{-12} - 10^{-9}. \qquad (2.12)$$

The subordinate role of the magnetic induction in electrorheological fluids is mathematically interpreted through the assumption that

$$\frac{E_0}{B_0} \frac{L_0}{c T_0} \sim 1, \qquad (2.13)$$

which is consistent with the assumption that the magnetic induction is only induced by oscillations of the electric field (cf. (2.28) and (2.35)).

Concerning the amount of free charges we shall assume that

$$\frac{q_0 L_0}{E_0} \sim 10^{-12}. \qquad (2.14)$$

Defining a small non-dimensional number ε through

$$\varepsilon \equiv 10^{-3} \qquad (2.15)$$

---

[11] One easily can extend the approach presented here to smaller Reynolds numbers with the effect that some convective terms should be neglected.

we can recast (2.10), (2.12), (2.13) and (2.14) into the form

$$\frac{L_0}{cT_0} = O(\varepsilon^2) - O(\varepsilon^4) \,, \tag{2.16}$$

$$\frac{V_0}{c} = O(\varepsilon^3) - O(\varepsilon^4) \,, \tag{2.17}$$

$$\frac{E_0}{B_0}\frac{L_0}{cT_0} = O(1) \,, \tag{2.18}$$

$$\frac{q_0 L_0}{E_0} = O(\varepsilon^4) \,. \tag{2.19}$$

We will approximate the system (1.43)–(1.52), written in non-dimensional quantities, by neglecting terms of order ε^3 and higher, while retaining terms up to order ε^2.

In preparation for this we shall discuss the role of $\mathcal{E}$ in the constitutive relations for $\mathbf{P}$ and $\mathbf{T}$. It follows from the definition of the electromotive intensity $\mathcal{E}$ (1.9) that

$$
\begin{aligned}
|\bar{\mathcal{E}}|^2 = \frac{|\mathcal{E}|^2}{E_0^2} &= \frac{1}{E_0^2}\left(|\mathbf{E}|^2 + \frac{2}{c}\mathbf{E}\cdot(\mathbf{v}\times\mathbf{B}) + \frac{1}{c^2}|\mathbf{v}\times\mathbf{B}|^2\right) \\
&= |\bar{\mathbf{E}}|^2 + 2\frac{V_0}{c}\frac{L_0}{cT_0}\bar{\mathbf{E}}\cdot(\bar{\mathbf{v}}\times\bar{\mathbf{B}}) + \frac{V_0^2}{c^2}\frac{L_0^2}{V_0^2 c^2}|\bar{\mathbf{v}}\times\bar{\mathbf{B}}|^2 \\
&= |\bar{\mathbf{E}}|^2 + O(\varepsilon^5) \,,
\end{aligned}
\tag{2.20}
$$

where we used (2.18), (2.16) and (2.17).

In view of (2.20), (2.5) and the regularity assumption of the material function ψ we obtain

$$\bar{\mathbf{P}}(\bar{\rho}, \bar{\theta}, \bar{\mathcal{E}}) = \bar{\mathbf{P}}(\bar{\rho}, \bar{\theta}, \bar{\mathbf{E}}) + O(\varepsilon^5) \,. \tag{2.21}$$

From the invariance requirements and $(1.41)_3$ it follows that $\mathbf{T}$ is a symmetric isotropic function of its arguments. Hence representation theorems (cf. Spencer [118]) yield that

$$
\begin{aligned}
\mathbf{T} = \mathbf{T}(\rho, \theta, \mathbf{D}, \mathcal{E}) \\
= \alpha_1\mathbf{I} + \alpha_2\mathcal{E}\otimes\mathcal{E} + \alpha_3\mathbf{D} + \alpha_4\mathbf{D}^2 + \alpha_5(\mathbf{D}\mathcal{E}\otimes\mathcal{E} + \mathcal{E}\otimes\mathbf{D}\mathcal{E}) \\
+ \alpha_6(\mathbf{D}^2\mathcal{E}\otimes\mathcal{E} + \mathcal{E}\otimes\mathbf{D}^2\mathcal{E}) \,,
\end{aligned}
\tag{2.22}
$$

where α_i, $i = 1, \ldots, 6$, are functions of the invariants

$$\rho\,, \theta\,, |\mathcal{E}|^2\,, \mathrm{tr}\,\mathbf{D}\,, \mathrm{tr}\,|\mathbf{D}|^2\,, \mathrm{tr}\,\mathbf{D}^3\,, \mathrm{tr}\,(\mathbf{D}\mathcal{E}\otimes\mathcal{E})\,, \mathrm{tr}\,(\mathbf{D}^2\mathcal{E}\otimes\mathcal{E}) \,. \tag{2.23}$$

From (2.20) we again deduce in a way similar to that for the polarization that

$$\bar{\alpha}_i(\bar{\rho}, \bar{\theta}, \bar{\mathbf{D}}, \bar{\mathcal{E}}) = \frac{\alpha_i(\rho, \theta, \mathbf{D}, \mathcal{E})}{\alpha_i^0} = \bar{\alpha}_i(\bar{\rho}, \bar{\theta}, \bar{\mathbf{D}}, \bar{\mathbf{E}}) + O(\varepsilon^5) \,, \tag{2.24}$$

Where α_i^0 are appropriate non-dimensional numbers.

Using (1.19), (1.23) and (2.7) we re-write (1.48)–(1.52) in terms of $\mathbf{E}, \mathbf{B}, \mathbf{P}$ and q_e only. Now we carry out a dimensional analysis of this re-written formulation of Maxwell's equations. Using (2.16)–(2.19) and (2.21) we obtain

$$\bar{\operatorname{div}} \bar{\mathbf{E}} + \bar{\operatorname{div}} \bar{\mathbf{P}} = \frac{q_0 L_0}{E_0} \bar{q}_e + O(\varepsilon^5) \,, \tag{2.25}$$

$$\bar{\operatorname{curl}} \bar{\mathbf{E}} + \frac{B_0}{E_0} \frac{L_0}{c T_0} \frac{\partial \bar{\mathbf{B}}}{\partial \bar{t}} = \mathbf{0} \,, \tag{2.26}$$

$$\bar{\operatorname{div}} \bar{\mathbf{B}} = 0 \,, \tag{2.27}$$

$$\bar{\operatorname{curl}} \bar{\mathbf{B}} + \frac{E_0}{B_0} \frac{V_0}{c} \bar{\operatorname{curl}} (\bar{\mathbf{v}} \times \bar{\mathbf{P}}) = \frac{E_0}{B_0} \frac{L_0}{c T_0} \frac{\partial}{\partial \bar{t}} (\bar{\mathbf{E}} + \bar{\mathbf{P}}) \tag{2.28}$$

$$+ \frac{q_0 L_0}{E_0} \frac{V_0}{c} \frac{E_0}{B_0} \bar{q}_e \bar{\mathbf{v}} + O(\varepsilon^5) \,,$$

$$\frac{\partial \bar{q}_e}{\partial \bar{t}} + \frac{V_0 T_0}{L_0} [\bar{\nabla} \bar{q}_e] \bar{\mathbf{v}} = - \frac{V_0 T_0}{L_0} \bar{q}_e \bar{\operatorname{div}} \bar{\mathbf{v}} \,. \tag{2.29}$$

Now using that (cf. (2.16)–(2.18))

$$\frac{V_0 T_0}{L_0} = \frac{V_0}{c} \frac{c T_0}{L_0} = O(\varepsilon^{-1}) - O(\varepsilon^2) \,, \tag{2.30}$$

and

$$\frac{B_0}{E_0} = O(\varepsilon^2) - O(\varepsilon^4) \,, \tag{2.31}$$

and neglecting terms of $O(\varepsilon^3)$ we can approximate (1.48)–(1.52) by

$$\operatorname{div} (\mathbf{E} + \mathbf{P}) = 0 \,, \tag{2.32}$$

$$\operatorname{curl} \mathbf{E} = \mathbf{0} \,, \tag{2.33}$$

$$\operatorname{div} \mathbf{B} = 0 \,, \tag{2.34}$$

$$\operatorname{curl} \mathbf{B} + \frac{1}{c} \operatorname{curl} (\mathbf{v} \times \mathbf{P}) = \frac{1}{c} \frac{\partial (\mathbf{E} + \mathbf{P})}{\partial t} \,, \tag{2.35}$$

$$\dot{q}_e + q_e \operatorname{div} \mathbf{v} = 0 \,, \tag{2.36}$$

where $\mathbf{P} = \mathbf{P}(\rho, \theta, \mathbf{E})$. It follows from (1.19), (1.23) and (2.7) that (2.32)–(2.35) can be re-written again in terms of $\mathbf{E}, \mathbf{B}, \mathbf{H}, \mathbf{D}_e$ only. Note, that (2.32)–(2.36) is exactly the electric quasi-static approximation of Maxwell's equations (cf. Romano [110]).

Now we turn to the approximation of the equations (1.43)–(1.45) and (1.47). The conservation of mass (1.43) remains un-effected. In the momentum equation (1.44) we use (2.22), (2.21) and (2.24), which leads to

$$
\begin{aligned}
& \frac{\rho_0 V_0 L_0}{E_0^2 T_0} \bar{\rho} \frac{\partial \bar{\mathbf{v}}}{\partial \bar{t}} + \frac{\rho_0 V_0 L_0}{E_0^2 T_0} \frac{T_0 V_0}{L_0} [\bar{\nabla} \bar{\mathbf{v}}] \bar{\mathbf{v}} \\
& - \{\frac{\alpha_1^0}{E_0^2}\} \bar{\nabla} \bar{\alpha}_1 - \{\alpha_2^0\} \overline{\operatorname{div}} (\bar{\alpha}_2 \bar{\mathbf{E}} \otimes \bar{\mathbf{E}}) \\
& - \{\frac{\alpha_3^0}{E_0^2 T_0}\} \frac{T_0 V_0}{L_0} \overline{\operatorname{div}} (\bar{\alpha}_3 \bar{\mathbf{D}}) - \{\frac{\alpha_4^0 V_0}{E_0^2 T_0 L_0}\} \frac{T_0 V_0}{L_0} \overline{\operatorname{div}} (\bar{\alpha}_4 \bar{\mathbf{D}}^2) \\
& - \{\frac{\alpha_5^0}{T_0}\} \frac{T_0 V_0}{L_0} \overline{\operatorname{div}} (\bar{\alpha}_5 (\bar{\mathbf{D}} \bar{\mathbf{E}} \otimes \bar{\mathbf{E}} + \bar{\mathbf{E}} \otimes \bar{\mathbf{D}} \bar{\mathbf{E}})) \\
& - \{\frac{\alpha_6^0 V_0}{T_0 L_0}\} \frac{T_0 V_0}{L_0} \overline{\operatorname{div}} (\bar{\alpha}_6 (\bar{\mathbf{D}}^2 \bar{\mathbf{E}} \otimes \bar{\mathbf{E}} + \bar{\mathbf{E}} \otimes \bar{\mathbf{D}}^2 \bar{\mathbf{E}})) \\
& = \rho_0 f_0 \frac{L_0}{E_0^2} \bar{\rho} \bar{\mathbf{f}} + \frac{q_0 L_0}{E_0} \bar{q}_e \bar{\mathbf{E}} + \frac{q_0 L_0}{E_0} \frac{B_0}{E_0} \bar{q}_e \bar{\mathbf{v}} \times \bar{\mathbf{B}} \\
& + \frac{L_0}{c T_0} \frac{B_0}{E_0} \frac{\partial \bar{\mathbf{P}}}{\partial \bar{t}} \times \bar{\mathbf{B}} + \frac{V_0}{c} \frac{B_0}{E_0} ([\bar{\nabla} \bar{\mathbf{P}}(\mathbf{E})] \bar{\mathbf{v}} + (\overline{\operatorname{div}} \bar{\mathbf{v}}) \mathbf{P}) \times \bar{\mathbf{B}} \\
& + \frac{V_0}{c} \frac{B_0}{E_0} \bar{\mathbf{v}} \times ([\bar{\nabla} \bar{\mathbf{B}}] \bar{\mathbf{P}}) + [\bar{\nabla} \bar{\mathbf{E}}] \bar{\mathbf{P}} + O(\varepsilon^5),
\end{aligned}
\tag{2.37}
$$

where in $O(\varepsilon^5)$ only terms arising from (2.21) and (2.24) are included. This form of the non-dimensionalization was chosen in order to evaluate the relative importance of the various terms that occur in the equation retaining the stress tensor in generality. Simplifications of the stress tensor, i.e. the forms of the material functions, will be discussed in the next sections. On using (2.16)–(2.19) and (2.31) we see that all terms on the right-hand side of (2.37) appearing in the expression for the electro-magnetic force except the term $[\bar{\nabla} \bar{\mathbf{E}}] \bar{\mathbf{P}}$ have to be neglected. We shall also retain the mechanical force term. Let us denote the terms in the squiggly brackets by R_i^{-1}, which are Reynolds number like quantities[12]. Therefore all terms arising from the stress tensor are to be kept in virtue of (2.30). Further, one easily computes that

$$
\frac{\rho_0 V_0 L_0}{E_0^2 T_0} = \begin{cases} O(1) - O(\varepsilon^2) & \text{if } E_0 \sim 10^2 \text{ statvolt cm}^{-1}, \\ O(\varepsilon^{-1}) - O(\varepsilon) & \text{if } E_0 \sim 1 \quad \text{statvolt cm}^{-1}, \\ O(\varepsilon^{-1}) - O(\varepsilon^{-3}) & \text{if } E_0 \sim 10^{-2} \text{ statvolt cm}^{-1}. \end{cases}
\tag{2.38}
$$

Hence the first term on the left-hand side of (2.37) has to be kept. The second term should be strictly speaking neglected if $E_0 \sim 10^2$ statvolt cm^{-1}, which is reasonable, because for such values of E_0 the fluid nearly solidifies. However, since we are interested in the response for the full range of values of E_0, especially when the electric field suddenly changes, we will retain it.

With regard to the approximation of the energy equation, we first re-write (1.45) using (1.16), (1.3), (1.41)$_2$, (1.54) and the definition of the specific heat

$$
c_v = -\theta \frac{\partial^2 \psi}{\partial \theta^2}
\tag{2.39}
$$

---

[12]Note, that these Reynolds number like quantities R_i are related to the corresponding Reynolds numbers Re_i, defined as in (2.11) by $R_i^{-1} = Re_i^{-1} \frac{\rho_0 V_0 L_0}{E_0^2 T_0}$.

as

$$c_v \rho \dot{\theta} - k \, \Delta\theta - \theta \Big(\rho \frac{\partial^2 \psi}{\partial\theta\partial\boldsymbol{\mathcal{E}}} \cdot \dot{\boldsymbol{\mathcal{E}}} - \rho^2 \frac{\partial^2 \psi}{\partial\theta\partial\rho} \operatorname{tr} \mathbf{D} \Big) = \mathbf{T} \cdot \mathbf{D} + \rho^2 \frac{\partial \psi}{\partial\rho} \operatorname{tr} \mathbf{D} + \rho r \,, \qquad (2.40)$$

where c_v and ψ are functions of $\rho, \theta, \boldsymbol{\mathcal{E}}$ and $\mathbf{T} = \mathbf{T}(\rho, \theta, \mathbf{D}, \boldsymbol{\mathcal{E}})$. Similar as in (2.21) and (2.24) we assume

$$\bar{\psi}(\bar{\rho}, \bar{\theta}, \bar{\boldsymbol{\mathcal{E}}}) = \bar{\psi}(\bar{\rho}, \bar{\theta}, \bar{\mathbf{E}}) + O(\varepsilon^5) \,, \qquad (2.41)$$

which together with $(1.41)_2$, (2.21) and (1.54) implies

$$\begin{aligned}
\bar{\mathbf{P}}(\bar{\rho}, \bar{\theta}, \bar{\boldsymbol{\mathcal{E}}}) &= -\bar{\rho} \frac{\partial \bar{\psi}}{\partial \mathbf{E}}(\bar{\rho}, \bar{\theta}, \bar{\mathbf{E}}) + O(\varepsilon^5) \,, \\
\bar{\phi}(\bar{\rho}, \bar{\theta}, \bar{\boldsymbol{\mathcal{E}}}) &= \bar{\rho}^2 \frac{\partial \bar{\psi}}{\partial \rho}(\bar{\rho}, \bar{\theta}, \bar{\mathbf{E}}) + O(\varepsilon^5) \,.
\end{aligned} \qquad (2.42)$$

Now we neglect only terms coming from (2.21), (2.24) and (2.42) in the non-dimensionalized energy equations (2.40), since we have no evidence how the derivatives of the free potential ψ behave. This leads to

$$c_v \rho \dot{\theta} - k \Delta\theta + \theta \Big(\frac{\partial \mathbf{P}}{\partial \theta} \cdot \dot{\mathbf{E}} - \frac{\partial \phi}{\partial \theta} \operatorname{tr} \mathbf{D} \Big) = (\mathbf{T} + \phi \mathbf{I}) \cdot \mathbf{D} + \rho r \,.$$

In the case of the Clausius-Duhem inequality we proceed similarly. Thus we have approximated the thermo-mechanical part of the system (1.43)–(1.52) by

$$\dot{\rho} + \rho \operatorname{div} \mathbf{v} = 0 \,, \qquad (2.43)$$

$$\rho \dot{\mathbf{v}} - \operatorname{div} \mathbf{T} = \rho \mathbf{f} + [\nabla \mathbf{E}] \mathbf{P} \,, \qquad (2.44)$$

$$c_v \rho \dot{\theta} - k \Delta\theta + \theta \Big(\frac{\partial \mathbf{P}}{\partial \theta} \cdot \dot{\mathbf{E}} - \frac{\partial \phi}{\partial \theta} \operatorname{tr} \mathbf{D} \Big) = (\mathbf{T} + \phi \mathbf{I}) \cdot \mathbf{D} + \rho r \,, \qquad (2.45)$$

$$(\mathbf{T} + \phi \mathbf{I}) \cdot \mathbf{D} + k \frac{|\nabla \theta|^2}{\theta} \geq 0 \,, \qquad (2.46)$$

where $\mathbf{P}$, c_v and ϕ are functions of $\rho, \theta, \mathbf{E}$ and $\mathbf{T} = \mathbf{T}(\rho, \theta, \mathbf{D}, \mathbf{E})$.

In the next sections, we shall discuss various simplified structures for the stress **T** with a view towards obtaining a simplified model that can capture the behaviour exhibited by electrorheological fluids.

1.3 Linear Models for the Stress Tensor T

Even a cursory glance at (2.22), (2.23), the representation for the stress[13] **T**, reveals that it has little practical utility or applicability in virtue of its generality, with as many as six material functions that can depend in an arbitrary manner on the invariants[13] (2.23). It would be futile to experimentally determine all these material functions and thus we are left with the task of trying to simplify the expression for

---

[13]Of course, we have to replace $\boldsymbol{\mathcal{E}}$ by **E**.

the stress without forsaking the possibility of obtaining a model that can reflect the behaviour observed for electrorheological fluids. This and the following section is devoted to a discussion of special constitutive models with a view towards developing a theoretical framework that is amenable to mathematical analysis. In this section we assume that the stress tensor is linear in $\mathbf{D}$ and has quadratic growth in $\mathbf{E}$, and then we obtain restrictions on the form of $\mathbf{T}$, which are posed by the Clausius-Duhem inequality in the case of a compressible, an incompressible and a mechanically incompressible but electrically compressible fluid. In the following section, we discuss the case, when the stress power satisfies a growth condition and the material is incompressible, which is analogous to shear dependent viscous fluids.

From the approximation described in the previous section we know that the stress tensor depends on $\rho, \theta, \mathbf{D}, \mathbf{E}$ and therefore has the form (Spencer [118])

$$
\begin{aligned}
\mathbf{T} = {} & \alpha_1 \mathbf{I} + \alpha_2 \mathbf{E} \otimes \mathbf{E} + \alpha_3 \mathbf{D} + \alpha_4 \mathbf{D}^2 + \alpha_5 (\mathbf{DE} \otimes \mathbf{E} + \mathbf{E} \otimes \mathbf{DE}) \\
& + \alpha_6 (\mathbf{D}^2 \mathbf{E} \otimes \mathbf{E} + \mathbf{E} \otimes \mathbf{D}^2 \mathbf{E}) \,,
\end{aligned}
\tag{3.1}
$$

where α_i, $i = 1, \ldots, 6$ are functions of

$$
\rho, \theta, |\mathbf{E}|^2, \operatorname{tr} \mathbf{D}, \operatorname{tr} \mathbf{D}^2, \operatorname{tr} \mathbf{D}^3, \operatorname{tr} (\mathbf{DE} \otimes \mathbf{E}), \operatorname{tr} (\mathbf{D}^2 \mathbf{E} \otimes \mathbf{E}) \,.
\tag{3.2}
$$

Assuming now that $\mathbf{T}$ is linear in $\mathbf{D}$ and has quadratic growth in $\mathbf{E}$, we observe

$$
\begin{aligned}
\alpha_1 &= \alpha_{11} + \alpha_{12} \operatorname{tr} \mathbf{D} + \alpha_{13} |\mathbf{E}|^2 + \alpha_{14} |\mathbf{E}|^2 \operatorname{tr} \mathbf{D} + \alpha_{15} \operatorname{tr} (\mathbf{DE} \otimes \mathbf{E}) \,, \\
\alpha_2 &= \alpha_{21} + \alpha_{22} \operatorname{tr} \mathbf{D} \,, \\
\alpha_3 &= \alpha_{31} + \alpha_{32} |\mathbf{E}|^2 \,, \\
\alpha_4 &= 0 \,, \\
\alpha_5 &= \alpha_{51} \,, \\
\alpha_6 &= 0 \,,
\end{aligned}
\tag{3.3}
$$

where α_{ij} are functions of ρ and θ only. The Clausius-Duhem inequality (2.46) reads

$$
(\mathbf{T} + \phi \mathbf{I}) \cdot \mathbf{D} + k \frac{|\nabla \theta|^2}{\theta} \geq 0 \,,
\tag{3.4}
$$

where the thermodynamic pressure ϕ is given by

$$
\hat{\phi}(\rho, \theta, |\mathbf{E}|^2) = \rho^2 \frac{\partial \hat{\psi}}{\partial \rho}(\rho, \theta, |\mathbf{E}|^2) \,.
\tag{3.5}
$$

Now, holding the temperature constant we obtain from (3.1), (3.3) and (3.4)

$$
\begin{aligned}
& (\alpha_{11} + \alpha_{13} |\mathbf{E}|^2 + \phi) \operatorname{tr} \mathbf{D} + (\alpha_{12} + \alpha_{14} |\mathbf{E}|^2)| \operatorname{tr} \mathbf{D}|^2 \\
& + (\alpha_{21} + (\alpha_{22} + \alpha_{15}) \operatorname{tr} \mathbf{D}) \operatorname{tr} (\mathbf{DE} \otimes \mathbf{E}) \\
& + (\alpha_{31} + \alpha_{32} |\mathbf{E}|^2) |\mathbf{D}|^2 + 2\alpha_{51} |\mathbf{DE}|^2 \geq 0 \,.
\end{aligned}
\tag{3.6}
$$

This inequality has to hold for all admissible $\mathbf{D}$ and $\mathbf{E}$. Due to the presence of external source terms in the thermo-mechanical balance equations one sees that the fields $\theta(t, \mathbf{x}) = \theta(t)$, $\mathbf{E}(t, \mathbf{x}) = \mathbf{E}(t)$, and $\mathbf{v}(t, \mathbf{x})$ with div $\mathbf{v}(t, \mathbf{x}) = \mathbf{a}(t)$, where $\mathbf{a}(t)$ is an arbitrary function of time, together with an appropriate density $\rho(t, \mathbf{x}) = \rho(t)$ and magnetic induction $\mathbf{B}(t, \mathbf{x})$ are solutions of (2.32)–(2.35) and (2.43)–(2.45). Therefore all $\mathbf{D}$ and $\mathbf{E}$ are admissible and by specifying and rescaling[14] their values we obtain restrictions on α_{ij}. In preparation for this it is useful to establish relations between the quantities appearing in (3.6)

Lemma 3.7. *Let* $\mathbf{D} \in \mathbb{R}^{3 \times 3}_{\text{sym}}$, *and* $\mathbf{E} \in \mathbb{R}^3$. *Then it holds*

$$| \operatorname{tr} \mathbf{D}|^2 \le 3|\mathbf{D}|^2. \tag{3.8}$$

Moreover, if $\operatorname{tr} \mathbf{D} = 0$ *we obtain*

$$|\mathbf{DE}|^2 \le \tfrac{2}{3}|\mathbf{D}|^2|\mathbf{E}|^2, \tag{3.9}$$

$$| \operatorname{tr}(\mathbf{DE} \otimes \mathbf{E})| \le \sqrt{\tfrac{2}{3}}\, |\mathbf{D}||\mathbf{E}|^2. \tag{3.10}$$

PROOF : Since $\mathbf{D}$ is symmetric it is always diagonalizable and (3.8) follows immediately by seeking the maximum of

$$(\operatorname{tr} \mathbf{D})^2 = \left(d_{11} + d_{22} + d_{33}\right)^2 \tag{3.11}$$

under the constraint $|\mathbf{D}|^2 = 1$. The maximum is 3 and it is attained for $\mathbf{D} = \mathbf{I}$ and (3.8) follows. All quantities in (3.9) are invariant under rotations and therefore we may chose such a basis that

$$\mathbf{E} = \begin{pmatrix} 1 \\ 0 \\ 0 \end{pmatrix}, \qquad \mathbf{D} = \begin{pmatrix} a & c & d \\ c & b & e \\ d & e & -(a+b) \end{pmatrix}. \tag{3.12}$$

Therefore we have to seek the maximum of

$$|\mathbf{DE}|^2 = a^2 + c^2 + d^2 \tag{3.13}$$

under the constraint $|\mathbf{D}|^2 = 1$. A straightforward computation shows that the maximum is $\tfrac{2}{3}$, which is attained for

$$a = \frac{2}{\sqrt{6}}, \qquad b = -\frac{1}{\sqrt{6}}, \qquad c = d = e = 0, \tag{3.14}$$

and (3.9) follows. Note, that $\operatorname{tr}(\mathbf{DE} \otimes \mathbf{E}) = \mathbf{DE} \cdot \mathbf{E}$ and (3.10) follows along the same line of arguments as (3.9). The maximum is attained again for (3.12), (3.14). ∎

Now we shall discuss the consequences of the Clausius-Duhem inequality in the case of a compressible, an incompressible and a mechanically incompressible but electrically compressible electrorheological fluid.

---

[14] A similar rescaling argument was used in a completely different context by Nečas, Šilhavý [94], Růžička [111].

1.3.1   Compressible Electrorheological Fluids

Let us first discuss the case of a compressible electrorheological fluid and fix ρ and θ in the coefficients α_{ij} and in the thermodynamic pressure, which will be denoted $\phi(|\mathbf{E}|^2)$. Setting $\mathbf{E} = 0$ in (3.6) we have

$$(\alpha_{11} + \phi(0)) \operatorname{tr} \mathbf{D} + \alpha_{12}|\operatorname{tr} \mathbf{D}|^2 + \alpha_{31}|\mathbf{D}|^2 \geq 0 , \tag{3.15}$$

and if we now replace $\mathbf{D}$ by $\gamma\mathbf{D}$, multiply (3.15) by γ^{-1} and let $\gamma \to 0$ we obtain

$$\alpha_{11} = -\phi(0) , \tag{3.16}$$

and the remaining part of (3.15) immediately gives using (3.8)

$$\alpha_{31} \geq 0 , \quad 3\alpha_{12} + \alpha_{31} \geq 0 , \tag{3.17}$$

which are the usual restrictions on the viscosities for a compressible viscous fluid. Further, rescaling (3.6) through $\mathbf{D} \to \gamma\mathbf{D}$, multiplying by γ^{-1} and letting $\gamma \to 0$ we get

$$\left(- \phi(0) + \alpha_{13}|\mathbf{E}|^2 + \phi(|\mathbf{E}|^2)\right) \operatorname{tr} \mathbf{D} + \alpha_{21} \operatorname{tr}(\mathbf{DE} \otimes \mathbf{E}) \geq 0 ,$$

which immediately leads to

$$\alpha_{21} = 0 , \tag{3.18}$$

by changing the sign of $\mathbf{D}$, and choosing $\operatorname{tr} \mathbf{D} = 0$. Moreover we note that

$$-\alpha_{13} = \frac{\phi(|\mathbf{E}|^2) - \phi(0)}{|\mathbf{E}|^2} \to \frac{\partial\phi(0)}{\partial|\mathbf{E}|^2}$$

as $|\mathbf{E}|^2 \to 0$, and thus we deduce the following formula for $\phi(|\mathbf{E}|^2)$

$$\phi(|\mathbf{E}|^2) = -\alpha_{11} - \alpha_{13}|\mathbf{E}|^2 = \phi(0) + |\mathbf{E}|^2 \frac{\partial\phi(0)}{\partial|\mathbf{E}|^2} ,$$

where we suppressed the dependence on ρ, θ. Next, choosing $\mathbf{D}$ in (3.6) such that $\operatorname{tr} \mathbf{D} = 0$ we obtain

$$(\alpha_{31} + \alpha_{32}|\mathbf{E}|^2)|\mathbf{D}|^2 + 2\alpha_{51}|\mathbf{DE}|^2 \geq 0 .$$

Setting $\mathbf{DE} = 0$, rescaling $\mathbf{E} \to \gamma\mathbf{E}$, multiplying by γ^{-2}, letting $\gamma \to \infty$ then yields

$$\alpha_{32} \geq 0 , \tag{3.19}$$

and (rescaling $\mathbf{E} \to \gamma\mathbf{E}$, multiplying by γ^{-2}, letting $\gamma \to \infty$)

$$\alpha_{32}|\mathbf{E}|^2|\mathbf{D}|^2 + 2\alpha_{51}|\mathbf{DE}|^2 \geq 0 .$$

The last inequality implies that (on choosing $\mathbf{E}, \mathbf{D}$ as in (3.12), (3.14))

$$\alpha_{32} + \tfrac{4}{3}\alpha_{51} \geq 0 . \tag{3.20}$$

Setting $\mathbf{DE} = \mathbf{0}$ reduces (3.6) to

$$(\alpha_{12} + \alpha_{14}|\mathbf{E}|^2)|\operatorname{tr}\mathbf{D}|^2 + (\alpha_{31} + \alpha_{32}|\mathbf{E}|^2)|\mathbf{D}|^2 \geq 0,$$

which gives (on rescaling $\mathbf{E} \to \gamma\mathbf{E}$, multiplying by γ^{-2}, letting $\gamma \to \infty$ and using (3.8))

$$3\alpha_{14} + \alpha_{32} \geq 0. \tag{3.21}$$

Next, we decompose $\mathbf{D}$ as

$$\mathbf{D} = \tfrac{1}{3}(\operatorname{tr}\mathbf{D})\mathbf{I} + \mathbf{G}, \qquad \operatorname{tr}\mathbf{G} = 0, \tag{3.22}$$

where now $\mathbf{G}$ and $\operatorname{tr}\mathbf{D}$ may be chosen independently. Inequality (3.6) can be re-written as (cf. (3.16), (3.18))

$$\begin{aligned}
\mathbf{T}\cdot\mathbf{D} &= |\operatorname{tr}\mathbf{D}|^2\{\alpha_{12} + \tfrac{1}{3}\alpha_{31} + |\mathbf{E}|^2(\alpha_{14} + \tfrac{1}{3}\alpha_{32} + \tfrac{1}{3}\alpha_{22} + \tfrac{1}{3}\alpha_{15} + \tfrac{2}{9}\alpha_{51})\} \\
&\quad + \operatorname{tr}\mathbf{D}\operatorname{tr}\mathbf{GE}\otimes\mathbf{E}(\alpha_{22} + \alpha_{15} + \tfrac{4}{3}\alpha_{51}) \\
&\quad + (\alpha_{31} + \alpha_{32}|\mathbf{E}|^2)|\mathbf{G}|^2 + 2\alpha_{51}|\mathbf{GE}|^2 \\
&\geq 0.
\end{aligned} \tag{3.23}$$

On rescaling $\mathbf{E} \to \gamma\mathbf{E}$, multiplying by γ^{-2} and letting $\gamma \to \infty$ we notice

$$\begin{aligned}
0 &\leq (\operatorname{tr}\mathbf{D})^2|\mathbf{E}|^2(\alpha_{14} + \tfrac{1}{3}(\alpha_{32} + \alpha_{22} + \alpha_{15}) + \tfrac{2}{9}\alpha_{51}) \\
&\quad + \operatorname{tr}\mathbf{D}\operatorname{tr}\mathbf{GE}\otimes\mathbf{E}(\alpha_{22} + \alpha_{15} + \tfrac{4}{3}\alpha_{51}) + \alpha_{32}|\mathbf{G}|^2|\mathbf{E}|^2 + 2\alpha_{51}|\mathbf{GE}|^2.
\end{aligned} \tag{3.24}$$

Choosing now $\mathbf{G} = 0$ provides

$$\alpha_{14} + \tfrac{1}{3}(\alpha_{32} + \alpha_{22} + \alpha_{15}) + \tfrac{2}{9}\alpha_{51} \geq 0. \tag{3.25}$$

The right-hand side of (3.24) is a polynomial of second order in $\operatorname{tr}\mathbf{D}$ and its non-negativity is equivalent to the condition

$$\begin{aligned}
&|\alpha_{22} + \alpha_{15} + \tfrac{4}{3}\alpha_{51}|^2|\mathbf{E}\cdot\mathbf{GE}|^2 \\
&\leq 4|\mathbf{E}|^2(\alpha_{14} + \tfrac{1}{3}(\alpha_{32} + \alpha_{22} + \alpha_{15}) + \tfrac{2}{9}\alpha_{51})(\alpha_{32}|\mathbf{E}|^2|\mathbf{G}|^2 + 2\alpha_{51}|\mathbf{GE}|^2).
\end{aligned} \tag{3.26}$$

Choosing now $\mathbf{E}$ and $\mathbf{G}$ as in (3.12), (3.14) we obtain

$$|\alpha_{22} + \alpha_{15} + \tfrac{4}{3}\alpha_{51}| \leq \sqrt{6}\sqrt{\alpha_{14} + \tfrac{1}{3}(\alpha_{32} + \alpha_{22} + \alpha_{15}) + \tfrac{2}{9}\alpha_{51}}\sqrt{\alpha_{32} + \tfrac{4}{3}\alpha_{51}}. \tag{3.27}$$

Conversely, starting with (3.27) we can recover (3.26). This can easily be seen by multiplying the square of (3.27) by $|\mathbf{GE}|^2|\mathbf{E}|^2$, using (3.9), $|\mathbf{GE}\cdot\mathbf{E}|^2 \leq |\mathbf{GE}|^2|\mathbf{E}|^2$ and (3.25), (3.20). But (3.26) is equivalent to (3.24). Adding to (3.24) (cf. (3.17))

$$0 \leq (\operatorname{tr}\mathbf{D})^2(\alpha_{12} + \tfrac{1}{3}\alpha_{31}) + \alpha_{31}|\mathbf{G}|^2$$

we restore (3.23), which is equivalent to the Clausius-Duhem inequality (3.6). Thus we have shown

Lemma 3.28. *The stress tensor* $\mathbf{T}$ *for a compressible electrorheological fluid given by* (3.1) *and* (3.3) *satisfies the Clausius-Duhem inequality if and only if* $\mathbf{T}$ *is of the form*

$$
\begin{aligned}
\mathbf{T} = \big(& -\phi(|\mathbf{E}|^2) + (\alpha_{12} + \alpha_{14}|\mathbf{E}|^2)\operatorname{tr}\mathbf{D} + \alpha_{15}\operatorname{tr}(\mathbf{DE}\otimes\mathbf{E})\big)\mathbf{I} \\
& + \alpha_{22}(\operatorname{tr}\mathbf{D})\mathbf{E}\otimes\mathbf{E} + (\alpha_{31} + \alpha_{32}|\mathbf{E}|^2)\mathbf{D} \\
& + \alpha_{51}(\mathbf{DE}\otimes\mathbf{E} + \mathbf{E}\otimes\mathbf{DE})\,,
\end{aligned}
\tag{3.29}
$$

where the coefficients and the thermodynamic pressure have to satisfy[15]

$$
\alpha_{31} \geq 0\,, \quad \alpha_{32} \geq 0\,,
$$

$$
3\alpha_{12} + \alpha_{31} \geq 0\,, \quad 3\alpha_{14} + \alpha_{32} \geq 0\,, \quad \alpha_{32} + \tfrac{4}{3}\alpha_{51} \geq 0\,,
$$

$$
\alpha_{14} + \tfrac{1}{3}\big(\alpha_{15} + \alpha_{22} + \alpha_{32} + \tfrac{2}{3}\alpha_{51}\big) \geq 0\,,
\tag{3.30}
$$

$$
\big|\alpha_{22} + \alpha_{15} + \tfrac{4}{3}\alpha_{51}\big| \leq \sqrt{6}\sqrt{\alpha_{14} + \tfrac{1}{3}(\alpha_{15} + \alpha_{22} + \alpha_{32} + \tfrac{2}{3}\alpha_{51})}\sqrt{\alpha_{32} + \tfrac{4}{3}\alpha_{51}}\,,
$$

$$
\phi(|\mathbf{E}|^2) = -\alpha_{11} - \alpha_{13}|\mathbf{E}|^2\,.
$$

1.3.2   Incompressible Electrorheological Fluids

An incompressible fluid is only capable of isochoric motions. Such an internal constraint can be expressed by

$$
\operatorname{div}\mathbf{v} = \operatorname{tr}\mathbf{D} = 0\,.
\tag{3.31}
$$

Thus, in the representation (3.1) for the stress tensor $\mathbf{T}$ we have $\alpha_1 = -\phi$, where ϕ is the indeterminate part of the stress due to the constraint (3.31). Assuming again that $\mathbf{T}$ is linear in $\mathbf{D}$ and has quadratic growth in $\mathbf{E}$ we obtain

$$
\begin{aligned}
\alpha_1 &= -\phi\,, \\
\alpha_2 &= \alpha_{21}\,, \\
\alpha_3 &= \alpha_{31} + \alpha_{32}|\mathbf{E}|^2\,, \\
\alpha_4 &= 0\,, \\
\alpha_5 &= \alpha_{51}\,, \\
\alpha_6 &= 0\,,
\end{aligned}
\tag{3.32}
$$

where α_{ij} are functions of θ only. The Clausius-Duhem inequality (2.46) in this situation reads

$$
\alpha_{21}\operatorname{tr}(\mathbf{DE}\otimes\mathbf{E}) + (\alpha_{31} + \alpha_{32}|\mathbf{E}|^2)|\mathbf{D}|^2 + 2\alpha_{51}|\mathbf{DE}|^2 \geq 0\,.
\tag{3.33}
$$

Proceeding similarly as in the case of compressible fluids we can show

---

[15]Note, that all coefficients α_{ij} and $\phi(|\mathbf{E}|^2)$ are functions of θ and ρ.

Lemma 3.34. *The stress tensor* **T** *for an incompressible electrorheological fluid given by* (3.1) *and* (3.32) *satisfies the Clausius-Duhem inequality if and only if* **T** *is of the form*

$$\mathbf{T} = -\phi\mathbf{I} + (\alpha_{31} + \alpha_{32}|\mathbf{E}|^2)\mathbf{D} + \alpha_{51}(\mathbf{DE} \otimes \mathbf{E} + \mathbf{E} \otimes \mathbf{DE}),\qquad(3.35)$$

where the coefficients, which are functions of θ, *satisfy*

$$\alpha_{31} \geq 0,\quad \alpha_{32} \geq 0,\quad \alpha_{32} + \tfrac{4}{3}\alpha_{51} \geq 0.\qquad(3.36)$$

1.3.3 Mechanically Incompressible but Electrically Compressible Electrorheological Fluids

Preliminary experiments (Vel, Rajagopal and Yalamanchili [125]) seem to indicate that there are electrorheological fluids which undergo only isochoric motions in processes where $\mathbf{E} = const.$, while they are compressible if the electric field $\mathbf{E}$ changes. This situation is similar to that in a viscous fluid which can undergo only isochoric motions if the temperature is constant, while it can sustain motions that are not isochoric due to changes in the temperature, which is at the heart of the celebrated Oberbeck-Boussinesq approximation, which was recently re-examined in Rajagopal, Růžička, Srinivasa [106]. We here will follow the approach outlined in [106] for treating internal constraints.

   The situation described above of a fluid that can undergo only isochoric motions if $\mathbf{E} = const.$, while being capable of motions that are not isochoric as $\mathbf{E}$ changes, is mathematically expressed by[16]

$$\det \mathbf{F} = f(|\mathbf{E}|^2),\qquad(3.37)$$

where **F** is the deformation gradient. Computing the total time derivative of (3.37) implies

$$\operatorname{div}\mathbf{v} = \operatorname{tr}\mathbf{D} = \beta(|\mathbf{E}|^2)\frac{d}{dt}|\mathbf{E}|^2,\qquad(3.38)$$

where

$$\beta(|\mathbf{E}|^2) = \frac{f'(|\mathbf{E}|^2)}{f(|\mathbf{E}|^2)}.\qquad(3.39)$$

From (3.38) and (2.43) it follows that

$$\frac{\dot{\rho}}{\rho} = -\operatorname{tr}\mathbf{D} = -\beta\frac{d}{dt}|\mathbf{E}|^2,\qquad(3.40)$$

i.e. the density is completely determined by the electric field $\mathbf{E}$. To ensure that the constraint (3.37) is satisfied, we shall decompose the stress tensor as

$$\mathbf{T} = -\phi\mathbf{I} + \tilde{\mathbf{T}},\qquad(3.41)$$

---

[16]Note, that det **F** is a measure for the change of the volume.

where

$$\phi = -\tfrac{1}{3}\operatorname{tr}\mathbf{T}, \quad \operatorname{tr}\tilde{\mathbf{T}} = 0. \tag{3.42}$$

ϕ is usually called the mechanical pressure. Inserting (3.42) into (2.44) we see

$$\rho\dot{\mathbf{v}} - \operatorname{div}\tilde{\mathbf{T}} + \nabla\phi = \rho\mathbf{f} + [\nabla\mathbf{E}]\mathbf{P}. \tag{3.43}$$

Taking the divergence of (3.43) we obtain an equation for ϕ, which reads

$$-\Delta\phi + \frac{1}{\rho}\nabla\rho\cdot\nabla\phi = -\rho\operatorname{div}\mathbf{f} - \operatorname{div}\left([\nabla\mathbf{E}]\mathbf{P}\right)$$
$$+ \frac{1}{\rho}\nabla\rho\cdot\left([\nabla\mathbf{E}]\mathbf{P} + \operatorname{div}\tilde{\mathbf{T}}\right) + \rho\frac{d}{dt}\left(\beta\frac{d}{dt}|\mathbf{E}|^2\right) \tag{3.44}$$
$$+ \rho\mathbf{L}\cdot\mathbf{L}^T - \operatorname{div}\left(\operatorname{div}\tilde{\mathbf{T}}\right).$$

We see from the above equation that no constitutive relation for ϕ is required, but once the electrical field $\mathbf{E}$ is known we can compute ρ from (3.40) and, $\mathbf{v}$ and ϕ are determined through (2.44) and (3.44). From our treatment of the constraint (3.37) it follows that we have to re-examine the consequences of the Clausius-Duhem inequality (1.17). As pointed out in Section 2 the appropriate form of (1.17) is

$$-\rho(\dot{\psi} + \eta\dot{\theta}) + \mathbf{T}\cdot\mathbf{D} + k\frac{|\nabla\theta|^2}{\theta} - \dot{\mathbf{E}}\cdot\mathbf{P} \geq 0, \tag{3.45}$$

where $\psi, \eta, \mathbf{T}$ and $\mathbf{P}$ are functions of $\mathbf{E}, \mathbf{D}$ and θ. By (3.40) and arguments similar to that, which led to (1.41), (1.42) we deduce that

$$\eta = -\frac{\partial\psi}{\partial\theta}, \quad \frac{\partial\psi}{\partial\mathbf{D}} = 0,$$
$$\mathbf{P} = -2\beta\phi\mathbf{E} - \rho\frac{\partial\psi}{\partial\mathbf{E}},$$

and (3.45) reduces to

$$\tilde{\mathbf{T}}\cdot\mathbf{D} + k\frac{|\nabla\theta|^2}{\theta} \geq 0. \tag{3.46}$$

Assuming now that $\tilde{\mathbf{T}}$ is linear in $\mathbf{D}$ and has quadratic growth in $\mathbf{E}$ we have the representation (3.1) for $\tilde{\mathbf{T}}$ with α_i satisfying (3.3), where α_{ij} are functions of θ only, because ρ is determined through $\mathbf{E}$.

Holding the temperature θ fixed we get from (3.46) inequality (3.6), where ϕ has to be omitted. Replacing now $\mathbf{E}$ by $\gamma\mathbf{E}$ and letting $\gamma \to 0$ we obtain

$$\alpha_{11}\operatorname{tr}\mathbf{D} + \alpha_{12}|\operatorname{tr}\mathbf{D}|^2 + \alpha_{31}|\mathbf{D}|^2 \geq 0, \tag{3.47}$$

and if we now re-scale $\mathbf{D}$ by $\gamma\mathbf{D}$, multiply by γ^{-1} and let $\gamma \to 0$ we observe

$$\alpha_{11} = 0, \tag{3.48}$$

and the remaining part of (3.47) yields again (3.17). Rescaling (3.6) through $\mathbf{D} \to \gamma \mathbf{D}$, multiplying by γ^{-1} and letting $\gamma \to 0$ we obtain

$$\alpha_{13}|\mathbf{E}|^2 \operatorname{tr} \mathbf{D} + \alpha_{21} \operatorname{tr}(\mathbf{DE} \otimes \mathbf{E}) \geq 0,$$

which immediately gives

$$\alpha_{13} = \alpha_{21} = 0, \tag{3.49}$$

by changing the sign of $\mathbf{D}$, and choosing $\operatorname{tr} \mathbf{D} = 0$. Further we proceed as in the compressible case and we arrive at (3.29) and (3.30)$_{1-4}$, where $\phi(|\mathbf{E}|^2)$ has to be replaced by $-\phi$. Using (3.42)$_2$ we get

$$\begin{aligned} 0 &= (3\alpha_{12} + \alpha_{31}) \operatorname{tr} \mathbf{D} + (3\alpha_{14} + \alpha_{22} + \alpha_{32})|\mathbf{E}|^2 \operatorname{tr} \mathbf{D} \\ &\quad + (2\alpha_{51} + 3\alpha_{15}) \operatorname{tr}(\mathbf{DE} \otimes \mathbf{E}), \end{aligned} \tag{3.50}$$

from which we easily deduce

$$\alpha_{15} = -\tfrac{2}{3}\alpha_{51}, \quad \alpha_{12} = -\tfrac{1}{3}\alpha_{31}, \quad \alpha_{14} = -\tfrac{1}{3}(\alpha_{22} + \alpha_{32}). \tag{3.51}$$

Furthermore, (3.30)$_7$ implies

$$\alpha_{22} = -\tfrac{2}{3}\alpha_{51}. \tag{3.52}$$

The restrictions (3.30) together with (3.52), (3.51) can be re-written as[17]

$$\alpha_{31} \geq 0, \quad \alpha_{32} \geq 0, \quad \alpha_{51} \geq 0. \tag{3.53}$$

$$\alpha_{15} = -\tfrac{2}{3}\alpha_{51}, \quad \alpha_{12} = -\tfrac{1}{3}\alpha_{31}, \quad \alpha_{14} = -\tfrac{1}{3}\alpha_{32} + \tfrac{2}{9}\alpha_{51}. \tag{3.54}$$

Conversely, using (3.54) and the splitting (3.22) we easily see that

$$\tilde{\mathbf{T}} \cdot \mathbf{D} = \alpha_{31}|\mathbf{G}| + \alpha_{32}|\mathbf{E}|^2|\mathbf{G}|^2 + 2\alpha_{51}|\mathbf{GE}|^2,$$

which is nonnegative by (3.53). Therefore we have proved

Lemma 3.55. *The stress tensor* **T** *for a mechanically incompressible but electrically compressible electrorheological fluid given by (3.1) and (3.3) satisfies the Clausius-Duhem inequality if and only if* **T** *is of the form*

$$\begin{aligned} \mathbf{T} &= -\phi\mathbf{I} - \tfrac{1}{3}\big(\alpha_{31} + (\alpha_{32} - \tfrac{2}{3}\alpha_{51})|\mathbf{E}|^2\big)(\operatorname{tr} \mathbf{D})\,\mathbf{I} \\ &\quad - \tfrac{2}{3}\alpha_{51}\operatorname{tr}(\mathbf{DE} \otimes \mathbf{E})\,\mathbf{I} - \tfrac{2}{3}\alpha_{51}(\operatorname{tr} \mathbf{D})\mathbf{E} \otimes \mathbf{E} \\ &\quad + \big(\alpha_{31} + \alpha_{32}|\mathbf{E}|^2\big)\mathbf{D} + \alpha_{51}\big(\mathbf{DE} \otimes \mathbf{E} + \mathbf{E} \otimes \mathbf{DE}\big), \end{aligned} \tag{3.56}$$

where $\alpha_{ij} = \alpha_{ij}(\theta)$ *satisfy*

$$\alpha_{31} \geq 0, \quad \alpha_{32} \geq 0, \quad \alpha_{51} \geq 0. \tag{3.57}$$

---

[17]Though the same symbols α_{ij} and ϕ are used for the material functions, they are different in each of the cases, i.e. the compressible, the incompressible and the mechanically incompressible but electrically compressible. Thus, as a consequence, we cannot directly compare the inequalities established for these material functions.

1.4   Incompressible Electrorheological Fluids with Shear Dependent Viscosities

Now we discuss in some detail how to obtain a model of an incompressible electrorheological fluid with shear dependent viscosities, which could be thought of as an analogue of simple viscous fluids with shear dependent viscosities.

Let us therefore consider the predictions of the general stress tensor given by (3.1) (with α_1 replaced by $-\phi$, due to the incompressibility of the fluid) for viscometric flows with constant temperature. It is known that viscometric flows are locally a simple shear flow (cf. Huilgol [45]) .

$$\mathbf{v} = \kappa x_2 \mathbf{e}_1, \qquad \kappa = \text{const.} \tag{4.1}$$

with an electric field given by

$$\mathbf{E} = E_1 \mathbf{e}_1 + E_2 \mathbf{e}_2 + E_3 \mathbf{e}_3 . \tag{4.2}$$

Then the components of the extra stress $\mathbf{S} = \mathbf{T} + \phi \mathbf{I}$ are found to be

$$
\begin{aligned}
S_{11} &= \alpha_2 E_1^2 + \frac{\alpha_4}{4}\kappa^2 + \alpha_5 \kappa E_1 E_2 + \frac{\alpha_6}{2}\kappa^2 E_1^2 \\
S_{22} &= \alpha_2 E_2^2 + \frac{\alpha_4}{4}\kappa^2 + \alpha_5 \kappa E_1 E_2 + \frac{\alpha_6}{2}\kappa^2 E_2^2 \\
S_{33} &= \alpha_2 E_3^2 \\
S_{12} &= \alpha_2 E_1 E_2 + \frac{\alpha_3}{2}\kappa + \frac{\alpha_5}{2}\kappa(E_1^2 + E_2^2) + \frac{\alpha_6}{2}\kappa^2 E_1 E_2 \\
S_{13} &= \alpha_2 E_1 E_3 + \frac{\alpha_5}{2}\kappa E_2 E_3 + \frac{\alpha_6}{4}\kappa^2 E_1 E_3 \\
S_{23} &= \alpha_2 E_2 E_3 + \frac{\alpha_5}{2}\kappa E_1 E_3 + \frac{\alpha_6}{4}\kappa^2 E_2 E_3 ,
\end{aligned}
\tag{4.3}
$$

where α_i are functions of the invariants

$$|\mathbf{E}|^2 , \tfrac{1}{2}\kappa^2 , \kappa E_1 E_2 , \tfrac{\kappa^2}{2}(E_1^2 + E_2^2) . \tag{4.4}$$

From (4.3) it follows that the presence of the electric field induces shear stress components not only in the plane of flow, $x_3 = \text{const.}$, but also in the $x_1 - x_3$ and $x_2 - x_3$ planes. The shear stress components S_{13} and S_{23} are induced by the electric field component E_3, which also contributes to the normal stress T_{33} . All other components of the extra stress $\mathbf{S}$ depend on E_3 only through the dependence of the α_i's on $|\mathbf{E}|^2$. Therefore we can conclude that though the stress tensor is an isotropic function of its arguments, the shear stresses S_{12}, S_{13} and S_{23} depend not only on the magnitude $|\mathbf{E}|^2$ of the electric field, but also on the direction of it.[18] In this sense one can say that the fluid possesses a preferred direction, that depends on the applied electric field. For a purely simple viscous fluid the stress in a viscometric flow is determined[19] by the

---

[18]This feature is to be expected from the understanding of the underlying microstructural mechanism, which causes the properties of the electrorheological fluid.

[19]In our situation here we need 5 viscometric functions to determine the stress in a viscometric flow, since we have non trivial shear stresses not only in the plane of flow.

shear stress function $\tau \equiv S_{12}$ and the normal stress differences $N_1 \equiv S_{11} - S_{22}$ and $N_2 = S_{22} - S_{33}$. In our situation these quantities are

$$\tau = \tau(\kappa, \mathbf{E}) = (\alpha_2 + \alpha_6 \frac{\kappa^2}{2}) E_1 E_2 + \left(\alpha_3 + \alpha_5 (E_1^2 + E_2^2)\right) \frac{\kappa}{2}$$

$$N_1 = N_1(\kappa, \mathbf{E}) = (\alpha_2 + \alpha_6 \frac{\kappa^2}{2})(E_1^2 - E_2^2) \tag{4.5}$$

$$N_2 = N_2(\kappa, \mathbf{E}) = (\alpha_2 + \alpha_6 \frac{\kappa^2}{2}) E_2^2 - \alpha_2 E_3^2 + \alpha_4 \frac{\kappa^2}{4} + \alpha_5 E_1 E_2 \kappa,$$

where $\alpha_i, i = 1, \ldots, 6$ are functions of the invariants (4.4). However, in the absence of an electric field, i. e. $\mathbf{E} = \mathbf{0}$, (4.5) reduces to

$$N_1 = 0, \qquad N_2 = \alpha_4 \frac{\kappa^2}{2}.$$

This behaviour is not supported by experimental data (see Huigol [45] and the discussion in Málek, Rajagopal, Růžička [74] and Málek, Nečas, Rokyta, Růžička [70]).[20] For that we shall assume, even in the presence of an electric field, that

$$\alpha_4 \equiv 0. \tag{4.6}$$

Note, that in the presence of an electric field we have no experimental guidance how the normal stress differences of a real electrorheological fluid behave. Moreover, as pointed out before the fluid has a preferred direction and thus the experience for purely viscous fluids can not simply be carried over to our situation. Therefore we shall not make any further assumptions for the material functions on the basis of the form of the normal stress differences.

Previous mathematical investigations of shear dependent viscous fluids (see Málek, Rajagopal, Růžička [74], Málek, Nečas, Rokyta, Růžička [70]) suggest that terms involving $\mathbf{D}^2$ can be treated from the mathematical point of view as a perturbation of the remaining terms (under some smallness assumptions). Therefore we shall also assume

$$\alpha_6 \equiv 0 \tag{4.7}$$

in order to keep the model simple, while still retaining the ability to explain the observed phenomena. After a better understanding of this model has been achieved we can also include the terms involving α_4 and α_6.

Note, that under the assumptions (4.6) and (4.7) the model still predicts both normal stress differences N_1, N_2 and has a shear stress function τ, which depends on the direction of the electric field. For example, if the electric field is in the plane of

---

[20]It is precisely the absence of such experimental data that led to the demise of the Reiner-Rivlin model, which did not predict in the case of non-Newtonian fluids (in the absence of electric fields) both normal stress differences in simple shear flow. However, there are several fluids that exhibit neither of the normal stress differences but whose viscosities are shear dependent and are called generalized Newtonian fluids. The power-law fluids are a subclass of them.

flow, i.e. $E_3 = 0$, we observe (note, that $S_{13} = S_{23} = S_{33} = 0$)

$$\tau = \alpha_2 E_1 E_2 + (\alpha_3 + \alpha_5 |\mathbf{E}|^2) \frac{\kappa}{2},$$
$$N_1 = \alpha_2 (E_1^2 - E_2^2),$$
$$N_2 = \alpha_2 E_2^2 + \alpha_5 E_1 E_2 \kappa. \qquad (4.8)$$

Concerning the remaining material functions α_2, α_3 and α_5 we assume that the material shows the following behaviour: in the absence and presence of an electric field the behaviour is that of a generalized Newtonian fluid with power p. The power p can depend on the magnitude of the electric field (cf. Halsey, Martin, Adolf [42], Bayer [116]) and all terms have the same growth behaviour. Moreover we restrict ourselves to the case when α_2, α_3 and α_5 are functions of the invariants $\theta, |\mathbf{D}|^2$ and $|\mathbf{E}|^2$ only, because we are merely interested in the growth of the material functions α_2, α_3 and α_5. The growth of the other invariants can be obtained as a combination of these considered. Therefore we shall assume that

$$\alpha_2 = \alpha_{21}(1 + |\mathbf{D}|^2)^{\frac{p-1}{2}} + \alpha_{20},$$
$$\alpha_3 = (\alpha_{31} + \alpha_{33}|\mathbf{E}|^2)(1 + |\mathbf{D}|^2)^{\frac{p-2}{2}} + \alpha_{30} + \alpha_{32}|\mathbf{E}|^2, \qquad (4.9)$$
$$\alpha_5 = \alpha_{51}(1 + |\mathbf{D}|^2)^{\frac{p-2}{2}} + \alpha_{50},$$

where α_{ij} are functions of θ, and in general

$$p = p(|\mathbf{E}|^2). \qquad (4.10)$$

We further assume that $p : [0, \infty) \to (1, \infty)$ is a smooth function of $|\mathbf{E}|^2$ and that

$$\lim_{|\mathbf{E}|^2 \to 0} p(|\mathbf{E}|^2) = p_0, \qquad \lim_{|\mathbf{E}|^2 \to \infty} p(|\mathbf{E}|^2) = p_\infty, \qquad (4.11)$$

where

$$1 < p_\infty \leq p(|\mathbf{E}|^2) \leq p_0. \qquad (4.12)$$

Alternatively we can replace $\left(1 + |\mathbf{D}|^2\right)^{\beta/2}$ in (4.9) by

$$|\mathbf{D}|^\beta, \qquad (4.13)$$

where $\beta = p - 1$ or $p - 2$.

The models (4.9) and (4.13) can exhibit a drastically different behaviour. This can be illustrated if we consider the generalized viscosity $\mu(\kappa, \mathbf{E})$ defined through

$$\mu(\kappa, \mathbf{E}) = \frac{\tau(\kappa, \mathbf{E})}{\kappa} \qquad (4.14)$$

for these models. The limits

$$\mu_0 = \lim_{\kappa \to 0} \mu(\kappa, \mathbf{E})$$

and

$$\mu_\infty = \lim_{\kappa \to \infty} \mu(\kappa, \mathbf{E}),$$

respectively, are usually called zero shear viscosity respectively infinite shear viscosity. Let us assume for a moment $p \in (1, 2)$. One easily checks for the model (4.9) ($\mathbf{E} = E_1 \mathbf{e}_1 + E_2 \mathbf{e}_2$)[21] that

$$\begin{aligned} \mu_0 &= \tfrac{1}{2}(\alpha_{30} + \alpha_{31}) + \tfrac{1}{2}(\alpha_{32} + \alpha_{33} + \alpha_{50} + \alpha_{51})|\mathbf{E}|^2, \\ \mu_\infty &= \tfrac{1}{2}\alpha_{30} + \tfrac{1}{2}(\alpha_{32} + \alpha_{50})|\mathbf{E}|^2, \end{aligned} \quad (4.15)$$

while for the model (4.13) we get

$$\begin{aligned} \mu_0 &= \infty, \\ \mu_\infty &= \tfrac{1}{2}\alpha_{30} + \tfrac{1}{2}(\alpha_{32} + \alpha_{50})|\mathbf{E}|^2. \end{aligned} \quad (4.16)$$

On the other hand if $p \in (2, \infty)$ we have for the model (4.9)

$$\begin{aligned} \mu_0 &= \tfrac{1}{2}(\alpha_{30} + \alpha_{31}) + \tfrac{1}{2}(\alpha_{32} + \alpha_{33} + \alpha_{50} + \alpha_{51})|\mathbf{E}|^2, \\ \mu_\infty &= \infty \end{aligned} \quad (4.17)$$

and for the model (4.13)[22]

$$\begin{aligned} \mu_0 &= \tfrac{1}{2}\alpha_{30} + \tfrac{1}{2}(\alpha_{32} + \alpha_{50})|\mathbf{E}|^2, \\ \mu_\infty &= \infty. \end{aligned} \quad (4.18)$$

Relations (4.15)–(4.18) show how different material properties are included in the above models (4.9) and (4.13). For a more detailed discussion for purely generalized Newtonian fluids we refer the reader to Málek, Nečas, Rokyta, Růžička [70] and Málek, Rajagopal, Růžička [74]. From the mathematical point of view the case when $\mu_\infty = 0$ for $p \in (1, 2)$ is more challenging and therefore we will assume

$$\alpha_{30} = \alpha_{32} = \alpha_{50} = 0. \quad (4.19)$$

It is worth noticing that under the assumption (4.19) the form of the viscometric functions (4.8) indicates that it might be possible to design experiments, with the possibility of changing the direction of the electric field, which allow to determine the constants in (4.9) and (4.13), respectively, for the material functions α_2, α_3 and α_5.

Before we discuss the consequences of the Clausius-Duhem inequality and related problems, we shall illustrate the features of the above described models by solving the boundary value problem of a flow between infinite parallel plates, which are a distance

---

[21]From the Clausius-Duhem inequality (3.4) it follows that $\alpha_{20} = -\alpha_{21}$ (cf. (4.33)), which is already used in the computation of the limits.

[22]From the Clausius-Duhem inequality (3.4) it follows that $\alpha_{20} = 0$, which is already used in the computation of the limits.

$2h$ apart. We assume that there is no external body force and that the electric field and the velocity are given by

$$\mathbf{E} = \begin{pmatrix} E_1 \\ E_2 \\ 0 \end{pmatrix}, \qquad \mathbf{v} = \begin{pmatrix} f(x_2) \\ 0 \\ 0 \end{pmatrix}. \tag{4.20}$$

We will solve this problem for the model (3.1), (4.6), (4.7), (4.13)[23], (4.19) and (3.31). Firstly, we observe that the extra stress $\mathbf{S}$ is a function of x_2 only. Thus the equations of motion read

$$\partial_2 S_{12} + \partial_1 \phi = 0 \,,$$
$$\partial_2 S_{22} + \partial_2 \phi = 0 \,, \tag{4.21}$$
$$\partial_3 \phi = 0 \,.$$

From the last equation follows that $\phi = \phi(x_1, x_2)$ and from the first two equations we conclude

$$\partial_1^2 \phi = 0 \,, \qquad \partial_1 \partial_2 \phi = 0 \,,$$

and therefore ϕ is an affine function of x_1 given by

$$\phi(x_1, x_2) = -Ax_1 + k(x_2) \,. \tag{4.22}$$

The constant A can be interpreted as the pressure drop, which maintains the flow. Inserting this into (4.21) and integrating we obtain

$$S_{12} = -Ax_2 + c_0 \,, \qquad c_0 = \text{const.} \tag{4.23}$$

This and the explicit form of the extra stress component S_{12} (cf. (4.8)) leads to an ordinary differential equation for $f(x_2)$, namely

$$-Ax_2 + c_0 = S_{12}(f'(x_2)) \tag{4.24}$$

to which we add no-slip boundary conditions

$$f(\pm h) = 0 \,. \tag{4.25}$$

Inserting (4.20) into (4.24) and using (4.8) we obtain

$$S_{12}(f') = \frac{1}{2^{p/2}} \left(\alpha_{21} \sqrt{2} E_1 E_2 + \alpha_{31} + (\alpha_{33} + \alpha_{51}) |\mathbf{E}|^2 \right) (f')^{p-1} \tag{4.26}$$
$$=: \gamma_0 (f')^{p-1}$$

if $f' \geq 0$ and

$$S_{12}(f') = \frac{-1}{2^{p/2}} \left(-\alpha_{21} \sqrt{2} E_1 E_2 + \alpha_{31} + (\alpha_{33} + \alpha_{51}) |\mathbf{E}|^2 \right) (-f')^{p-1} \tag{4.27}$$
$$=: -\gamma_1 (-f')^{p-1}$$

---

[23] We have chosen this model since the problem is solvable explicitly. For the flow considered there is no essential difference to the model (4.9), but the resulting ordinary differential equation must be solved numerically.

if $f' \leq 0$. Note, that both constants γ_0 and γ_1 are positive due to the restrictions on α_{ij} from the Clausius-Duhem inequality. Furthermore, one expects the flow to take place in direction of $\mathbf{e}_1$ and thus we get $f'(-h) > 0$, $f'(h) < 0$. From (4.24) it is clear that there is exactly one point $x^* = \frac{c_0}{A}$, where $f' = 0$ and where one has to switch from (4.26) to (4.27). One easily checks that

$$
f(x_2) =
\begin{cases}
\left(\dfrac{A}{\gamma_0}\right)^{\frac{1}{p-1}} \dfrac{p-1}{p}\left[(h\gamma + h)^{\frac{p}{p-1}} - (h\gamma - x_2)^{\frac{p}{p-1}}\right] & x_2 \in [-h, h\gamma], \\[3mm]
\left(\dfrac{A}{\gamma_1}\right)^{\frac{1}{p-1}} \dfrac{p-1}{p}\left[(h - h\gamma)^{\frac{p}{p-1}} - (x_2 - h\gamma)^{\frac{p}{p-1}}\right] & x_2 \in [h\gamma, h],
\end{cases}
\tag{4.28}
$$

where

$$
\gamma = \frac{\gamma_0^{1/p} - \gamma_1^{1/p}}{\gamma_0^{1/p} + \gamma_1^{1/p}},
\tag{4.29}
$$

is a solution of (4.24), (4.25). The constant c_0 can be computed as $c_0 = Ah\gamma$. The flow profile (4.28) clearly shows how the solution of the system (4.21) depends not only on the magnitude of the electric field but also on the direction of it. From (4.28) one can compute all components of the extra stress $\mathbf{S}$ and the pressure ϕ. We see that the pressure ϕ is given by

$$
\phi(x_1, x_2) = -Ax_1 + S_{22}(x_2) + c_3,
$$

where

$$
S_{22}(x_2) =
\begin{cases}
\dfrac{A}{\gamma_0} 2^{\frac{-p}{2}} (h\gamma - x_2)\left(\sqrt{2}\,\alpha_{21} E_2^2 + 2\,\alpha_{51} E_1 E_2\right) & x_2 \in [-h, h\gamma], \\[3mm]
\dfrac{A}{\gamma_1} 2^{\frac{-p}{2}} (x_2 - h\gamma)\left(\sqrt{2}\,\alpha_{21} E_2^2 - 2\,\alpha_{51} E_1 E_2\right) & x_2 \in [h\gamma, h],
\end{cases}
$$

while the effective force in direction x_2 is given by

$$
T_{22}(x_1) = Ax_1 - c_3.
\tag{4.30}
$$

Note, that this force does not depend neither on the electric field nor on the value of p and in fact is the same as for a Navier-Stokes fluid.

To illustrate the described solution we will plot the velocity profile for the following situations: The electric field is given as

$$
\mathbf{E} = \beta \begin{pmatrix} \alpha \\ \sqrt{1 - \alpha^2} \\ 0 \end{pmatrix},
$$

the material constants are chosen as $\alpha_{31} = 1$, $\alpha_{33} = 1$, $\alpha_{51} = 3/4$, $\alpha_{51} = \sqrt{6}$, the distance between the plates is $h = 1$. In Figure 1 and 2 we have chosen $p = 2$ and the pressure drop $A = 10$. Figure 1 shows the effect of an increasing magnitude of the electric field. Figure 2 illustrates the effect of different directions of the electric field with constant magnitude.

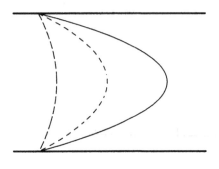

Fig. 1: $p = 2$, $\beta = 0, 1/2, 4/3$

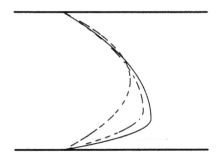

Fig. 2: $p = 2$, $\alpha = 0, \frac{\sqrt{2}}{2}, \frac{\sqrt{3}}{2}$

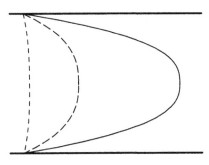

Fig. 3: $p = 1.5$, $\beta = 0, 1/2, 4/3$

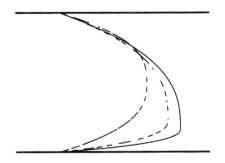

Fig. 4: $p = 1.5$, $\alpha = 0, \frac{\sqrt{2}}{2}, \frac{\sqrt{3}}{2}$

One clearly sees that the velocity profile is asymmetric if $\mathbf{E}$ is not perpendicular to the plates and that the maximal velocity depends on the value of α. Note, that this effect is maximal for $\alpha = \sqrt{2}/2$. In Figure 3 and 4 the same situation is depicted for $p = 3/2$, the pressure drop A is normalized such that the flow rate is the same for $\alpha = 1$, $\beta = 4/3$ as in the case $p = 2$. Between Figure 2 and 4 there is no scaling involved, Figure 3 is scaled $1/6$ times with respect to Figure 1. Figure 1 is scaled $1/5$ times with respect to Figure 2.

Summarizing we can say that the above pictures show that an increasing electric field increases the viscosity and thus the velocity profile becomes more flat. This effect is most significant if $\mathbf{E} = (0, E_2, 0)^\top$, in all other cases, i.e. $E_1 E_2 \neq 0$, the maximal velocity is larger than in the case $E_1 E_2 = 0$ and the velocity profile becomes asymmetric.

Now let us discuss in some detail the consequences of the Clausius-Duhem inequality and related questions for the model (4.9). Here we restrict ourselves to this case since in Chapter 2 we will deal only with this model or the model (3.35) and refer the reader to Rajagopal, Růžička [104], [105] for more details about the model (4.13) and related ones. Holding the temperature θ fixed we obtain from (3.1), (4.6), (4.7),

(3.31) and (3.4)

$$(\alpha_{20} + \alpha_{21}(1 + |\mathbf{D}|^2)^{\frac{p-1}{2}})\mathbf{E} \cdot \mathbf{D}\mathbf{E}$$
$$+ (\alpha_{31} + \alpha_{33}|\mathbf{E}|^2)(1 + |\mathbf{D}|^2)^{\frac{p-2}{2}}|\mathbf{D}|^2 \tag{4.31}$$
$$+ 2\alpha_{51}(1 + |\mathbf{D}|^2)^{\frac{p-2}{2}}|\mathbf{D}\mathbf{E}|^2 \geq 0.$$

Setting $\mathbf{E} = 0$ inequality (4.31) reduces to

$$\alpha_{31} \geq 0, \tag{4.32}$$

and rescaling $\mathbf{D} \to \gamma\mathbf{D}$, multiplying by γ^{-1}, letting $\gamma \to 0$ and changing $\mathbf{D} \to -\mathbf{D}$ we obtain

$$\alpha_{20} = -\alpha_{21}. \tag{4.33}$$

Changing in (4.31) $\mathbf{D}$ by $-\mathbf{D}$ and adding the results we eliminate the terms with α_{20} and α_{21}. This inequality is multiplied by $(1 + |\mathbf{D}|^2)^{-\frac{p-2}{2}}$ and then we re-scale $\mathbf{E} \to \gamma\mathbf{E}$, multiply by γ^{-2} and let $\gamma \to \infty$, which gives

$$\alpha_{33}|\mathbf{E}|^2|\mathbf{D}|^2 + 2\alpha_{51}|\mathbf{D}\mathbf{E}|^2 \geq 0. \tag{4.34}$$

Choosing $\mathbf{D}, \mathbf{E} \neq 0$ such that $\mathbf{D}\mathbf{E} = 0$ we obtain

$$\alpha_{33} \geq 0, \tag{4.35}$$

and from (4.34) and Lemma 3.7 we deduce

$$\alpha_{33} + \tfrac{4}{3}\alpha_{51} \geq 0. \tag{4.36}$$

Now we would like to derive some condition for α_{21}. Using (4.33) we can re-write (4.31) as (also using that $\mathbf{D}$ can be replaced by $-\mathbf{D}$)

$$|\alpha_{21}||\mathbf{E} \cdot \mathbf{D}\mathbf{E}||\mathbf{D}|\frac{(1 + |\mathbf{D}|^2)^{\frac{p-1}{2}} - 1}{(1 + |\mathbf{D}|^2)^{\frac{p-2}{2}}|\mathbf{D}|} \tag{4.37}$$
$$\leq (\alpha_{31} + \alpha_{33}|\mathbf{E}|^2)|\mathbf{D}|^2 + 2\alpha_{51}|\mathbf{D}\mathbf{E}|^2.$$

Choosing $\mathbf{D}, \mathbf{E} \neq 0$ such that $|\mathbf{E} \cdot \mathbf{D}\mathbf{E}| = |\mathbf{E}||\mathbf{D}\mathbf{E}|$, and using (3.9) we get from (4.37) that

$$|\alpha_{21}|\frac{(1 + |\mathbf{D}|^2)^{\frac{p-1}{2}} - 1}{(1 + |\mathbf{D}|^2)^{\frac{p-2}{2}}|\mathbf{D}|} \leq \frac{\alpha_{31}}{|\mathbf{E}|^2\gamma} + \frac{\alpha_{33}}{\gamma} + 2\gamma\alpha_{51} \tag{4.38}$$

must hold for all $|\mathbf{E}|, |\mathbf{D}| \geq 0$ and all $\gamma \in [0, \sqrt{\frac{2}{3}}]$. Note, that the left-hand side of (4.38) depends on $|\mathbf{D}|$ and $|\mathbf{E}|^2$ only, since $p = p(|\mathbf{E}|^2)$, while the right-hand side depends on γ and $|\mathbf{E}|^2$.

One can show that the function

$$f(x) = \frac{(1+x^2)^{\frac{p-1}{2}} - 1}{(1+x^2)^{\frac{p-2}{2}} x} \tag{4.39}$$

is bounded on the interval $[0, \infty]$. More precisely it holds

$$\sup_{x \in [0,\infty]} f(x) = k_0(p), \tag{4.40}$$

where $k_0(p) = 1$ if $p \in (1,3]$ and $k_0(p) > 1$ if $p > 3$. Note, that $k_0(p)$ is an increasing function of p for $p \geq 3$. For $p \geq 3$ the supremum in (4.40) is a maximum which is attained at x_0, where x_0 solves the equation

$$(p-1)x^2 + 1 = (1+x^2)^{\frac{p-1}{2}}. \tag{4.41}$$

For $p \in (1,3]$ the supremum is approached for $x \to \infty$. Taking now the supremum of (4.38) with respect to $|\mathbf{D}|$ and using (4.40), we see that

$$|\alpha_{21}| \leq \frac{1}{k_0(p(|\mathbf{E}|^2))} \left(\frac{\alpha_{31}}{|\mathbf{E}|^2 \gamma} + \frac{\alpha_{33}}{\gamma} + 2\alpha_{51}\gamma \right) \tag{4.42}$$

must hold for all $|\mathbf{E}| \geq 0, \gamma \in [0, \sqrt{\frac{2}{3}}]$. Now we would like to find the infimum of the right-hand side of (4.42), but this is in general impossible, since we have no explicit formula for the solution x_0 of (4.41) and the function $p(|\mathbf{E}|^2)$ might behave very different for various fluids. But letting $|\mathbf{E}|^2 \to \infty$ in (4.42) we arrive at

$$|\alpha_{21}| \leq \frac{1}{k_0(p_\infty)} \left(\frac{\alpha_{33}}{\gamma} + 2\alpha_{51}\gamma \right), \tag{4.43}$$

which holds for all $\gamma \in [0, \sqrt{\frac{2}{3}}]$. An easy calculation shows that this implies

$$k_0(p_\infty)|\alpha_{21}| \leq \begin{cases} 2\sqrt{\alpha_{33}}\sqrt{2\alpha_{51}} & \text{if } \alpha_{33} \leq \frac{4}{3}\alpha_{51} \\[2mm] \sqrt{\frac{3}{2}}(\alpha_{33} + \frac{4}{3}\alpha_{51}) & \text{if } \frac{4}{3}|\alpha_{51}| \leq \alpha_{33}. \end{cases} \tag{4.44}$$

In other words, condition (4.44) ensures that

$$k(p_\infty)|\alpha_{21}||\mathbf{D}||\mathbf{E}|\,|\mathbf{D}\mathbf{E}| \leq \alpha_{33}|\mathbf{D}|^2|\mathbf{E}|^2 + 2\alpha_{51}|\mathbf{D}\mathbf{E}|^2 \tag{4.45}$$

is satisfied for all $\mathbf{E} \in \mathbb{R}^3$ and $\mathbf{D} \in X \equiv \{\mathbf{D} \in \mathbb{R}^{3\times3}_{\text{sym}}, \text{tr } \mathbf{D} = 0\}$. Thus we have shown

Lemma 4.46. *If the stress tensor* $\mathbf{T}$ *given by* (3.1), (4.6), (4.7), (4.9), (4.19) *and* (3.31) *satisfies the Clausius-Duhem inequality, then* $\mathbf{T}$ *is of the form*

$$\mathbf{T} = -\phi\mathbf{I} + \alpha_{21}\left((1+|\mathbf{D}|^2)^{\frac{p-1}{2}} - 1 \right)\mathbf{E} \otimes \mathbf{E}$$

$$+ (\alpha_{31} + \alpha_{33}|\mathbf{E}|^2)(1+|\mathbf{D}|^2)^{\frac{p-2}{2}}\mathbf{D} \tag{4.47}$$

$$+ \alpha_{51}(1+|\mathbf{D}|^2)^{\frac{p-2}{2}}(\mathbf{D}\mathbf{E} \otimes \mathbf{E} + \mathbf{E} \otimes \mathbf{D}\mathbf{E}),$$

where $p = p(|\mathbf{E}|^2)$ *and the coefficients* $\alpha_{ij} = \alpha_{ij}(\theta)$ *satisfy* (4.32), (4.35), (4.36), *and* (4.44).

Remark 4.48. As already indicated the condition (4.44) is in general not sufficient for the Clausius-Duhem inequality to hold. However if p does not depend on $|\mathbf{E}|^2$ or if $p_0 \geq 3$ the condition (4.44) is also sufficient. One easily checks that the relation

$$k_0(p_0)|\alpha_{21}| \leq \begin{cases} 2\sqrt{\alpha_{33}}\sqrt{2\alpha_{51}} & \text{if } \alpha_{33} \leq \frac{4}{3}\alpha_{51} \\ \sqrt{\frac{3}{2}}(\alpha_{33} + \frac{4}{3}\alpha_{51}) & \text{if } \frac{4}{3}|\alpha_{51}| \leq \alpha_{33} . \end{cases} \tag{4.49}$$

together with (4.32), (4.35) and (4.36) are sufficient for the Clausius-Duhem inequality.

The Clausius-Duhem inequality is closely related to the coercivity of the operator induced by $-\operatorname{div}\mathbf{T}$. In fact, if we assume strict inequalities in (4.32), (4.35), (4.36) and (4.49), we get that there is a constant $C_0 > 0$, such that for all $\mathbf{D} \in X$, $\mathbf{E} \in \mathbb{R}^3$

$$\mathbf{T}(\mathbf{D}, \mathbf{E}) \cdot \mathbf{D} \geq C_0(1 + |\mathbf{E}|^2)\big(1 + |\mathbf{D}|^2\big)^{\frac{p(|\mathbf{E}|^2)-2}{2}}|\mathbf{D}|^2 . \tag{4.50}$$

For the mathematical investigations in the next section it is important to know, under which conditions on the coefficients α_{ij} the operator induced by $-\operatorname{div}\mathbf{T}$ is monotone. One easily checks that the conditions (4.32), (4.35), (4.36) and (4.49) with sharp inequalities are not sufficient for the monotonicity. It is well known (cf. Gajewski, Gröger, Zacharias [35], p. 64) that for differentiable operators the monotonicity is equivalent to the condition

$$\frac{\partial T_{ij}(\mathbf{D}, \mathbf{E})}{\partial D_{kl}} B_{ij}B_{kl} \geq 0, \qquad \forall\, \mathbf{B}, \mathbf{D} \in X, \ \forall\, \mathbf{E} \in \mathbb{R}^3 . \tag{4.51}$$

An easy calculation shows, for $\mathbf{T}$ given by (4.47), that

$$\begin{aligned} &\frac{\partial T_{ij}(\mathbf{D}, \mathbf{E})}{\partial D_{kl}} B_{ij}B_{kl} \\ &= \big(1 + |\mathbf{D}|^2\big)^{\frac{p-2}{2}} \Big\{\alpha_{21}(p-1)\frac{(\mathbf{B} \cdot \mathbf{D})(\mathbf{BE} \cdot \mathbf{E})}{(1 + |\mathbf{D}|^2)^{1/2}} \\ &\quad + (\alpha_{31} + \alpha_{33}|\mathbf{E}|^2)\Big(|\mathbf{B}|^2 + (p-2)\frac{(\mathbf{B} \cdot \mathbf{D})^2}{1 + |\mathbf{D}|^2}\Big) \\ &\quad + 2\,\alpha_{51}\Big(|\mathbf{BE}|^2 + (p-2)\frac{(\mathbf{B} \cdot \mathbf{D})(\mathbf{BE} \cdot \mathbf{DE})}{1 + |\mathbf{D}|^2}\Big)\Big\} . \end{aligned} \tag{4.52}$$

Note, that the second term in the squiggly brackets is always non-negative, while the first and the third one can change their signs. Replacing $\mathbf{D}$ by $-\mathbf{D}$ implies that the term in (4.52) with α_{21} should be replaced by

$$-|\alpha_{21}|(p-1)\frac{|\mathbf{B} \cdot \mathbf{D}|\,|\mathbf{BE} \cdot \mathbf{E}|}{(1 + |\mathbf{D}|^2)^{1/2}} . \tag{4.53}$$

The terms with α_{51} are much more involved, since α_{51} has no sign and moreover, it is complicated to find $\mathbf{B}, \mathbf{D}, \mathbf{E}$, such that the infimum of

$$|\mathbf{BE}|^2 + (p-2)\frac{(\mathbf{B} \cdot \mathbf{D})(\mathbf{BE} \cdot \mathbf{DE})}{1 + |\mathbf{D}|^2} \tag{4.54}$$

is attained. The situation becomes even more difficult if (4.53) is taken into account. Even for diagonal matrixes $\mathbf{B}, \mathbf{D}$ and $p = $ const. it is not possible to get handy necessary conditions on α_{21} and α_{51} for (4.51) to hold. For that, let us find sufficient conditions for (4.51). The term with α_{31} is handled easily. For all $\mathbf{B}, \mathbf{D} \in X$, $\mathbf{E} \in \mathbb{R}^3$

$$
\alpha_{31}(1 + |\mathbf{D}|^2)^{\frac{p-2}{2}} \left(|\mathbf{B}|^2 + (p-2)\frac{(\mathbf{B} \cdot \mathbf{D})^2}{1 + |\mathbf{D}|^2} \right)
$$

$$
\geq \alpha_{31}\gamma(p_\infty)(1 + |\mathbf{D}|^2)^{\frac{p-2}{2}}|\mathbf{B}|^2 \equiv C_1(1 + |\mathbf{D}|^2)^{\frac{p-2}{2}}|\mathbf{B}|^2
\tag{4.55}
$$

is valid, where C_1 is independent of $\mathbf{B}, \mathbf{D}$ and $\mathbf{E}$ and $\gamma(p) = p-1$, if $p \in (1,2)$, whereas $\gamma(p) = 1$ if $p \geq 2$. For the remaining terms we proceed as follows. The terms with coefficients α_{21}, α_{33} and α_{51} in the squiggly brackets in (4.52) are bounded from below by (cf. (3.9))

$$
\alpha_{33}\gamma(p)|\mathbf{E}|^2|\mathbf{B}|^2 + 2\,\alpha_{51}|\mathbf{BE}|^2
$$

$$
- \left(|\alpha_{21}|(p-1) + 2\sqrt{\tfrac{2}{3}}\,|\alpha_{51}(p-2)| \right)|\mathbf{BE}|\,|\mathbf{B}||\mathbf{E}| .
\tag{4.56}
$$

Introducing the following new variables

$$
x = |\mathbf{B}||\mathbf{E}|, \qquad y = |\mathbf{BE}|,
$$

we can re-write (4.56) as

$$
f(x,y) = \alpha_{33}\gamma(p)\,x^2 + 2\,\alpha_{51}y^2 - \left(|\alpha_{21}|(p-1) + 2\sqrt{\tfrac{2}{3}}\,|\alpha_{51}(p-2)| \right)xy ,
\tag{4.57}
$$

where $p = p(|\mathbf{E}|^2)$. We would like to find conditions on p and α_{21} ensuring that there exists a constant C_2 independent of x, y such that

$$
f(x,y) \geq C_2\,x^2
\tag{4.58}
$$

for all $x \in [0, \infty)$ and $y \in [0, \sqrt{\tfrac{2}{3}}x]$. Using a similar method as in the discussion of the Clausius-Duhem inequality we obtain that

$$
|\alpha_{21}|(p-1) + 2\sqrt{\tfrac{2}{3}}\,|\alpha_{51}(p-2)| <
\begin{cases}
2\sqrt{\gamma(p)\alpha_{33}}\,\sqrt{2\alpha_{51}} & \text{if } \gamma(p) \leq \frac{4\,\alpha_{51}}{3\,\alpha_{33}}, \\[2ex]
\sqrt{\tfrac{3}{2}}\left(\gamma(p)\alpha_{33} + \tfrac{4}{3}\alpha_{51}\right) & \text{if } \left|\frac{4\,\alpha_{51}}{3\,\alpha_{33}}\right| \leq \gamma(p),
\end{cases}
$$

must be satisfied for all $p \in [p_\infty, p_0]$. The above condition looks very compact, but both sides of these inequalities and the conditions depend on $p \in [p_\infty, p_0]$ and it must be ensured that the right-hand side is larger than the left-hand side. Let us fix α_{33} and α_{51} and regard the above conditions as an inequality of two functions depending on p. One easily sees that the left-hand side is a linear function in the intervals $[p_\infty, 2]$ and $[2, p_0]$, which is increasing in the latter one. The right-hand side is a concave function in the interval $[p_\infty, 2]$ and constant in the interval $[2, p_0]$. Therefore it is enough to check the above conditions for p_∞ and p_0 to ensure that they hold on the whole interval in-between. Let us denote

$$r = \frac{4\,\alpha_{51}}{3\,\alpha_{33}},$$

and

$$\gamma_1(p) = \frac{1}{p-1}\left(\gamma(p) + r\left(1 + |p-2|\right)\right),$$
$$\gamma_2(p) = \frac{1}{p-1}\left(2\sqrt{r\,\gamma(p)} - r|p-2|\right),$$
$$\gamma_3(p) = \frac{1}{p-1}\left(\gamma(p) + r\left(1 - |p-2|\right)\right).$$

Straightforward calculations lead to the following requirements on α_{21}, p_∞ and p_0 to ensure that (4.58) holds:

(1) if $r \in (-1, 0]$

$$|\alpha_{21}| < \sqrt{\tfrac{3}{2}}\,\alpha_{33}\,\min\{\gamma_1(p_\infty), \gamma_1(p_0)\},$$
$$\frac{1-3r}{1-r} < p_\infty \leq p_0 < 1 - \frac{1}{r},$$
(4.59)

(2) if $r \in (0,1)$ and $r < \gamma(p_\infty)$

$$|\alpha_{21}| < \sqrt{\tfrac{3}{2}}\,\alpha_{33}\,\gamma_3(p_0),$$
$$r + 1 < p_\infty \leq p_0 < 3 + \frac{1}{r},$$
(4.60)

(3) if $r \in (0,1)$ and $r \in (\gamma(p_0), 1)$

$$|\alpha_{21}| < \sqrt{\tfrac{3}{2}}\,\alpha_{33}\,\gamma_2(p_\infty),$$
$$2\frac{r+1}{r}\left(1 - \frac{1}{\sqrt{r+1}}\right) < p_\infty \leq p_0 < 1 + r,$$
(4.61)

(4) if $r \in (0,1)$ and $r \in [\gamma(p_\infty), \gamma(p_0)]$

$$|\alpha_{21}| < \sqrt{\tfrac{3}{2}}\,\alpha_{33}\,\min\{\gamma_2(p_\infty), \gamma_3(p_0)\},$$
$$2\frac{r+1}{r}\left(1 - \frac{1}{\sqrt{r+1}}\right) < p_\infty \leq p_0 < 2 + \frac{1}{r},$$
(4.62)

(5) if $r \in [1, \infty)$

$$|\alpha_{21}| < \sqrt{\tfrac{3}{2}}\,\alpha_{33}\,\min\{\gamma_2(p_\infty), \gamma_2(p_0)\},$$
$$2\frac{r+1}{r}\left(1 - \frac{1}{\sqrt{r+1}}\right) < p_\infty \leq p_0 < 2\left(1 + \frac{1}{\sqrt{r}}\right).$$
(4.63)

In the picture below we have depicted the restrictions on the largest possible range of p_∞ and p_0 in dependence on r.

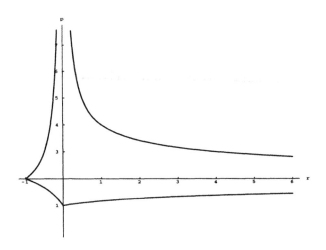

Figure 5

Therefore we have shown

Lemma 4.64. *Let* **T** *be given by* (4.47). *Then the operator induced by* $-\operatorname{div}$ **T** *satisfies for all* **B**, **D** $\in X$ *the inequality*

$$\frac{\partial T_{ij}(\mathbf{D},\mathbf{E})}{\partial D_{kl}} B_{ij} B_{kl} \geq \left(C_1 + C_2 |\mathbf{E}|^2\right)\left(1 + |\mathbf{D}|^2\right)^{\frac{p(|\mathbf{E}|^2)-2}{2}} |\mathbf{B}|^2 \qquad (4.65)$$

if (4.32), (4.35), (4.36) *and* (4.49) *hold with sharp inequalities and if* (4.59)–(4.63) *are satisfied.*

Remark 4.66. One could obtain slightly less restrictive sufficient conditions for the inequality (4.65) if one works instead of (4.56) with the better lower bound

$$\alpha_{33}|\mathbf{E}|^2\left(|\mathbf{B}|^2 + (p-2)\frac{(\mathbf{B}\cdot\mathbf{D})^2}{1+|\mathbf{D}|^2}\right) + 2\alpha_{51}|\mathbf{BE}|^2$$

$$- \left(|\alpha_{21}|(p-1) + 2\sqrt{\tfrac{2}{3}}\,|\alpha_{51}(p-2)|\right)\frac{|\mathbf{B}\cdot\mathbf{D}|}{(1+|\mathbf{D}|^2)^{1/2}}|\mathbf{BE}|\,|\mathbf{E}|\,.$$

The resulting conditions are even less handy than (4.59)–(4.63) and moreover they are not sufficient in the case $\alpha_{51} < 0$ to ensure that also the approximative stress tensor $\mathbf{S}^A$ (cf. (3.3.13)) is monotone.

The last constitutive relation we should discuss is that for the polarization $\mathbf{P}$. For many materials it is reasonable to assume a linear relation between $\mathbf{P}$ and $\mathbf{E}$ and therefore we shall assume that

$$\mathbf{P} = \chi^E(\rho, \theta)\, \mathbf{E}, \tag{4.67}$$

where χ^E is the dielectric susceptibility. In the next chapter we are only concerned with processes for incompressible shear dependent electrorheological fluids. In this situation χ^E depends on θ only.

2 Mathematical Framework

2.1 Setting of the Problem and Introduction

In the following chapters we are mainly concerned with the existence theory for steady and unsteady flows of an incompressible electrorheological fluid with shear dependent viscosities. It follows from the treatment in Chapter 1 that the motion is governed by the system (1.2.32)–(1.2.36), (1.2.44), (1.3.31) and (1.2.45), where the stress tensor $\mathbf{T}$ is given by either (1.3.35) or (1.4.47) and the polarization $\mathbf{P}$ is fixed by (1.4.67). Here we shall consider the isothermal case only, i.e. the temperature θ is constant. This assumption in fact means that we have at our disposal a heat source r, which holds the temperature constant (see below (1.5)). As we are dealing with an incompressible fluid this assumption also implies that all material functions α_{ij}, χ^E are constant. Therefore we have to deal with the system[1]

$$\operatorname{div} \mathbf{E} = 0,$$
$$\operatorname{curl} \mathbf{E} = \mathbf{0}, \tag{1.1}$$

$$\rho_0 \frac{\partial \mathbf{v}}{\partial t} - \operatorname{div} \mathbf{S} + \rho_0 [\nabla \mathbf{v}]\mathbf{v} + \nabla \phi = \rho_0 \mathbf{f} + \chi^E [\nabla \mathbf{E}]\mathbf{E},$$
$$\operatorname{div} \mathbf{v} = 0, \tag{1.2}$$

$$\operatorname{div} \mathbf{B} = 0,$$
$$\operatorname{curl} \mathbf{B} + \frac{\chi^E}{c} \operatorname{curl} (\mathbf{v} \times \mathbf{E}) = \frac{1 + \chi^E}{c} \frac{\partial \mathbf{E}}{\partial t}, \tag{1.3}$$

$$\frac{\partial q_e}{\partial t} + \operatorname{div} (q_e \mathbf{v}) = 0, \tag{1.4}$$

$$\mathbf{S} \cdot \mathbf{D} + \rho r = 0, \tag{1.5}$$

where $\mathbf{S} = \mathbf{T} + \phi \mathbf{I}$. The system (1.1)–(1.5) is separated, so we can first solve the quasi-static Maxwell's equations (1.1) for the electric field and then seek the velocity field by solving (1.2). Having at our disposal $\mathbf{E}$ and $\mathbf{v}$ we can solve (1.3), (1.4) and (1.5). Note that equation (1.5) has to be interpreted as an equation for the heat source r. It was already pointed out in Section 1.2 that the magnetic field $\mathbf{B}$ is of secondary

---

[1]Recall, that $[\nabla \mathbf{v}]\mathbf{v} = v_j \partial_j \mathbf{v}$.

importance, which is reflected by the structure of the above system. Moreover, for this first mathematical investigation of electrorheological fluids we are mainly interested in the electric field $\mathbf{E}$ and the velocity field $\mathbf{v}$, and not in the free charges q_e and the heat source r. Therefore we shall only consider (1.1) and (1.2), to which appropriate initial and boundary conditions should be added. In particular, we will consider the quasi-static Maxwell's equations (1.1) with the boundary condition

$$\mathbf{E} \cdot \mathbf{n} = \mathbf{E}_0 \cdot \mathbf{n} \qquad \text{on } \partial\Omega, \qquad (1.6)$$

where $\mathbf{E}_0$ is a given electric field and $\mathbf{n}$ is the outward normal; this boundary condition describes the contact between a dielectric and a conductor. The equation of motion (1.2) will be completed with Dirichlet boundary conditions

$$\mathbf{v} = \mathbf{0} \qquad \text{on } \partial\Omega \times (0, T), \qquad (1.7)$$

and an initial condition

$$\mathbf{v}(0) = \mathbf{v}_0 \qquad \text{in } \Omega, \qquad (1.8)$$

if the unsteady system is treated. Here and in the following, if not otherwise stated, Ω denotes a three-dimensional bounded smooth domain[2] and $T > 0$ a given length of the time interval, which will be denoted by $I = (0, T)$.

We would like to make some comments about the structure of the system (1.1), (1.2). The quasi-static Maxwell's equations (1.1) are widely studied in the literature (cf. the overview article Milani, Picard [83]) and well understood. In Section 2.3 we will give precise formulations and quotations of the for us relevant existence, uniqueness and regularity results. The situation is quite different for the system (1.2), when $\mathbf{S} = \mathbf{T} + \phi\mathbf{I}$ is given by (1.4.47). The properties of the stress tensor have already been discussed in some detail in Section 1.4. There we have given necessary and sufficient conditions for the coercivity and sufficient conditions for the monotonicity of the operator induced by $-\operatorname{div} \mathbf{S}$. Since the material function p depends in general on the magnitude of the electric field $|\mathbf{E}|^2$ we have to deal with an elliptic or parabolic system of partial differential equations with non-standard growth conditions, i.e. (cf. (1.4.47), (1.4.50), (1.4.10)–(1.4.12))

$$\mathbf{S}(\mathbf{D}, \mathbf{E}) \cdot \mathbf{D} \geq C_0(1 + |\mathbf{E}|^2)(1 + |\mathbf{D}|^2)^{\frac{p_\infty - 2}{2}}|\mathbf{D}|^2, \qquad (1.9)$$

$$|\mathbf{S}(\mathbf{D}, \mathbf{E})| \leq C_3(1 + |\mathbf{D}|^2)^{\frac{p_0 - 1}{2}}|\mathbf{E}|^2, \qquad (1.10)$$

where in general $1 < p_\infty \leq p_0$.[3] The system is further complicated by the internal constraint $(1.2)_2$ and the dependence of the elliptic operator on the modulus of the symmetric part of the velocity gradient.

In recent years the interest in the study of elliptic equations and systems with non-standard growth conditions has increased. Most investigations are concerned with the

---

[2] Here we restrict ourselves to this case in order to make the computations not more complicated, but it is clear that the methods can also be applied in the case of two-dimensional domains with appropriate changes.

[3] Of course in the case $p_\infty = p_0$ we have standard growth conditions.

regularity of minimizers of variational integrals and solutions of elliptic and parabolic equations and systems. Let us give a brief overview of some of the results.

The first investigations of equations and scalar variational integrals with non-standard growth conditions go back to Zhikov [135], Giaquinta [37] and Marcellini [76]. In Zhikov [135] the variational integral

$$\int_\Omega (1 + |\nabla u|^2)^{\alpha(x)} \, dx \tag{1.11}$$

is studied, and it is shown that non-regular minimizers do not exist if α is Hölder continuous. In Marcellini [76] and Giaquinta [37] anisotropic variational integrals of the type

$$\int_\Omega f(\nabla u) \, dx \tag{1.12}$$

have been considered, where the convex function f satisfies for all $\boldsymbol{\xi} \in \mathbb{R}^d$

$$m|\boldsymbol{\xi}|^p \le f(\boldsymbol{\xi}) \le M(1 + |\boldsymbol{\xi}|)^q, \tag{1.13}$$

for $M \ge m > 0$ and $1 < p < q$. It is shown that there exist unbounded minimizers if the difference $q - p$ is large enough in dependence on the dimension d of the domain Ω; e.g. Giaquinta [37] considers the case $p = 2, q = 4$ and $d \ge 6$ (see also Marcellini [77], [78], Hong [84]). On the other hand there is a variety of papers dealing with conditions on p and q, which ensure regularity of minimizers. For example, in Marcellini [77], [78] and Hong [84] it is shown that minimizers of (1.12), (1.13) belonging to the space $W_{\mathrm{loc}}^{1,q}(\Omega)$ are locally Lipschitz continuous if $2 \le p$ and

$$p \le q < p\frac{d}{d-2}, \tag{1.14}$$

and Moscariello, Nania [89] prove that minimizers of (1.12), (1.13) belonging to the space $W_{\mathrm{loc}}^{1,1}(\Omega)$ are bounded if

$$1 < p \le q < p\frac{d}{d-p}. \tag{1.15}$$

Moreover they obtain that locally bounded minimizers of (1.12), (1.13) with $f(\boldsymbol{\xi}) = f(|\boldsymbol{\xi}|)$ are Hölder continuous without restrictions on p and q. In Boccardo, Marcellini, Sbordone [13] and Fusco, Sbordone [32] it is shown that minimizers of (1.12) with a special anisotropic structure, i.e. f satisfies

$$\sum_{i=1}^d |\xi_i|^{q_i} \le f(\boldsymbol{\xi}) \le c\Big(1 + \sum_{i=1}^d |\xi_i|^{q_i}\Big), \tag{1.15}$$

are locally bounded if

$$1 \le q_i \le \bar{q}\frac{d}{d-\bar{q}}, \qquad \frac{1}{\bar{q}} = \frac{1}{d} \sum_{i=1}^d \frac{1}{q_i}. \tag{1.16}$$

These results have been extended to the case

$$\int_\Omega f(x, u, \nabla u)\, dx\,,$$

e.g. by Fusco, Sbordone [33], Mascolo, Papi [81], Chiadò Piat, Coscia [99] and others.
    Similar results also hold for weak solutions $u \in W^{1,q}_{\mathrm{loc}}(\Omega) \cap L^\infty_{\mathrm{loc}}(\Omega)$ of elliptic equations

$$\mathrm{div}\ \mathbf{a}(x, u, \nabla u) = b(x, u, \nabla u)\,, \tag{1.17}$$

where for almost all $x \in \Omega$, and for all $s \in \mathbb{R}$, $\boldsymbol{\xi}, \boldsymbol{\lambda} \in \mathbb{R}^n$

$$\frac{\partial a_i(x, s, \boldsymbol{\xi})}{\partial \xi_j} \lambda_i \lambda_j \geq m\big(1 + |\boldsymbol{\xi}|^2\big)^{\frac{p-2}{2}} |\boldsymbol{\lambda}|^2\,, $$
$$\left|\frac{\partial \mathbf{a}}{\partial \boldsymbol{\xi}}(x, s, \boldsymbol{\xi})\right| \leq M\big(1 + |\boldsymbol{\xi}|^2\big)^{\frac{q-2}{2}}\,, \tag{1.18}$$

is satisfied for positive constants m, M and exponents $1 < p \leq q$. It is shown in
Marcellini [78], [79] that weak solutions belong to the space $W^{1,\infty}_{\mathrm{loc}}(\Omega) \cap W^{2,2}_{\mathrm{loc}}(\Omega)$
if (1.14) and some structure conditions on $\mathbf{a}(x, u, \nabla u)$ and $b(x, u, \nabla u)$ are satisfied
(cf. Lieberman [67], where gradient estimates for smooth solutions are proved). More-
over, existence of weak solutions to (1.17), (1.18) in the case $b(x) \in L^{p'}(\Omega) \cap L^\infty_{\mathrm{loc}}(\Omega)$
and $2 \leq p \leq q < \frac{d+2}{d}p$ is shown in Marcellini [78], while the case $b(x) \in L^1(\Omega)$ is
treated in Boccardo, Gallouet, Marcellini [12], which to the knowledge of the author
are the only papers dealing with the problem of existence of weak solutions to (1.17).
    Finally, Lieberman [67] obtains gradient estimates under some structure conditions
also for the parabolic version of (1.17) (cf. Mkrtychyan [86], [87]).

    In the vectorial case only minimizer of variational integrals

$$\int_\Omega \mathbf{F}(x, \nabla \mathbf{u})\, dx \tag{1.19}$$

and the corresponding Euler equation

$$\mathrm{div}\ \mathbf{A}(x, \nabla \mathbf{u}) = \mathbf{0} \tag{1.20}$$

are investigated. For example, in Leonetti [62] and Bhattacharya, Leonetti [10] it
is shown via a new Poincaré inequality that $\mathbf{F}(|\nabla \mathbf{u}|)$ belongs to the Morrey space
$L^{1,\lambda}_{\mathrm{loc}}(\Omega)$. Leonetti [63] proves that weak solutions $\mathbf{u} \in W^{1,q}(\Omega)$ of (1.20), where
$\mathbf{A}(x, \nabla \mathbf{u})$ satisfies some structure conditions and

$$\frac{\partial A_{ij}}{\partial \xi_{kl}}(x, \boldsymbol{\xi}) \lambda_{ij} \lambda_{kl} \geq m(1 + |\boldsymbol{\xi}|^2)^{\frac{p-2}{2}} |\boldsymbol{\lambda}|^2\,, $$
$$\left|\frac{\partial \mathbf{A}}{\partial \boldsymbol{\xi}}(x, \boldsymbol{\xi})\right| \leq M(1 + |\boldsymbol{\xi}|^2)^{\frac{q-2}{2}}\,, \tag{1.21}$$

have higher differentiability properties, namely

$$(1 + |\nabla \mathbf{u}|^2)^{p/4} \in W_{\text{loc}}^{1,2}(\Omega),\tag{1.22}$$

and for $p \geq 2$ one gets $\mathbf{u} \in W_{\text{loc}}^{2,2}(\Omega)$. Note, that no restriction on the ratio q/p is imposed, but that also no higher integrability is proved. Choe [18] proves that locally bounded minimizers of (1.19) with $\mathbf{F}(|\nabla \mathbf{u}|)$, where $\mathbf{F}_{\xi_{ij}\xi_{kl}}(|\boldsymbol{\xi}|)$ satisfies an appropriate version of (1.21), belong to the space $C_{\text{loc}}^{1,\alpha}(\Omega)$ if

$$1 < p \leq q < p+1,\tag{1.23}$$

and that minimizers $\mathbf{u} \in W^{1,p}(\Omega)$ are locally bounded if

$$1 < p \leq q < p\left(1 + \frac{1}{d}\right).$$

Coscia, Mingione [20] have proved $C_{\text{loc}}^{1,\alpha}(\Omega)$-regularity for local minimizers of (1.19) provided that $\mathbf{F}(x, \nabla \mathbf{u}) = |\nabla \mathbf{u}|^{p(x)}$ under the only assumption that p is locally Hölder continuous. Acerbi, Fusco [4] show for anisotropic functionals satisfying an appropriate version of (1.15) partial regularity of minimizers, namely $\mathbf{u} \in C^{0,\alpha}(\Omega_0)$, $|\Omega - \Omega_0| = 0$, under the condition (1.16). Bhattacharya, Leonetti [11] treat a very special anisotropic structure and in the case $p = 2, q < \frac{2d}{d-2}$, they obtain that

$$\mathbf{u} \in W_{\text{loc}}^{1,\infty}(\Omega) \cap W_{\text{loc}}^{2,2}(\Omega).$$

See also Leonetti [64], where the existence of second order derivatives for some anisotropic functionals is established. The above results are generalized in Leonetti [65].

   To the knowledge of the author there are no results for elliptic systems with a right-hand side depending on $\mathbf{u}$, $\nabla \mathbf{u}$ or for parabolic systems with non-standard growth conditions. Our treatment of the system (1.2) is built on techniques developed in previous investigations of generalized Newtonian fluids, which had been initiated by Nečas [93], and were further developed and extended in Málek, Nečas, Růžička [71], [72], Bellout, Bloom, Nečas [9], Málek, Nečas, Rokyta, Růžička [70] (cf. Málek, Rajagopal, Růžička [74], Málek, Růžička, Thäter [75]).

2.2   Function Spaces

Now we introduce some notation and useful function spaces for the treatment of the system (1.1), (1.2), (1.6)–(1.8). Let $\Omega \subset \mathbb{R}^d, d \geq 2$, be a bounded smooth domain and $T > 0$ be the given length of the time interval $I = (0, T)$. Sometimes we write Q_T instead of $I \times \Omega$.

   Let $1 \leq p, q \leq \infty$ and $k \in \mathbb{N}$, then $(L^q(\Omega), \|\cdot\|_q)$ denotes the usual Lebesgue space and $(W^{k,q}(\Omega), \|\cdot\|_{k,q})$ is used for the standard Sobolev space. $L_0^q(\Omega)$ is the subspace of $L^q(\Omega)$ consisting of functions having mean value zero. We denote by $\mathcal{D}(\Omega)$ the space of smooth functions with compact support in Ω and define $W_0^{k,q}(\Omega)$ as the completion of $\mathcal{D}(\Omega)$ in the norm $\|\cdot\|_{k,q}$. The Bochner space $L^q(I; X(\Omega))$, where $X(\Omega)$ is some function space over Ω, will be equipped with the norm $\left(\int_0^T \|\cdot\|_{X(\Omega)}^q \, dt\right)^{1/q}$.

We shall also work with weighted Lebesgue and Sobolev spaces. Let $0 < \rho \in L^1(\Omega)$ be the density of a Radon measure ν with respect to the Lebesgue measure, i.e. for all Lebesgue measurable sets E the measure $\nu(E)$ is defined by

$$\nu(E) = \int_E \rho \, dx \,.$$

Then the space $L^q(\Omega, \nu)$, $1 < q < \infty$, is a reflexive, separable Banach space with the norm

$$\|f\|_{q,\nu} = \left(\int_\Omega |f(x)|^q \, \rho(x) \, dx \right)^{1/q} .$$

The weighted Sobolev space $(W^{k,q}(\Omega, \nu), \| \cdot \|_{k,q,\nu})$ is defined in an obvious analogous manner. The space $W_0^{k,q}(\Omega, \nu)$ is the completion of $\mathcal{D}(\Omega)$ in the norm $\| \cdot \|_{k,q,\nu}$. We set

$$
\begin{aligned}
\mathcal{V} &\equiv \{ \mathbf{u} \in \mathcal{D}(\Omega), \operatorname{div} \mathbf{u} = 0 \} \,, \\
H &\equiv \text{closure of } \mathcal{V} \text{ in } \| \cdot \|_2\text{-norm} \,, \\
V_q &\equiv \text{closure of } \mathcal{V} \text{ in } \| \nabla \cdot \|_q\text{-norm} \,, \\
V^s &\equiv \text{closure of } \mathcal{V} \text{ in } \| \cdot \|_{s,2}\text{-norm} \,, \\
H(\operatorname{div}) &\equiv \{ \mathbf{u} \in L^2(\Omega); \operatorname{div} \mathbf{u} \in L^2(\Omega) \} \,, \\
H(\operatorname{curl}) &\equiv \{ \mathbf{u} \in L^2(\Omega); \operatorname{curl} \mathbf{u} \in L^2(\Omega) \} \,.
\end{aligned}
\tag{2.1}
$$

The properties of the above spaces are well known (cf. Galdi [36], Girault, Raviart [38]). We just mention that V_q, $1 < q < \infty$, is a separable, reflexive Banach space and that H, V^s and $H(\operatorname{curl}), H(\operatorname{div})$ are Hilbert spaces.

We shall also work with the spaces $L^{p(x)}(\Omega)$ and $W^{k,p(x)}(\Omega)$, respectively, which are called, respectively, generalized Lebesgue and generalized Sobolev spaces. They have been studied by Hudzik [44], Musielak [91] and Ková čik, Rákosník [55]. Since they are not so well known and since we shall also need a weighted version of these spaces we shall discuss their properties in some detail.[4] For given $p(x) \in L^\infty(\Omega), 1 < p_\infty \le p(x) \le p_0 < \infty$, we fix

$$|f|_{p(x)} = \int_\Omega |f(x)|^{p(x)} \, dx \,. \tag{2.2}$$

It is easy to see that $| \cdot |_{p(x)}$ defines a convex modular (cf. Musielak [91]), which preserves ordering, i.e. $f \le g$ implies $|f|_{p(x)} \le |g|_{p(x)}$. Similarly to the Luxemburg norm in Orlicz spaces we can define

$$\|f\|_{p(x)} = \inf\{ \lambda > 0; \; \left| \frac{f}{\lambda} \right|_{p(x)} \le 1 \} \,, \tag{2.3}$$

---

[4]We refer the reader to Ková čik, Rákosník [55] for more details and proofs.

which is a norm on the space

$$L^{p(x)}(\Omega) \equiv \{f \in L^1(\Omega); \ |\lambda f|_{p(x)} < \infty \quad \text{for some } \lambda > 0\}. \qquad (2.4)$$

We define the dual function $p'(x)$ pointwise as the dual exponent to $p(x)$, i.e. $p'(x) = \frac{p(x)}{p(x)-1}$ and the number $r_p = r_{p(x)} = \frac{1}{p_\infty} + \frac{1}{p_0}$. Then we deduce by a pointwise application of Young's inequality that the generalized Hölder inequality

$$\int_\Omega |f(x)g(x)| \, dx \leq r_p \|f\|_{p(x)} \|g\|_{p'(x)} \qquad (2.5)$$

holds for all functions $f \in L^{p(x)}(\Omega)$, $g \in L^{p'(x)}(\Omega)$. We can also define an equivalent norm on $L^{p(x)}(\Omega)$ by

$$\|f\|_{p(x)} = \sup_{|g|_{p'(x)} \leq 1} \left| \int_\Omega f(x)g(x) \, dx \right|, \qquad (2.6)$$

which is the analogue of the Orlicz norm in Orlicz spaces. From (2.6) and the pointwise application of Young's inequality we obtain that

$$\|f\|_{p(x)} \leq \|f\|_{p(x)} \leq \frac{1}{(p_0)'} + \frac{1}{p_\infty} |f|_{p(x)}. \qquad (2.7)$$

Moreover, norm convergence in (2.3) or (2.6) is equivalent to the modular convergence in (2.2). It can be shown that the space $(L^{p(x)}(\Omega), \|\cdot\|_{p(x)})$ is complete and that the dual space can be characterized as $L^{p'(x)}(\Omega)$. In fact, for all elements $G \in (L^{p(x)}(\Omega))^*$ there exists a function $g \in L^{p'(x)}(\Omega)$ such that for all $f \in L^{p(x)}(\Omega)$

$$G(f) = \int_\Omega f(x)\, g(x) \, dx. \qquad (2.8)$$

The proof[5] follows the same lines as the one for the usual Lebesgue space $L^p(\Omega)$. Namely, defining for all Lebesgue-measurable sets F

$$\mu(F) = G(\chi_F),$$

where χ_F is the characteristic function of the set F, we obtain that μ is a σ-additive, absolutely continuous with respect to the Lebesgue measure set function. Therefore the Radon-Nikodým theorem yields the existence of a function $g \in L^1(\Omega)$ such that (2.8) holds for all simple functions. Since all functions f in $L^{p(x)}(\Omega)$ can be approximated by simple functions s_n

$$s_n(x) \nearrow f(x) \qquad \text{a.e. } x \text{ in } \Omega,$$

the Lebesgue monotone convergence theorem, the Vitali-Hahn-Saks theorem and the Vitali theorem lead to (2.8) for all $f \in L^{p(x)}(\Omega)$. From (2.5) and the first inequality in (2.7) we get

$$\|g\|_{p'(x)} \leq \|G\|_{(L^{p(x)}(\Omega))^*} \leq r_p \|g\|_{p'(x)}, \qquad (2.9)$$

which concludes the proof.

---

[5]For more details in the case $L^p(\Omega)$ see e.g. Kufner, John, Fučik [57] or Dunford, Schwartz [23].

From the characterization of the dual space $(L^{p(x)}(\Omega))^*$ we immediately obtain that $L^{p(x)}(\Omega)$ is a reflexive Banach space. Moreover, one can show that $\mathcal{D}(\Omega)$ is dense in $L^{p(x)}(\Omega)$ and that $L^{p(x)}(\Omega)$ is separable. Indeed, let $f \in L^{p(x)}(\Omega)$, then we see by truncation that there exists a bounded function g such that $\|f - g\|_{p(x)} \leq \varepsilon$. Luzin's theorem provides the existence of $h \in C(\Omega)$ such that $h = g$ on a large closed set and $\|g - h\|_{p(x)} \leq \varepsilon$. The absolute continuity of the integral then gives the existence of an open set G such that $\|h - h\chi_G\|_{p(x)} \leq c$, and moreover there exists a polynomial m such that also $\|h\chi_G - m\chi_G\|_{p(x)} \leq \varepsilon$. Finally we find a function $\varphi \in C_0^\infty(\Omega)$, $0 \leq \varphi \leq 1$, such that $\|m\chi_G - m\varphi\|_{p(x)} \leq \varepsilon$. This shows that $\mathcal{D}(\Omega)$ is dense in $L^{p(x)}(\Omega)$.

But in contrast to Lebesgue and Orlicz spaces the generalized Lebesgue space $L^{p(x)}(\Omega)$ is not $p(x)$-mean continuous. More precisely, let the ball $B_r(x_0) \subset \Omega$ and let p be continuous and non-constant on $B_r(x_0)$. Then, there exists a function $f \in L^{p(x)}(\Omega)$ and a sequence $h_n \to 0$ such that $f_n(x) \equiv f(x+h_n)$ do not belong to the space $L^{p(x)}(\Omega)$.

The generalized Sobolev space $W^{k,p(x)}(\Omega)$, $k \in \mathbb{N}$, is defined as the set[6]

$$W^{k,p(x)}(\Omega) = \{f \in L^{p(x)}; \; \partial^\alpha f \in L^{p(x)}(\Omega) \quad \forall |\alpha| \leq k\}, \qquad (2.10)$$

which is endowed with the norm

$$\|f\|_{k,p(x)} = \sum_{|\alpha| \leq k} \|\partial^\alpha f\|_{p(x)}. \qquad (2.11)$$

The space $W_0^{k,p(x)}(\Omega)$ is defined as the completion of $\mathcal{D}(\Omega)$ in the norm (2.11). Using standard arguments one can derive from the properties of the space $L^{p(x)}(\Omega)$ that $W^{k,p(x)}(\Omega)$ and $W_0^{k,p(x)}(\Omega)$ are separable, reflexive Banach spaces and that every element G of the dual of $W_0^{k,p(x)}(\Omega)$ is characterized as

$$G(f) = \sum_{|\alpha| \leq k} \int_\Omega \partial^\alpha f(x)\, g_\alpha(x)\, dx, \qquad (2.12)$$

where $g_\alpha \in L^{p'(x)}(\Omega)$. However, note that in general it is not known if smooth functions are dense in $W^{1,p(x)}(\Omega)$. The only result in this direction is due to Edmunds, Rákosník [25], who prove the density of smooth functions for a special class of functions $p(x)$.

Let us now discuss the embedding properties of the generalized Lebesgue and Sobolev spaces. Firstly, we know that

$$L^{p(x)}(\Omega) \hookrightarrow L^{q(x)}(\Omega), \qquad (2.13)$$

if and only if

$$q(x) \leq p(x) \qquad \text{a.e. in } \Omega. \qquad (2.14)$$

Beside this trivial embedding one would like to have an embedding of the Sobolev type, namely

$$W_0^{1,p(x)}(\Omega) \hookrightarrow L^{q(x)}(\Omega), \qquad (2.15)$$

---

[6]Here α is a multi-index and $\partial^\alpha f$ denotes the generalized derivative of f.

where

$$\frac{1}{q(x)} = \frac{1}{p^*(x)} \equiv \frac{1}{p(x)} - \frac{1}{d}. \tag{2.16}$$

Counterexamples show that in general this is not possible. However, if p is continuous on $\bar{\Omega}$ and if $p_0 < d$ one disposes of the following embedding: for all $\varepsilon \in (0, 1/d)$ there exists a constant $c > 0$ such that for all $f \in W_0^{1,p(x)}(\Omega)$

$$\|f\|_{q(x)} \le c\|f\|_{1,p(x)}, \tag{2.17}$$

where

$$1 \le q(x) \le p^*(x) - \varepsilon. \tag{2.18}$$

This means that one "loses" an arbitrarily small power in the desired result (2.15), (2.16). In fact, the embedding $W_0^{1,p(x)} \hookrightarrow L^{q(x)}(\Omega)$, where q satisfies (2.18), is not only continuous but compact. This in turn implies that $\|\nabla f\|_{p(x)}$ is an equivalent norm on the space $W_0^{1,p(x)}(\Omega)$. Note, that one can get embeddings of similar type as (2.17), (2.18) under weaker assumptions on p. On the other hand if we assume more about p, we can improve the integrability in the above result and "lose" only a logarithmic factor. More precisely we have

Proposition 2.19. *Let p be a measurable function such that $p \in [p_\infty, p_0]$, $p_0 < d$ and let the sets*

$$\Omega_q = \{x \in \Omega; \, p(x) \ge q\} \tag{2.20}$$

have a Lipschitz continuous boundary for all $q \in [p_\infty, p_0]$. Moreover, assume that

$$\int_{p_\infty}^{p_0} c_q^{q^*} \, dq \le c_0 < \infty, \tag{2.21}$$

where c_q are the continuity constants of the embedding $W^{1,q}(\Omega_q) \hookrightarrow L^{q^*}(\Omega_q)$, i.e.

$$\|f\|_{L^{q^*}(\Omega_q)} \le c_q \|f\|_{W^{1,q}(\Omega_q)}. \tag{2.22}$$

Then there exists a constant c depending on p_0, p_∞, d and $|\Omega|$ such that

$$\int_\Omega \frac{|f(x)|^{p^*(x)}}{\ln(2 + |f(x)|)} \, dx \le c \left(1 + \left(\int_\Omega |f(x)|^{p(x)} + |\nabla f(x)|^{p(x)} \, dx\right)^{p_0^*/p_0}\right) \tag{2.23}$$

is satisfied for all $f \in W^{1,p(x)}(\Omega)$.

PROOF : From the definition of the set Ω_q it is clear that functions belonging to the space $W^{1,p(x)}(\Omega)$ also belong to the space $W^{1,q}(\Omega_q)$, for all $q \in [p_\infty, p_0]$. Therefore we obtain from (2.22) raised to the power q^* and integrated over (p_∞, p_0)

$$\int\limits_{p_\infty}^{p_0}\int\limits_{\Omega_q} |f|^{q^*}\, dx\, dq \le \int\limits_{p_\infty}^{p_0} c_q^{q^*}\Big(\int\limits_{\Omega_q} |f|^q + |\nabla f|^q\, dx\Big)^{q^*/q}\, dq$$

$$\le \int\limits_{p_\infty}^{p_0} c_q^{q^*}\Big(\int\limits_{\Omega_q} |f|^{p(x)} + |\nabla f|^{p(x)}\, dx\Big)^{q^*/q}\, dq \qquad (2.24)$$

$$\le \int\limits_{p_\infty}^{p_0} c_q^{q^*}\, dq\Big(\int\limits_{\Omega} |f|^{p(x)} + |\nabla f|^{p(x)} + |\Omega|\, dx\Big)^{p_0^*/p_0}$$

$$\le c\Big(1 + \Big(\int\limits_{\Omega} |f|^{p(x)} + |\nabla f|^{p(x)} + |\Omega|\, dx\Big)^{p_0^*/p_0}\Big).$$

Now, the left-hand side of (2.24) can be written as

$$\int\limits_{p_\infty}^{p_0}\int\limits_{\Omega} |f(x)|^{q^*}\, \chi_q(x)\, dx\, dq, \qquad (2.25)$$

where χ_q is the characteristic function of the set Ω_q. From Fubini's theorem we get that (2.25) is equal to

$$\int\limits_{\Omega}\int\limits_{p_\infty}^{p(x)} |f(x)|^{q^*}\, dq\, dx \ge \int\limits_{\{x\in\Omega,|f(x)|\ge 2\}}\int\limits_{p_\infty}^{p(x)} |f(x)|^{q^*}\, dq\, dx. \qquad (2.26)$$

Using that the integrand can be written as

$$\exp\Big(\frac{qd}{d-q}\, \ln|f(x)|\Big) \qquad (2.27)$$

we can compute the inner integral in (2.26) as

$$\int\limits_{p_\infty}^{p(x)} |f(x)|^{\frac{qd}{d-q}}\, dq = -(d-p(x))|f(x)|^{\frac{p(x)d}{d-p(x)}} + d^2\, \frac{\ln|f(x)|}{|f(x)|^d}\, \ln\frac{d^2}{d-p(x)}$$

$$+ d^2\, \frac{\ln|f(x)|}{|f(x)|^d}\, \sum_{k=1}^{\infty} \frac{(\ln|f(x)|\frac{d^2}{d-p(x)})^k}{k\, k!} \qquad (2.28)$$

$$+ (d-p_\infty)|f(x)|^{\frac{p_\infty d}{d-p_\infty}} - d^2\, \frac{\ln|f(x)|}{|f(x)|^d}\, \ln\frac{d^2}{d-p_\infty}$$

$$- d^2\, \frac{\ln|f(x)|}{|f(x)|^d}\, \sum_{k=1}^{\infty} \frac{(\ln|f(x)|\frac{d^2}{d-p_\infty})^k}{k\, k!}.$$

Using (2.27) and the power series for $\exp y$ in the first term on the right-hand side; the first and the third term in (2.28) can be written as

$$
(d - p(x)) \exp(-d \ln |f(x)|) \left\{ \sum_{k=1}^{\infty} \frac{k+1}{k} \frac{\left(\ln |f(x)| \frac{d^2}{d-p(x)} \right)^{k+1}}{(k+1)!} \right.
$$
$$
\left. - \sum_{k=0}^{\infty} \frac{\left(\ln |f(x)| \frac{d^2}{d-p(x)} \right)^{k}}{k!} \right\}.
$$

(2.29)

Furthermore the difference in the squiggly bracket is equal to

$$
-1 - \ln |f(x)| \frac{d^2}{d - p(x)} + \sum_{k=2}^{\infty} \frac{(\ln |f(x)| \frac{d^2}{d-p(x)})^k}{k!(k-1)}
$$
$$
\geq -2 - \frac{3}{2} \ln |f(x)| \frac{d^2}{d - p(x)} - \frac{d - p(x)}{d^2 \ln |f(x)|}
$$
$$
+ \frac{d - p(x)}{d^2 \ln |f(x)|} \exp \left(\frac{d^2}{d-p(x)} \ln |f(x)| \right)
$$

and therefore the first and third term in (2.28) are bounded from below by

$$
\left(\frac{d - p(x)}{d} \right)^2 \frac{|f(x)|^{p^*(x)}}{\ln |f(x)|} - \frac{2(d - p(x))}{|f(x)|^d}
$$
$$
- \frac{3}{2} d^2 \frac{\ln |f(x)|}{|f(x)|^d} - \left(\frac{d - p(x)}{d} \right)^2 \frac{1}{\ln |f(x)| |f(x)|^d}.
$$

(2.30)

Similarly we can treat the corresponding terms with p_∞ instead of $p(x)$ in (2.28) and we get that they are bounded from below by

$$
-3 \left(\frac{d - p_\infty}{d} \right)^2 \frac{|f(x)|^{p_\infty^*}}{\ln |f(x)|} + \frac{4(d - p_\infty)}{|f(x)|^d}
$$
$$
+ \frac{5}{2} d^2 \frac{\ln |f(x)|}{|f(x)|^d} + \left(\frac{d - p_\infty}{d} \right)^2 \frac{3}{\ln |f(x)| |f(x)|^d}.
$$

(2.31)

One easily sees that the modulus of the last three terms in (2.30) and (2.31) is bounded by some constant depending on p_0, p_∞ and d if one uses that $|f(x)| \geq 2$. The modulus of the first term in (2.31) is bounded by

$$
3 \left(\frac{d - p_\infty}{d} \right)^2 \frac{|f(x)|^{p_\infty^*}}{\ln 2},
$$

(2.32)

which after integration with respect to x can be dominated by (2.24) with $q = p_\infty$. The first term in (2.30), which remains on the left-hand side, is bounded from below by

$$
\left(\frac{d - p_0}{d} \right)^2 \frac{|f(x)|^{\frac{dp(x)}{d-p(x)}}}{\ln(1 + |f(x)|)}.
$$

(2.33)

Putting all computations together we arrive at

$$\int_\Omega \frac{|f(x)|^{p^*(x)}}{\ln(2+|f(x)|)} \, dx \leq c(c_0, p_0, p_\infty, d, |\Omega|)\big(1 + |f|_{p(x)} + |\nabla f|_{p(x)}\big)^{p_0^*/(p_0)},$$

where we have added on both sides of the inequality the integrand integrated over the set $\{x \in \Omega, |f(x)| \leq 2\}$. The last inequality immediately yields the assertion.   ∎

Remark 2.34. 1) In the same way as we have introduced generalized Lebesgue spaces $L^{p(x)}(\Omega)$ one can introduce generalized Orlicz spaces $L_{M(x)}(\Omega)$ (cf. Hudzik [48] and the literature quoted there). In this sense the integral on the left-hand side of (2.23) defines a modular on the generalized Orlicz space $L_{M(x)}(\Omega)$, where $M(x,t) = t^{p^*(x)} \ln^{-1}(2+t)$. Therefore inequality (2.23) states that the embedding of the space $W^{1,p(x)}(\Omega)$ into the space $L_{M(x)}(\Omega)$ holds true.

2) Note, that condition (2.21) is fulfilled if the measure of the set Ω_{p_0} is positive.

3) Motivated by our studies of electrorheological fluids [113] Edmunds, Rákosník [26] have proved the optimal embedding $W_0^{1,p(x)}(\Omega) \hookrightarrow L^{p^*(x)}(\Omega)$ if p is Lipschitz continuous and $\sup_\Omega p(x) < d$.

The underlying idea of the previous proposition is that one could use the classical L^q information on the level sets Ω_q in order to get something for $L^{p(x)}(\Omega)$. This idea cannot only be used for embeddings, but also in many other situations, where the L^q-theory is well developed, as e.g. theory of singular operators, regularity theory of elliptic equations and systems to name a few.

The following counterexample of Kirchheim [53] shows that in general it is impossible to recover from the $L^q(\Omega_q)$ estimates an $L^{p(x)}(\Omega)$ information.

Proposition 2.35. *Let $\Omega = [0,1]$ and let $p(x) = 2 + x$. Then there exists a function g such that $g \notin L^{p(x)}(0,1)$, but for all $q \in [2,3]$*

$$\int_{q-2}^1 g(x)^q \, dx \leq c_0, \qquad (2.36)$$

where c_0 is independent of q.

PROOF : Let $b_k \in (0,1)$ be arbitrary. Then we have

$$\lim_{x \to b_k^-} (b_k - x)^{-\frac{x}{x+2}} = \infty,$$

and hence there exists $a_k \in (0, b_k)$ such that

$$(b_k - a_k)^{-\frac{a_k}{a_k+2}} = 2^{k+1}. \qquad (2.37)$$

Now, given $0 < a_k < b_k < 1$, we define $b_{k+1} \in (0, a_k)$ by

$$(b_k - a_k)^{\frac{b_{k+1}-a_k}{a_k+2}} = 2^k. \qquad (2.38)$$

Such b_{k+1} exists since the function $F(x) = (b_k - a_k)^{\frac{x-a_k}{a_k+2}}$ is continuous and decreasing and satisfies $F(0^+) = 2^{k+1}$ and $F(a_k^-) = 1$. Put

$$\lambda_k = (b_k - a_k)^{-\frac{1}{a_k+2}} \tag{2.39}$$

and hence (cf. (2.38))

$$\lambda_k^{b_{k+1}-a_k} = 2^{-k} . \tag{2.40}$$

Now we define $g(x)$ through

$$g(x) = \sum_{k=1}^{\infty} \lambda_k \chi_{(a_k,b_k)}(x) .$$

One immediately finds that $g \notin L^{p(x)}(0,1)$ since

$$\int_0^1 g(x)^{p(x)} \, dx = \sum_{k=1}^{\infty} \int_{a_k}^{b_k} \lambda_k^{2+x} \, dx$$

$$\geq \sum_{k=1}^{\infty} (b_k - a_k) \lambda_k^{a_k+2} = \sum_{k=1}^{\infty} 1 = \infty ,$$

where we used (2.39). It remains to show (2.36). Let $q > 2$ and let k_0 be such that $b_{k_0+1} \leq q - 2 < b_{k_0}$. Then we have

$$\int_{q-2}^1 g(x)^q \, dx = \sum_{k=1}^{k_0-1} \lambda_k^q (b_k - a_k) + \int_{q-2}^{b_{k_0}} \lambda_{k_0}^q \chi_{(a_{k_0},b_{k_0})}(x) \, dx = I_1 + I_2 .$$

Clearly, as $q \leq b_{k_0} + 2$ and $\lambda_k > 1$ we get

$$I_1 \leq \sum_{k=1}^{k_0-1} \lambda_k^{b_{k_0}+2}(b_k - a_k)$$

$$\leq \sum_{k=1}^{k_0-1} \lambda_k^{b_{k+1}-a_k} \lambda_k^{a_k+2}(b_k - a_k) \leq \sum_{k=1}^{\infty} 2^{-k} = 1 ,$$

where we used that the sequence b_k is decreasing and (2.39), (2.40). Further we see, again using (2.39), that

$$I_2 \leq \lambda_{k_0}^{b_{k_0}+2}(b_{k_0} - a_{k_0})$$

$$\leq \sup_k \lambda_k^{b_k-a_k} \lambda_k^{a_k+2}(b_k - a_k) = \sup_k (b_k - a_k)^{-\frac{b_k-a_k}{a_k+2}}$$

$$\leq \sup_k (b_k - a_k)^{-\frac{b_k-a_k}{2}} \leq \sup_{x \in (0,1)} x^{-\frac{x}{2}} = e^{\frac{1}{2e}} .$$

If $q = 2$ we observe

$$\int\limits_0^1 g(x)^2\, dx = \sum_{k=1}^{\infty} \lambda_k^2 (b_k - a_k)$$

$$= \sum_{k=1}^{\infty} (b_k - a_k)^{\frac{a_k}{a_k+2}} = \sum_{k=1}^{\infty} 2^{-(k+1)} = \frac{1}{2}\,.$$

All together we showed that

$$\int\limits_{q-2}^1 g(x)^q\, dx \le 1 + e^{\frac{1}{2e}}$$

is satisfied for all $q \in [2,3]$ and the proposition is proved.   ∎

Unfortunately we have not been able to construct a counterexample to the boundedness of the maximal operator and of singular integral operator in generalized Lebesgue spaces $L^{p(x)}(\Omega)$. However, we conjecture that there will be no analogue of such fundamental theorems, as e.g. boundedness of singular integral operators (cf. Calderon, Zygmund [17]), boundedness of the maximal function (cf. Stein [119]), the negative norm theorem (cf. Nečas [92]) and Korn's inequality (cf. Nečas [92]), in generalized Lebesgue spaces $L^{p(x)}(\Omega)$ even for very nice functions $p(x)$.

Let us finish this section by introducing some more function spaces, necessary notation and some auxiliary results which we shall need in the sequel. We define

$$\begin{aligned}
V_{p(x)} &\equiv \text{closure of } \mathcal{V} \text{ in } \|\,.\,\|_{1,p(x)}\text{-norm}\,,\\
E_{p(x)} &\equiv \text{closure of } \mathcal{V} \text{ in } \|\mathbf{D}(\,.\,)\|_{p(x)}\text{-norm}\,,
\end{aligned} \tag{2.41}$$

where $\mathbf{D}(.)$ denotes the symmetric part of the gradient. Clearly $V_{p(x)}$ is a Banach space, which is a closed subspace of $W_0^{1,p(x)}(\Omega)$. Therefore $V_{p(x)}$ is also separable and reflexive and the embedding

$$V_{p(x)} \hookrightarrow E_{p(x)}$$

holds true. However, since we do not have Korn's inequality, it is not clear if the spaces $V_{p(x)}$ and $E_{p(x)}$ coincide, even for $p \in C(\bar{\Omega})$. But $E_{p(x)}$ is also a separable reflexive Banach space. From the embedding (2.13) and Korn's inequality in $W_0^{1,p_\infty}(\Omega)$ we conclude

$$\|\mathbf{D}(\mathbf{u})\|_{p(x)} \ge c\,\|\mathbf{D}(\mathbf{u})\|_{p_\infty} \ge \tilde{c}\,\|\mathbf{u}\|_{1,p_\infty}\,, \tag{2.42}$$

which shows that $(E_{p(x)}, \|\mathbf{D}(.)\|_{p(x)})$ is a Banach space. Moreover $E_{p(x)}$ is reflexive and separable. This can easily be seen by considering the map

$$P : E_{p(x)} \to L_{N(d)}^{p(x)}(\Omega) : \mathbf{u} \to (D_{ij}(\mathbf{u}))_{1 \le i \le j \le d}\,,$$

where $N(d) = \frac{1}{2}d(d+1)$ and $L^{p(x)}_{N(d)}(\Omega) \equiv \prod\limits_{j=1}^{N(d)} L^{p(x)}(\Omega)$. If we endow the space

$L^{p(x)}_{N(d)}(\Omega)$ with the norm $\sum\limits_{j=1}^{N(d)} \|u_j\|_{p(x)}$ we obtain from the corresponding properties

of $L^{p(x)}(\Omega)$ that $L^{p(x)}_{N(d)}(\Omega)$ is a separable, reflexive Banach space. Using the embedding $E_{p(x)} \hookrightarrow V_{p_\infty}$ we get that P is injective. From the definitions of the norms in $E_{p(x)}$ and $L^{p(x)}_{N(d)}(\Omega)$ it is clear that P is an isometric isomorphism of $E_{p(x)}$ onto the subspace $W = P(E_{p(x)}) \subset L^{p(x)}_{N(d)}(\Omega)$, and since $E_{p(x)}$ is complete, W is a closed subspace of $L^{p(x)}_{N(d)}(\Omega)$. The reflexibility and separability of $E_{p(x)}$ follow immediately, since closed subspaces inherit these properties from $L^{p(x)}_{N(d)}(\Omega)$. Finally, we also need a weighted version of the space $E_{p(x)}$. Let $0 < \rho \in L^1(\Omega)$ be the density of a Radon measure ν with respect to the Lebesgue measure. Then we define the modular $|.|_{p(x),\nu}$ by

$$|f|_{p(x),\nu} \equiv \int\limits_{\Omega} |f(x)|^{p(x)} \, d\nu = \int\limits_{\Omega} |f(x)|^{p(x)} \rho(x) \, dx \qquad (2.43)$$

and

$$\|f\|_{p(x),\nu} = \inf\{\lambda > 0, \, |\tfrac{f}{\lambda}|_{p(x),\nu} \leq 1\}, \qquad (2.44)$$

which is a norm on the space

$$L^{p(x)}(\Omega, \nu) \equiv \{f \in L^1(\Omega, \nu); \, |\lambda f|_{p(x),\nu} < \infty \quad \text{for some } \lambda > 0\}. \qquad (2.45)$$

One easily checks that all properties stated for $L^{p(x)}(\Omega)$ also hold for $L^{p(x)}(\Omega, \nu)$, since no particular property of the Lebesgue measure which is not shared by a Radon measure was used in the proofs (cf. Dunford, Schwartz [23] for the corresponding theorems).

In particular, $L^{p(x)}(\Omega, \nu)$ is a separable, reflexive Banach space in which $\mathcal{D}(\Omega)$ is dense. The duality pairing $\langle \cdot, \cdot \rangle_{p(x),\nu}$ between $L^{p(x)}(\Omega, \nu)$ and $L^{p'(x)}(\Omega, \nu)$ is for all $f \in L^{p(x)}(\Omega, \nu)$, $g \in L^{p'(x)}(\Omega, \nu)$ given by

$$\langle f, g \rangle_{p(x),\nu} = \int\limits_{\Omega} f(x)g(x)\rho(x) \, dx.$$

Moreover, if $0 < c_0 \leq \rho$, then we note that

$$L^{p(x)}(\Omega, \nu) \hookrightarrow L^{p(x)}(\Omega), \qquad (2.46)$$

since one easily sees that

$$c_0 \int\limits_{\Omega} |f|^{p(x)} \, dx \leq \int\limits_{\Omega} |f|^{p(x)} \rho(x) \, dx$$

and this modular inequality immediately implies the above continuous embedding (cf. Pick [103]). For such ρ we define

$$E_{p(x),\nu} \equiv \text{closure of } \mathcal{V} \text{ with respect to the } \|\mathbf{D}(.)\|_{p(x),\nu}\text{-norm}. \qquad (2.47)$$

Similarly as for $E_{p(x)}$ one can show that $E_{p(x),\nu}$ is a separable, reflexive Banach space and that

$$E_{p(x),\nu} \hookrightarrow E_{p(x)} \,,$$
$$E_{p(x),\nu} \hookrightarrow E_{p_\infty,\nu} \hookrightarrow V_{p_\infty} \,,$$

(2.48)

and using a similar argument as for (2.7),

$$\|f\|_{p(x),\nu} \leq c \left(1 + |f|_{p(x),\nu}\right) \,.$$

(2.49)

2.3   Maxwell's Equations

In this section we will present existence, uniqueness and regularity results for the quasi-static Maxwell's equations

$$\begin{aligned} \operatorname{div} \mathbf{E} &= k \\ \operatorname{curl} \mathbf{E} &= \mathbf{h} \end{aligned} \quad \text{in } \Omega \,,$$
$$\mathbf{E} \cdot \mathbf{n} = \mathbf{E}_0 \cdot \mathbf{n} \qquad \text{on } \partial\Omega \,,$$

(3.1)

where $k, \mathbf{h}$ and $\mathbf{E}_0$ are given and where $\Omega \subseteq \mathbb{R}^3$ is a bounded domain. The system (3.1) is intensively studied in the literature and here we will give only the precise formulation of the results that we will need in the sequel. For more details we refer in the context of integral equation methods to Müller [90], Martensen [80], Kress [56] and Werner [128] and in the context of Hilbert space methods to Gaffney [34], Friedrichs [31], Leis [60], [61], Weck [127], Weber [126], Picard [102], [100], [101] and Milani, Picard [83].

Beside the spaces $H(\operatorname{div}), H(\operatorname{curl})$ introduced in the previous section we need some more spaces and notations:

$$\begin{aligned} \overset{\circ}{H}(\operatorname{div}) &= \{\mathbf{u} \in H(\operatorname{div}); \mathbf{n} \cdot \mathbf{u} = 0\} \,, \\ \overset{\circ}{H}(\operatorname{curl}) &= \{\mathbf{u} \in H(\operatorname{curl}); \mathbf{n} \times \mathbf{u} = 0\} \,, \\ H_0(\operatorname{div}) &= \{\mathbf{u} \in H(\operatorname{div}); \operatorname{div} \mathbf{u} = 0\} \,, \\ H_0(\operatorname{curl}) &= \{\mathbf{u} \in H(\operatorname{curl}); \operatorname{curl} \mathbf{u} = 0\} \,, \\ \overset{\circ}{H}_0(\operatorname{div}) &= \overset{\circ}{H}(\operatorname{div}) \cap H_0(\operatorname{div}) \,, \\ \overset{\circ}{H}_0(\operatorname{curl}) &= \overset{\circ}{H}(\operatorname{curl}) \cap H_0(\operatorname{curl}) \,, \\ H_N(\Omega) &= \overset{\circ}{H}_0(\operatorname{div}) \cap H_0(\operatorname{curl}) \,, \\ H_D(\Omega) &= \overset{\circ}{H}_0(\operatorname{curl}) \cap H_0(\operatorname{div}) \,. \end{aligned}$$

(3.2)

All these spaces are Hilbert spaces if they are equipped with the canonical scalar product. Note, that $\overset{\circ}{H}_0(\operatorname{div})$ and H defined in (2.1) are identical. For the investigation of Maxwell's equations it is crucial to have at disposal the compact embedding

$$\overset{\circ}{H}(\operatorname{div}) \cap H(\operatorname{curl}) \hookrightarrow\hookrightarrow L^2(\Omega)$$

(3.3)

and appropriate decompositions of $L^2(\Omega)$ into direct sums, as e.g.

$$L^2(\Omega) = H_N \oplus \operatorname{grad} W^{1,2}(\Omega) \oplus \operatorname{curl} \overset{\circ}{H}(\operatorname{curl}). \qquad (3.4)$$

The compact embedding is well known for smooth domains, where the proof is based on Gaffney's inequality and Rellich's compactness theorem. However, there are examples of non-smooth domains where Gaffney's inequality does not hold and thus the above arguments do not work. Nevertheless the following result characterizes a suitable class of domains.

Theorem 3.5. *Let Ω be a bounded domain with $\partial\Omega \in C^{0,1}$. Then the compact embedding (3.3) holds true.*

PROOF : This result is proved in Picard [102] and we will not explain the details here. Earlier results in this direction can be found in Weck [127]. The same result holds under even weaker assumptions, cf. Witsch [132]. ∎

As a consequence of Theorem 3.5 we obtain that various subspaces of $L^2(\Omega)$ are closed, which in turn enables us to prove decomposition results. In particular we have

Lemma 3.6. *Let Ω be a bounded domain with $\partial\Omega \in C^{0,1}$. Then*

$$\begin{aligned}
\operatorname{curl} H(\operatorname{curl}) &= \operatorname{curl}\left(\operatorname{curl} \overset{\circ}{H}(\operatorname{curl}) \cap H(\operatorname{curl})\right), \\
\operatorname{curl} \overset{\circ}{H}(\operatorname{curl}) &= \operatorname{curl}\left(\operatorname{curl} H(\operatorname{curl}) \cap \overset{\circ}{H}(\operatorname{curl})\right)
\end{aligned} \qquad (3.7)$$

are closed subspaces of $L^2(\Omega)$.

PROOF : The characterization of the spaces on the left-hand side is based on the decomposition theorem (cf. Leis [61]). The closedness follows from Theorem 3.5 (cf. Milani, Picard [83]). ∎

Lemma 3.8. *Under the assumption of Lemma 3.6 the following decompositions*

$$\begin{aligned}
L^2(\Omega) &= H_N \oplus \operatorname{grad} W^{1,2}(\Omega) \oplus \operatorname{curl} \overset{\circ}{H}(\operatorname{curl}), & (3.9) \\
H_0(\operatorname{div}) &= H_D \oplus \operatorname{curl} H(\operatorname{curl}), & (3.10) \\
H(\operatorname{div}) &= (\operatorname{grad} W^{1,2}(\Omega) \cap H(\operatorname{div})) \oplus H_N \oplus \overset{\circ}{H}(\operatorname{curl}) & (3.11)
\end{aligned}$$

hold true.

PROOF : see e.g. Milani, Picard [83]. ∎

Another important tool is the characterization of H_N and H_D, respectively, the spaces of harmonic vector fields with vanishing normal and tangential component, respectively. For this we need the following regularity result for harmonic vector fields.

Lemma 3.12. *Let Ω be a bounded domain with $\partial\Omega \in \mathcal{C}^{0,1}$. Then every vector field $\mathbf{v} \in H_0(\mathrm{div}\,) \cap H_0(\mathrm{curl}\,)$ belongs to the space $C(\Omega)$.*

PROOF :   see e.g. Morrey [88] p. 166.   ∎

Theorem 3.13. *The dimension of $H_N(\Omega)$ and $H_D(\Omega)$ is finite and given by*

$$\dim H_N = p, \qquad \dim H_D = m - 1 \tag{3.14}$$

where p is the first Betti number, i.e. the number of handles, and m the second Betti number, i.e. the number of connected components.

PROOF :   The proof can be found in Picard [101] and is based on Lemma 3.12 and a common procedure to construct a basis of $H_N(\Omega)$ and $H_D(\Omega)$, resp., (cf. Werner [128]).   ∎

Lemma 3.15. *Let Ω be a bounded domain with $\partial\Omega \in \mathcal{C}^{0,1}$. Then there exists a constant such that*

$$\|\mathbf{E}\|_2 \le c\Big(\|\operatorname{curl}\mathbf{E}\|_2 + \|\operatorname{div}\mathbf{E}\|_2 + \sum_{i=1}^{p} \big| \int_{\Omega} \boldsymbol{\omega}_i \cdot \mathbf{E}\, dx \big| \Big) \tag{3.16}$$

holds for all $\mathbf{E} \in \overset{\circ}{H}(\mathrm{div}\,) \cap H(\mathrm{curl}\,)$, where $\boldsymbol{\omega}_i$, $i = 1,\ldots,p$, is a basis of $H_N(\Omega)$.

PROOF :   see e.g. Milani, Picard [83].   ∎

Let us now discuss the solvability of (3.1). We assume that Ω is bounded and has a Lipschitz continuous boundary. First we re-formulate the boundary condition. It is well known that the mapping

$$\gamma_n : \mathbf{u} \to \mathbf{u} \cdot \mathbf{n} \tag{3.17}$$

defined in $\mathcal{D}(\bar{\Omega})$ can be continuously extended to a mapping, still denoted γ_n,

$$\gamma_n : H(\mathrm{div}\,) \to H^{-1/2}(\partial\Omega), \tag{3.18}$$

where $H^{-1/2}(\partial\Omega)$ is the dual space to $H^{1/2}(\partial\Omega)$. It can be shown that (cf. Girault, Raviart [38])

$$\|\gamma_n(\mathbf{u})\|_{-1/2,\partial\Omega} \le \|\mathbf{u}\|_{H(\mathrm{div}\,)},$$
$$\mathrm{Range}\,(\gamma_n) = H^{-1/2}(\partial\Omega), \tag{3.19}$$
$$\mathrm{Ker}\,(\gamma_n) = \overset{\circ}{H}(\mathrm{div}\,).$$

Therefore we can re-write the boundary condition $(3.1)_3$ as

$$\mathbf{E} - \mathbf{E}_0 \in \overset{\circ}{H}(\mathrm{div}\,) \tag{3.20}$$

where $\mathbf{E}_0 \in H(\mathrm{div}\,)$ is given. Now we can formulate an existence and uniqueness result for the Problem (3.1).

Theorem 3.21. *Let Ω be a bounded domain with $\partial\Omega \in \mathcal{C}^{0,1}$. Then there exists a solution $\mathbf{E} \in H(\mathrm{curl}) \cap H(\mathrm{div})$ of the problem* $(3.1)_{1,2}$, (3.20) *if and only if*

(i) $\mathbf{E}_0 \in H(\mathrm{div})$,

(ii) $\mathbf{h} \in H_0(\mathrm{div})$ *and* $\mathbf{h} \perp H_D(\Omega)$,

(iii) $k \in L^2(\Omega)$ *and* $k - \mathrm{div}\, \mathbf{E}_0 \perp \mathrm{const.}$

This solution satisfies the estimate

$$\|\mathbf{E}\|_2 \leq c\big(\|k\|_2 + \|\mathbf{E}_0\|_{-1/2,\partial\Omega} + \|\mathbf{h}\|_2\big)\,. \tag{3.22}$$

Moreover, the solution is unique if we require $\mathbf{E} \perp H_N(\Omega)$.

PROOF :   This theorem is a consequence of the results proved in Picard [100], [102]. We will only sketch the basic steps. One easily checks that a solution of $(3.1)_{1,2}$, (3.20) satisfies the necessary conditions (i)–(iii). Let us show that they are also sufficient. Due to (iii) there is exactly one $p \in W^{1,2}(\Omega)/\mathbb{R}$ solving the Neumann problem

$$\begin{aligned} -\Delta p &= k &&\text{in } \Omega\,, \\ \frac{\partial p}{\partial \mathbf{n}} &= \mathbf{E}_0 \cdot \mathbf{n} &&\text{on } \partial\Omega\,, \end{aligned} \tag{3.23}$$

and satisfying the estimate

$$\|\nabla p\|_2 \leq c\big(\|k\|_2 + \|\mathbf{E}_0\|_{-1/2,\partial\Omega}\big)\,. \tag{3.24}$$

From this and (3.11) we easily deduce that

$$\begin{aligned} &\nabla p \in H(\mathrm{div}) \cap H_0(\mathrm{curl})\,, \\ &\nabla p \perp H_N(\Omega)\,, \ \nabla p \perp \mathrm{curl}\,\overset{\circ}{H}(\mathrm{curl})\,. \end{aligned} \tag{3.25}$$

From (ii) an (3.10) we obtain that

$$h \in \mathrm{curl}\, H(\mathrm{curl}) \tag{3.26}$$

which together with $(3.7)_1$ implies that

$$h = \mathrm{curl}\,\phi \tag{3.27}$$

with

$$\phi \in \mathrm{curl}\,\overset{\circ}{H}(\mathrm{curl})\,. \tag{3.28}$$

This and (3.11), (3.9) imply that

$$\begin{aligned} &\phi \in \overset{\circ}{H}_0(\mathrm{div})\,, \\ &\phi \perp H_N(\Omega)\,, \phi \perp \mathrm{grad}\, W^{1,2}\,. \end{aligned} \tag{3.29}$$

Now, it is easy to check that

$$\mathbf{E} \equiv \nabla p + \phi \tag{3.30}$$

solves $(3.1)_{1,2}$, (3.20). The estimate (3.22) follows from (3.30), (3.24), (3.16), $(3.1)_2$ and $(3.29)_2$, namely

$$\|\mathbf{E}\|_2 \le \|\nabla p\|_2 + \|\phi\|_2$$
$$\le c\big(\|k\|_2 + \|\mathbf{E}_0\|_{-1/2,\partial\Omega} + \|\operatorname{curl}\phi\|_2\big)$$
$$\le c\big(\|k\|_2 + \|\mathbf{E}_0\|_{-1/2,\partial\Omega} + \|\mathbf{h}\|_2\big)\,.$$

Let $\mathbf{E}_1, \mathbf{E}_2$ be two solutions such that $\mathbf{E}_1, \mathbf{E}_2 \perp H_N(\Omega)$. From $(3.1)_{1,2}$, (3.20) we obtain $\mathbf{E}_1 - \mathbf{E}_2 \in H_N(\Omega)$ and from the orthogonality to $H_N(\Omega)$ we deduce $\mathbf{E}_1 - \mathbf{E}_2 = 0$. This completes the proof.   ∎

The regularity of the problem (3.1) within the context of Sobolev spaces has been studied in detail in Schwarz [115] in the context of manifolds with boundaries. In the last chapter he describes the consequences of his results in the case of three-dimensional vector fields.

Theorem 3.31. *Let Ω be a bounded domain with $\partial\Omega \in C^{2+l,1}, l \in \mathbb{N}$. Let $\mathbf{h}$, $k \in W^{l,p}(\Omega)$ and $\mathbf{E}_0 \in W^{l+1-\frac{1}{p},p}(\partial\Omega)$ be given, where $1 < p < \infty$. Then the solution of the problem $(3.1)_{1,2}$, (3.20), ensured by Theorem 3.21, satisfies the estimate*

$$\|\mathbf{E}\|_{l+1,p} \le c\big(\|k\|_{l,p} + \|\mathbf{h}\|_{l,p} + \|\mathbf{E}_0\|_{l+1-1/p,p,\partial\Omega}\big)\,. \tag{3.32}$$

PROOF :   This theorem is the contents of Corollary 3.2.6 and Lemma 3.5.4 in Schwarz [115].   ∎

Let us finish this section by discussing the time dependent problem

$$
\begin{aligned}
\operatorname{div}\mathbf{E} &= k & & \\
\operatorname{curl}\mathbf{E} &= \mathbf{h} & &\text{in } Q_T\,, \\
\mathbf{E}\cdot\mathbf{n} &= \mathbf{E}_0\cdot\mathbf{n} & &\text{on } I \times \partial\Omega\,,
\end{aligned}
\tag{3.33}
$$

where $k, \mathbf{h} \in C^1(\bar{I}, W^{l,p}(\Omega))$, $\mathbf{E}_0 \in C^1(\bar{I}, W^{l+1,p}(\Omega))$ are given[7]. We will denote by $\frac{\partial f}{\partial t}$ the time derivative of a function with values in a Banach space X if

$$\lim_{\substack{\varepsilon\to 0 \\ t+\varepsilon\in\bar{I}}} \left\| \frac{f(t+\varepsilon) - f(t)}{\varepsilon} - \frac{\partial f}{\partial t}(t) \right\|_X = 0\,. \tag{3.34}$$

---

[7]For the definition and properties of differentiable functions with values in Banach spaces we refer the reader to Gajewski, Gröger, Zacharias [35], Chapter 4.

Proposition 3.35. *Let Ω be a bounded domain with $\partial\Omega \in C^{2+l,1}, l \in \mathbb{N}$ and let $T > 0$ be given. Moreover, assume that $k, \mathbf{h} \in C^1(\bar{I}, W^{l,p}(\Omega))$, $\mathbf{E}_0 \in C^1(\bar{I}, W^{l+1,p}(\Omega))$, $1 < p < \infty$. Then there exists a solution of the problem (3.33) if $k, \mathbf{h}, \mathbf{E}_0$ and $\frac{\partial k}{\partial t}$, $\frac{\partial \mathbf{h}}{\partial t}, \frac{\partial \mathbf{E}_0}{\partial t}$ fulfill the conditions of Theorem 3.21 for all $t \in \bar{I}$. This solution satisfies the estimate*

$$\|\mathbf{E}\|_{C^1(\bar{I}, W^{l+1,p}(\Omega))} \leq c\left(\|k\|_{C^1(\bar{I}, W^{l,p}(\Omega))} + \|\mathbf{h}\|_{C^1(\bar{I}, W^{l,p}(\Omega))} + \|\mathbf{E}_0\|_{C^1(\bar{I}, W^{l+1,p}(\Omega))}\right) \quad (3.36)$$

and is unique if we require that $\mathbf{E}(t) \perp H_N(\Omega)$ for all $t \in I$.

PROOF : From the assumption of this proposition and the Theorem 3.31 it follows that for all $t \in \bar{I}$ the system (3.33) possesses a solution $\mathbf{E}(t) \in W^{l+1,p}(\Omega)$ satisfying the estimate (3.32) and that also the system

$$\text{div } \mathbf{G}(t) = \frac{\partial k}{\partial t}(t)$$

$$\text{curl } \mathbf{G}(t) = \frac{\partial \mathbf{h}}{\partial t}(t) \qquad \text{in } \Omega, \qquad (3.37)$$

$$\mathbf{G}(t) \cdot \mathbf{n} = \frac{\partial \mathbf{E}_0}{\partial t}(t) \cdot \mathbf{n} \qquad \text{on } \partial\Omega,$$

possesses a solution $\mathbf{G}(t) \in W^{l+1,p}(\Omega)$ satisfying for all $t \in \bar{I}$ the estimate

$$\|\mathbf{G}(t)\|_{l+1,p} \leq c\left(\left\|\frac{\partial k}{\partial t}(t)\right\|_{l,p} + \left\|\frac{\partial \mathbf{h}}{\partial t}(t)\right\|_{l,p} + \left\|\frac{\partial \mathbf{E}_0}{\partial t}(t)\right\|_{l+1,p}\right).$$

From these estimates, the linearity of the systems and (3.34) one easily deduces that $\mathbf{G}(t) = \frac{\partial \mathbf{E}}{\partial t}(t)$ for all $t \in \bar{I}$. The estimate (3.36) follows from the above considerations. For the uniqueness we proceed as in the proof of Theorem 3.21 and the proof is complete. ∎

3 Electrorheological Fluids with Shear Dependent Viscosities: Steady Flows

3.1 Introduction

In this chapter we shall deal with steady flows of incompressible electrorheological fluids in a bounded domain $\Omega \subseteq \mathbb{R}^3$. This motion is governed by the boundary value problem[1]

$$
\begin{aligned}
\text{div } \mathbf{E} &= 0 \\
\text{curl } \mathbf{E} &= 0 \qquad \text{in } \Omega, \\
\mathbf{E} \cdot \mathbf{n} &= \mathbf{E}_0 \cdot \mathbf{n} \qquad \text{on } \partial\Omega,
\end{aligned}
\tag{1.1}
$$

$$
\begin{aligned}
-\text{div } \mathbf{S}(\mathbf{D}(\mathbf{v}), \mathbf{E}) + [\nabla\mathbf{v}]\mathbf{v} + \nabla\phi &= \mathbf{f} + \chi^E[\nabla\mathbf{E}]\mathbf{E}, \\
\text{div } \mathbf{v} &= 0 \qquad \text{in } \Omega, \\
\mathbf{v} &= 0 \qquad \text{on } \partial\Omega,
\end{aligned}
\tag{1.2}
$$

where the data $\mathbf{f}$ and $\mathbf{E}_0$ are given. We shall assume that the extra stress $\mathbf{S} = \mathbf{T} + \phi\mathbf{I}$ is given by

$$
\begin{aligned}
\mathbf{S} = \alpha_{21}\big((1 + |\mathbf{D}|^2)^{\frac{p-1}{2}} - 1\big)\mathbf{E} \otimes \mathbf{E} \\
+ (\alpha_{31} + \alpha_{33}|\mathbf{E}|^2)(1 + |\mathbf{D}|^2)^{\frac{p-2}{2}}\mathbf{D} \\
+ \alpha_{51}(1 + |\mathbf{D}|^2)^{\frac{p-2}{2}}(\mathbf{DE} \otimes \mathbf{E} + \mathbf{E} \otimes \mathbf{DE}),
\end{aligned}
\tag{1.3}
$$

where $p = p(|\mathbf{E}|^2)$ is a C^1-function such that (cf. (1.4.11))

$$
1 < p_\infty \leq p(|\mathbf{E}|^2) \leq p_0.
\tag{1.4}
$$

We further require that the coefficients α_{ij} and the function p are such that (cf. Lemma 1.4.46)

$$
\alpha_{31} > 0, \qquad \alpha_{33} > 0, \qquad \alpha_{33} + \tfrac{4}{3}\alpha_{51} > 0,
\tag{1.5}
$$

---

[1] Note, that in (1.2) we have divided the steady version of (2.1.2) by the constant density ρ_0 and adapted the notation appropriately.

$$k(p_0)|\alpha_{21}| < \begin{cases} 2\sqrt{\alpha_{33}}\sqrt{2\alpha_{51}} & \text{if } \alpha_{33} \leq \tfrac{4}{3}\alpha_{51}\,, \\[2mm] \sqrt{\tfrac{3}{2}}(\alpha_{33} + \tfrac{4}{3}\alpha_{51}) & \text{if } \tfrac{4}{3}|\alpha_{51}| \leq \alpha_{33}\,, \end{cases} \tag{1.6}$$

and that for all $\mathbf{E} \in \mathbb{R}^3$, and all $\mathbf{B}, \mathbf{D} \in X = \{\mathbf{D} \in \mathbb{R}^{3\times3}_{\text{sym}}, \text{tr } \mathbf{D} = 0\}$

$$\frac{\partial S_{ij}(\mathbf{D}, \mathbf{E})}{\partial D_{kl}} B_{ij} B_{kl} \geq 0\,. \tag{1.7}$$

The last condition (1.7) ensures that for all $\mathbf{E}$ the operator $\mathbf{S}(\,.\,, \mathbf{E})$ is monotone. The conditions (1.5) and (1.6) imply (cf. Lemma 1.4.46 ff.) that there is a constant C_0 such that

$$\mathbf{S}(\mathbf{D}, \mathbf{E}) \cdot \mathbf{D} \geq C_0(1 + |\mathbf{E}|^2)(1 + |\mathbf{D}|^2)^{\frac{p(|\mathbf{E}|^2)-2}{2}} |\mathbf{D}|^2\,. \tag{1.8}$$

From (1.3) and (1.4.39) follows that

$$|\mathbf{S}(\mathbf{D}, \mathbf{E})| \leq C_3(1 + |\mathbf{E}|^2)(1 + |\mathbf{D}|^2)^{\frac{p(|\mathbf{E}|^2)-1}{2}}\,. \tag{1.9}$$

Let us define the Radon measure ν by its density $\rho = 1 + |\mathbf{E}|^2$ with respect to the Lebesgue measure, i.e. for all Lebesgue measurable sets $A \subseteq \Omega$, the measure ν is defined by

$$\nu(A) = \int_A \left(1 + |\mathbf{E}|^2\right) dx\,. \tag{1.10}$$

For given $\mathbf{E}$ we shall often write p or $p(x)$ instead of $p(|\mathbf{E}|^2)$ if there is no danger of confusion. Similarly we shall use E_p and $E_{p,\nu}$ instead of $E_{p(|\mathbf{E}|^2)}$ and $E_{p(|\mathbf{E}|^2),\nu}$ respectively. For the classical Lebesgue and classical Sobolev spaces we shall use the integrability exponents q, p_0, p_∞, but never p.

3.2 Weak Solutions

Let us start with the definition of weak solutions.

Definition 2.1. *The couple $(\mathbf{E}, \mathbf{v})$ is said to be a weak solution of the problem (1.1), (1.2) if and only if*

$$\begin{aligned} \mathbf{E} &\in H(\text{div}\,) \cap H(\text{curl})\,, \\ \mathbf{v} &\in E_{p(|\mathbf{E}|^2),\nu}\,, \end{aligned} \tag{2.2}$$

the system (1.1) is satisfied almost everywhere in Ω, $\mathbf{E} - \mathbf{E}_0 \in \overset{\circ}{H}(\text{div}\,)$ and the weak formulation of (1.2)

$$\begin{aligned} \int_\Omega \mathbf{S}(\mathbf{D}(\mathbf{v}), \mathbf{E}) \cdot \mathbf{D}(\boldsymbol{\varphi})\, dx &+ \int_\Omega [\nabla \mathbf{v}]\mathbf{v} \cdot \boldsymbol{\varphi}\, dx \\ &= \langle \mathbf{f}, \boldsymbol{\varphi} \rangle_{1,p_\infty} - \chi^E \int_\Omega \mathbf{E} \otimes \mathbf{E} \cdot \mathbf{D}(\boldsymbol{\varphi})\, dx \end{aligned} \tag{2.3}$$

is fulfilled for all $\boldsymbol{\varphi} \in E_{p(|\mathbf{E}|^2),\nu}$.

Now we formulate our main result concerning weak solutions.

Theorem 2.4. *Let Ω be such that $\partial\Omega \in C^{0,1}$ and let $\mathbf{E}_0 \in H(\mathrm{div})$, $\mathbf{f} \in (W_0^{1,p_\infty}(\Omega))^*$ be given. Then there exists a weak solution $(\mathbf{E}, \mathbf{v})$ of the problem (1.1), (1.2) whenever*

$$p_\infty > 9/5. \tag{2.5}$$

Remark 2.6. 1) From the proof of Theorem 2.4 it will become clear that the theorem also holds in the d-dimensional case, if (2.5) is replaced by

$$p_\infty > \frac{3d}{d+2}. \tag{2.7}$$

2) Note, that the lower bound (2.5) respectively (2.7) is the same as in the case $\mathbf{E} \equiv \mathbf{0}$ (i.e. $p_\infty = p_0 = p = \mathrm{const.}$) obtained by Lions [68] with the theory of monotone operators and compactness arguments. Recently Frehse, Málek, Steinhauer [30] and Růžička [112] obtained the existence of weak solutions in the case when $\mathbf{E} \equiv \mathbf{0}$ for $p_\infty > \frac{2d}{d+1}$. Both methods use results which are not available for generalized Lebesgue and generalized Sobolev spaces (cf. conjecture in Section 2.2).
3) Theorem 2.4 is formulated such that condition (2.5) restricts the class of possible materials independently of the process. We can re-formulate it also depending on the process. Namely, let $\mathbf{E} \in L^\infty(\Omega)$. Denoting

$$\tilde{p}_\infty \equiv p(\|\mathbf{E}\|_\infty^2)$$

we can replace condition (2.5) by

$$\tilde{p}_\infty > 9/5$$

and the theorem remains valid.

PROOF of Theorem 2.4: For $\mathbf{E}_0 \in H(\mathrm{div})$ it is shown in Theorem 2.3.21 that there exists a solution $\mathbf{E} \in H(\mathrm{div}) \cap H(\mathrm{curl})$ satisfying (1.1) and $\mathbf{E} - \mathbf{E}_0 \in \overset{\circ}{H}(\mathrm{div})$. Let us fix one such solution. From Lemma 2.3.12 and (2.3.22) follows that

$$\begin{aligned} \|\mathbf{E}\|_2 &\leq c\,\|\mathbf{E}_0\|_{-1/2,\partial\Omega}\,, \\ \mathbf{E} &\in C^0(\Omega)\,. \end{aligned} \tag{2.8}$$

We shall show the existence of a solution to (1.2) via a Galerkin approximation, Minty's trick and a compactness argument. Let $\boldsymbol{\omega}^j$, $j = 1, 2, \ldots$, be a Schauder basis of $E_{p,\nu}$ and let us denote the closed linear hull of $\boldsymbol{\omega}^1, \ldots, \boldsymbol{\omega}^n$ by X_n. We define the Galerkin approximation $\mathbf{v}^n$ by

$$\mathbf{v}^n = \sum_{j=1}^n a_j \boldsymbol{\omega}^j \tag{2.9}$$

and an operator $\mathbf{A} : \mathbb{R}^n \to \mathbb{R}^n : (a_1, \ldots, a_n) \to (\alpha_1, \ldots, \alpha_n)$, where α_i, $i = 1, \ldots, n$, is given by

$$
\begin{aligned}
\alpha_i = & \int_\Omega \mathbf{S}(\mathbf{D}(\mathbf{v}^n), \mathbf{E}) \cdot \mathbf{D}(\boldsymbol{\omega}^i)\, dx + \int_\Omega [\nabla \mathbf{v}^n] \mathbf{v}^n \cdot \boldsymbol{\omega}^i\, dx \\
& + \chi^E \int_\Omega \mathbf{E} \otimes \mathbf{E} \cdot \mathbf{D}(\boldsymbol{\omega}^i)\, dx - \langle \mathbf{f}, \boldsymbol{\omega}^i \rangle_{1, p_\infty}.
\end{aligned}
\tag{2.10}
$$

Obviously $\mathbf{A}$ is continuous. Multiplying (2.10) by a_i and summing up, using (1.8) and

$$
\int_\Omega [\nabla \mathbf{v}^n] \mathbf{v}^n \cdot \mathbf{v}^n\, dx = 0
\tag{2.11}
$$

we obtain

$$
\begin{aligned}
\mathbf{A}\mathbf{a} \cdot \mathbf{a} \geq & \, C_0 \int_\Omega (1 + |\mathbf{E}|^2)(1 + |\mathbf{D}(\mathbf{v}^n)|^2)^{\frac{p-2}{2}} |\mathbf{D}(\mathbf{v}^n)|^2\, dx \\
& - |\chi^E| \int_\Omega |\mathbf{E}|^2 |\mathbf{D}(\mathbf{v}^n)|\, dx - |\langle \mathbf{f}, \mathbf{v}^n \rangle_{1, p_\infty}| \\
= & \, I_1 - I_2 - I_3.
\end{aligned}
\tag{2.12}
$$

Using

$$
(1 + x^2)^{\frac{p(|\mathbf{E}|^2)-2}{2}} \geq (1 + x^2)^{\frac{p_\infty - 2}{2}}
\tag{2.13}
$$

and the pointwise inequality $(1 + x^2)^{\frac{p_\infty - 2}{2}} x^2 \geq c(x^{p_\infty} - 1)$ where $c = c(p_\infty, p_0)$ we deduce

$$
I_1 \geq c \int_\Omega (1 + |\mathbf{E}|^2)(|\mathbf{D}(\mathbf{v}^n)|^{p_\infty} + |\mathbf{D}(\mathbf{v}^n)|^p)\, dx - c.
\tag{2.14}
$$

From Young's and Korn's inequalities we obtain

$$
\begin{aligned}
I_2 &\leq \frac{c}{2} \int_\Omega |\mathbf{E}|^2 |\mathbf{D}(\mathbf{v}^n)|^p\, dx + \tilde{c} \int_\Omega |\mathbf{E}|^2\, dx, \\
I_3 &\leq \frac{c}{2} \int_\Omega |\mathbf{D}(\mathbf{v}^n)|^{p_\infty}\, dx + \tilde{c} \|\mathbf{f}\|_{-1, p_\infty'}^{p_\infty'}.
\end{aligned}
\tag{2.15}
$$

The inequalities (2.14) and (2.15) together with (2.12) lead to

$$
\begin{aligned}
\mathbf{A}\mathbf{a} \cdot \mathbf{a} \geq & \, \frac{c}{2} \Big(|\mathbf{D}(\mathbf{v}^n)|_{p, \nu} + \|\mathbf{D}(\mathbf{v}^n)\|_{p_\infty, \nu}^{p_\infty} \Big) \\
& - \tilde{c} \Big(1 + \|\mathbf{E}\|_2^2 + \|\mathbf{f}\|_{-1, p_\infty'}^{p_\infty'} \Big).
\end{aligned}
$$

This estimate shows on the one hand by a version of Brouwer's fixed point theorem (cf. Gajewski, Gröger, Zacharias [35], Lemma 2.1) the solvability of the Galerkin system

$$
\int_\Omega \mathbf{S}(\mathbf{D}(\mathbf{v}^n), \mathbf{E}) \cdot \mathbf{D}(\boldsymbol{\varphi}) \, dx + \int_\Omega [\nabla \mathbf{v}^n] \mathbf{v}^n \cdot \boldsymbol{\varphi} \, dx
$$
$$
= \langle \mathbf{f}, \boldsymbol{\varphi} \rangle_{1, p_\infty} - \chi^E \int_\Omega \mathbf{E} \otimes \mathbf{E} \cdot \mathbf{D}(\boldsymbol{\varphi}) \, dx \qquad \forall \boldsymbol{\varphi} \in X_n \tag{2.16}
$$

and on the other hand it provides, using (2.2.49) and Korn's inequality, the apriori estimate

$$
|\mathbf{D}(\mathbf{v}^n)|_{p,\nu} + \|\mathbf{D}(\mathbf{v}^n)\|_{p,\nu} + \|\nabla \mathbf{v}^n\|_{p_\infty}^{p_\infty} \le c_0 \big(1 + \|\mathbf{E}\|_2^2 + \|\mathbf{f}\|_{-1, p_\infty'}^{p_\infty'}\big). \tag{2.17}
$$

This estimate implies that we can choose a subsequence such that[2]

$$
\begin{aligned}
\mathbf{v}^n &\rightharpoonup \mathbf{v} && \text{weakly in } E_{p,\nu} \cap V_{p_\infty}, \\
\mathbf{v}^n &\to \mathbf{v} && \text{strongly in } L^q(\Omega),
\end{aligned} \tag{2.18}
$$

where $1 \le q < p_\infty^* = \frac{3p_\infty}{3 - p_\infty}$. Denoting

$$
\tilde{\mathbf{S}}(\mathbf{D}(\mathbf{u}), \mathbf{E}) = \frac{\mathbf{S}(\mathbf{D}(\mathbf{u}), \mathbf{E})}{1 + |\mathbf{E}|^2}, \tag{2.19}
$$

it follows from (2.17) and (1.9) that the sequence $\tilde{\mathbf{S}}(\mathbf{D}(\mathbf{v}^n), \mathbf{E})$ is bounded in $L^{p'(x)}(\Omega, \nu)$ and thus

$$
\tilde{\mathbf{S}}(\mathbf{D}(\mathbf{v}^n), \mathbf{E}) \rightharpoonup \boldsymbol{\chi} \quad \text{in} \quad L^{p'(x)}(\Omega, \nu). \tag{2.20}
$$

From (2.18) we obtain for all $\boldsymbol{\varphi} \in V_{p_\infty}$

$$
\int_\Omega [\nabla \mathbf{v}^n] \mathbf{v}^n \cdot \boldsymbol{\varphi} \, dx \to \int_\Omega [\nabla \mathbf{v}] \mathbf{v} \cdot \boldsymbol{\varphi} \, dx \tag{2.21}
$$

as long as $p_\infty > 9/5$. Re-writing the Galerkin system (2.16) by using (2.19), the limit $n \to \infty$ together with (2.20) and (2.21) leads to

$$
\int_\Omega \boldsymbol{\chi} \cdot \mathbf{D}(\boldsymbol{\varphi}) \, d\nu = - \int_\Omega [\nabla \mathbf{v}] \mathbf{v} \cdot \boldsymbol{\varphi} \, dx + \langle \mathbf{f}, \boldsymbol{\varphi} \rangle_{1, p_\infty}
$$
$$
- \chi^E \int_\Omega \mathbf{E} \otimes \mathbf{E} \cdot \mathbf{D}(\boldsymbol{\varphi}) \, dx, \tag{2.22}
$$

---

[2]Note, that the weak limits in $E_{p,\nu}$ and V_{p_∞} coincide. This follows easily by considering

$$
F_{\mathbf{f}}(\boldsymbol{\varphi}) = \int_\Omega \mathbf{D}(\mathbf{f}) \cdot \boldsymbol{\varphi} \, dx = \int_\Omega \mathbf{D}(\mathbf{f}) \cdot \frac{\boldsymbol{\varphi}}{1 + |\mathbf{E}|^2} \, d\nu \qquad \boldsymbol{\varphi} \in \mathcal{D}(\Omega),
$$

which defines elements from $V_{p_\infty^*}^*$ and $(E_{p,\nu})^*$.

which is satisfied for all $\varphi \in \bigcup\limits_{n=1}^{\infty} X_n$ and thus for all $\varphi \in E_{p,\nu}$. Observe that the last integral in (2.22) can be re-written as

$$-\chi^E \int\limits_{\Omega} \frac{\mathbf{E} \otimes \mathbf{E}}{1 + |\mathbf{E}|^2} \cdot \mathbf{D}(\varphi)\, d\nu\,,$$

which defines an element from $(E_{p,\nu})^*$. Therefore the Galerkin system (2.16) with $\varphi = \mathbf{v}^n$ yields

$$\int\limits_{\Omega} \tilde{\mathbf{S}}(\mathbf{D}(\mathbf{v}^n), \mathbf{E}) \cdot \mathbf{D}(\mathbf{v}^n)\, d\nu \to \langle \mathbf{f}, \mathbf{v} \rangle_{1,p_\infty} - \chi^E \int\limits_{\Omega} \mathbf{E} \otimes \mathbf{E} \cdot \mathbf{D}(\mathbf{v})\, dx \qquad (2.23)$$

if we take into account (2.18), (2.11) and (2.19). The monotonicity of $\mathbf{S}(\mathbf{D}, \mathbf{E})$ implies that for all $\varphi \in E_{p,\nu}$

$$0 \leq \int\limits_{\Omega} \left(\tilde{\mathbf{S}}(\mathbf{D}(\mathbf{v}^n), \mathbf{E}) - \tilde{\mathbf{S}}(\mathbf{D}(\varphi), \mathbf{E}) \right) \cdot \mathbf{D}(\mathbf{v}^n - \varphi)\, d\nu\,.$$

As $n \to \infty$ this relation, (2.23), (2.20) and (2.18) imply

$$0 \leq \langle \mathbf{f}, \mathbf{v} \rangle_{1,p_\infty} - \chi^E \int\limits_{\Omega} \mathbf{E} \otimes \mathbf{E} \cdot \mathbf{D}(\mathbf{v})\, dx - \int\limits_{\Omega} \boldsymbol{\chi} \cdot \mathbf{D}(\varphi)\, d\nu$$

$$- \int\limits_{\Omega} \tilde{\mathbf{S}}(\mathbf{D}(\varphi), \mathbf{E}) \cdot \mathbf{D}(\mathbf{v} - \varphi)\, dx\,,$$

which together with (2.22) for $\varphi = \mathbf{v}$ gives that

$$0 \leq \int\limits_{\Omega} \left(\boldsymbol{\chi} - \tilde{\mathbf{S}}(\mathbf{D}(\varphi), \mathbf{E}) \right) \cdot \mathbf{D}(\mathbf{v} - \varphi)\, d\nu$$

holds for all $\varphi \in E_{p,\nu}$. Choosing now $\varphi = \mathbf{v} \pm t\psi$, $\psi \in E_{p,\nu}$ we get (cf. (2.19))

$$\int\limits_{\Omega} \boldsymbol{\chi} \cdot \mathbf{D}(\psi)\, d\nu = \int\limits_{\Omega} \mathbf{S}(\mathbf{D}(\mathbf{v}), \mathbf{E}) \cdot \mathbf{D}(\psi)\, dx\,,$$

which together with (2.22) shows that $\mathbf{v}$ is a weak solution of (1.2). This concludes the proof.   ∎

If we require that $\mathbf{S}$ is not only monotone but uniformly monotone, i.e. for all $\mathbf{D}, \mathbf{B} \in X$

$$\frac{\partial S_{ij}(\mathbf{D}, \mathbf{E})}{\partial D_{kl}} B_{ij} B_{kl} \geq (C_1 + C_2 |\mathbf{E}|^2)(1 + |\mathbf{D}|^2)^{\frac{p(|\mathbf{E}|^2) - 2}{2}} |\mathbf{B}|^2 \qquad (2.24)$$

is valid, then we can also show that weak solutions are unique for small data[3]. To this end we need an apriori estimate similar to (2.17) but without an additive constant on the right-hand side. Thus let us establish a different coercivity property for $\mathbf{S}$.

---

[3]Note, that we have given sufficient conditions for (2.24) to hold in Lemma 1.4.64.

Lemma 2.25. *Let* **S** *be given by (1.3), satisfying (1.4)–(1.6) and (2.24). Then there exists a constant C_4 such that*

$$\mathbf{S}(\mathbf{D}, \mathbf{E}) \cdot \mathbf{D} \geq C_4(1 + |\mathbf{E}|^2)|\mathbf{D}|^{p_\infty} \tag{2.26}$$

is satisfied for $p_\infty \geq 2$ and

$$\mathbf{S}(\mathbf{D}, \mathbf{E}) \cdot \mathbf{D} \geq C_4(1 + |\mathbf{E}|^2) \begin{cases} |\mathbf{D}|^2 & |\mathbf{D}| \leq 1, \\ |\mathbf{D}|^{p_\infty} & |\mathbf{D}| \geq 1, \end{cases} \tag{2.27}$$

holds if $1 < p_\infty < 2$.

PROOF : Inequality (2.26) follows immediately from (1.8), (2.13), $2 \leq p_\infty$ and

$$\frac{1}{2} \leq \sup_{x \in [0,\infty)} \frac{(1+x)^\alpha}{1+x^\alpha} \leq \max(1, 2^{\alpha-1}), \qquad \alpha > 0. \tag{2.28}$$

In the case $p_\infty < 2$ we have

$$\mathbf{S}(\mathbf{D}, \mathbf{E}) \cdot \mathbf{D} = \int_0^1 \frac{d}{ds} \mathbf{S}(s\,\mathbf{D}, \mathbf{E})\, ds \cdot \mathbf{D}$$
$$\geq (C_1 + C_2|\mathbf{E}|^2) \int_0^1 \left(1 + |s\,\mathbf{D}|^2\right)^{\frac{p_\infty-2}{2}} ds\, |\mathbf{D}|^2, \tag{2.29}$$

where we used (2.24) and (2.13). Using (2.28) for $\alpha = 1/2$, the last integral in (2.29) can be estimated from below by

$$\int_0^1 \left(1 + s\,|\mathbf{D}|\right)^{p_\infty-2} ds = \frac{1}{p_\infty - 1} \frac{1}{|\mathbf{D}|} \left[(1 + s|\mathbf{D}|)^{p_\infty-1}\right]_0^1$$
$$= \frac{1}{p_\infty - 1} \frac{1}{|\mathbf{D}|} \left((1 + |\mathbf{D}|)^{p_\infty-1} - 1\right)$$

and thus we conclude that

$$\mathbf{S}(\mathbf{D}, \mathbf{E}) \cdot \mathbf{D} \geq \frac{C_1 + C_2|\mathbf{E}|^2}{p_\infty - 1} |\mathbf{D}| \left((1 + |\mathbf{D}|)^{p_\infty-1} - 1\right). \tag{2.30}$$

Since there exists a constant C such that (cf. Málek, Nečas, Rokyta, Růžička [70] formula (5.1.31))

$$\left((1 + |\mathbf{D}|)^{p_\infty-1} - 1\right) \geq C \begin{cases} |\mathbf{D}| & |\mathbf{D}| \leq 1, \\ |\mathbf{D}|^{p_\infty-1} & |\mathbf{D}| \geq 1, \end{cases} \tag{2.31}$$

we obtain (2.27) from (2.30) and (2.31). ∎

Corollary 2.32. *Let* $\mathbf{E}_0 \in H(\mathrm{div})$ *and* $\mathbf{f} \in (W_0^{1,p_\infty}(\Omega))^*$ *be given. Then any weak solution* $(\mathbf{E}, \mathbf{v})$ *of the system* (1.1), (1.2) *satisfies*

$$\|\mathbf{E}\|_2 \le C_5 \|\mathbf{E}_0\|_{H(\mathrm{div})}, \tag{2.33}$$

$$\|\mathbf{D}(\mathbf{v})\|_{p_\infty,\nu}^{p_\infty} \le C_6(\mathbf{f}, \mathbf{E}_0), \tag{2.34}$$

where $C_6(\mathbf{f}, \mathbf{E}_0)$ *is for* $p_\infty < 2$ *and* $p_\infty \ge 2$, *respectively, given by*

$$C_6(\mathbf{f}, \mathbf{E}_0) = \begin{cases} c \left(\|\mathbf{f}\|_{-1,p_\infty'}^2 + \|\mathbf{f}\|_{-1,p_\infty'}^{p_\infty'} + \|\mathbf{E}_0\|_{H(\mathrm{div})}^2 \right)^{\frac{p_\infty}{2}} \times \\ \quad \times \left(|\Omega| + \|\mathbf{E}_0\|_{H(\mathrm{div})}^2 + \|\mathbf{f}\|_{-1,p_\infty'}^2 + \|\mathbf{f}\|_{-1,p_\infty'}^{p_\infty'} \right)^{1-\frac{p_\infty}{2}}, \\ \\ 2c \|\mathbf{f}\|_{-1,p_\infty'}^{p_\infty'} + 2\|\mathbf{E}_0\|_{H(\mathrm{div})}^2. \end{cases} \tag{2.35}$$

PROOF : Inequality (2.33) follows from $(2.8)_1$ and $(2.3.19)_1$. Using $\mathbf{v}$ as a test function in (2.3) we obtain

$$\int_\Omega \mathbf{S}(\mathbf{D}(\mathbf{v}), \mathbf{E}) \cdot \mathbf{D}(\mathbf{v})\, dx = \langle \mathbf{f}, \mathbf{v} \rangle_{1,p_\infty} - \chi^E \int_\Omega \mathbf{E} \otimes \mathbf{E} \cdot \mathbf{D}(\mathbf{v})\, dx. \tag{2.36}$$

For $p_\infty \ge 2$ we use (2.26), Korn's and Young's inequalities to derive from (2.36)

$$C_4 \int_\Omega |\mathbf{D}(\mathbf{v})|^{p_\infty} (1 + |\mathbf{E}|^2)\, dx$$
$$\le c \|\mathbf{f}\|_{-1,p_\infty'}^{p_\infty'} + \|\mathbf{E}\|_2^2 + \frac{C_4}{2} \int_\Omega |\mathbf{D}(\mathbf{v})|^{p_\infty}\, dx + \frac{C_4}{2} \int_\Omega |\mathbf{D}(\mathbf{v})|^{p_\infty} |\mathbf{E}|^2\, dx,$$

which together with (2.33) implies (2.34), (2.35). For the case $p_\infty < 2$ we denote $\Omega_1 = \{x \in \Omega; |\mathbf{D}(\mathbf{v}(x))| \le 1\}$. Using (2.27), Korn's and Young's inequalities yield

$$C_4 \int_{\Omega_1} |\mathbf{D}(\mathbf{v})|^2 (1 + |\mathbf{E}|^2)\, dx + C_4 \int_{\Omega \setminus \Omega_1} |\mathbf{D}(\mathbf{v})|^{p_\infty} (1 + |\mathbf{E}|^2)\, dx$$
$$\le c \|\mathbf{f}\|_{-1,p_\infty'} \left(\|\mathbf{D}\|_{p_\infty,\Omega_1} + \|\mathbf{D}\|_{p_\infty,\Omega \setminus \Omega_1} \right) + c\|\mathbf{E}\|_2^2$$
$$\quad + \frac{C_4}{2} \int_{\Omega_1} |\mathbf{D}(\mathbf{v})|^2 |\mathbf{E}|^2\, dx + \frac{C_4}{2} \int_{\Omega \setminus \Omega_1} |\mathbf{D}(\mathbf{v})|^{p_\infty} |\mathbf{E}|^2\, dx$$
$$\le c \left(\|\mathbf{f}\|_{-1,p_\infty'}^{p_\infty'} + \|\mathbf{f}\|_{-1,p_\infty'}^2 + \|\mathbf{E}\|_2^2 \right) + \frac{C_4}{2} \int_{\Omega_1} |\mathbf{D}(\mathbf{v})|^2 (1 + |\mathbf{E}|^2)\, dx$$
$$\quad + \frac{C_4}{2} \int_{\Omega \setminus \Omega_1} |\mathbf{D}(\mathbf{v})|^{p_\infty} (1 + |\mathbf{E}|^2)\, dx,$$

which implies

$$
\int\limits_{\Omega_1} |\mathbf{D}(\mathbf{v})|^2 (1 + |\mathbf{E}|^2) \, dx + C_4 \int\limits_{\Omega \setminus \Omega_1} |\mathbf{D}(\mathbf{v})|^{p_\infty} (1 + |\mathbf{E}|^2) \, dx
$$
$$
\leq c \left(\|\mathbf{f}\|_{-1,p'_\infty}^{p'_\infty} + \|\mathbf{f}\|_{-1,p'_\infty}^2 + \|\mathbf{E}_0\|_{H(\mathrm{div})}^2 \right).
\tag{2.37}
$$

Since $p_\infty < 2$ we have, using (2.37),

$$
\int\limits_\Omega |\mathbf{D}(\mathbf{v})|^{p_\infty} (1 + |\mathbf{E}|^2) \, dx
$$
$$
\leq \int\limits_{\Omega \setminus \Omega_1} |\mathbf{D}(\mathbf{v})|^{p_\infty} (1 + |\mathbf{E}|^2) \, dx + \left(\int\limits_{\Omega_1} |\mathbf{D}(\mathbf{v})|^2 (1 + |\mathbf{E}|^2) \, dx \right)^{\frac{p_\infty}{2}} \left(\int\limits_{\Omega_1} 1 + |\mathbf{E}|^2 \, dx \right)^{1 - \frac{p_\infty}{2}}
$$
$$
\leq c \left(\|\mathbf{f}\|_{-1,p'_\infty}^{p'_\infty} + \|\mathbf{f}\|_{-1,p'_\infty}^2 + \|\mathbf{E}_0\|_{H(\mathrm{div})}^2 \right)
$$
$$
+ c \left(\|\mathbf{f}\|_{-1,p'_\infty}^{p'_\infty} + \|\mathbf{f}\|_{-1,p'_\infty}^2 + \|\mathbf{E}_0\|_{H(\mathrm{div})}^2 \right)^{\frac{p_\infty}{2}} \left(|\Omega| + \|\mathbf{E}_0\|_{H(\mathrm{div})}^2 \right)^{1 - \frac{p_\infty}{2}},
$$

which implies (2.34), (2.35) in the case $p_\infty < 2$. ∎

Proposition 2.38. *Let the solution* $\mathbf{E}$ *of* (1.1) *be orthogonal to* $H_N(\Omega)$. *Moreover, assume that* $\mathbf{S}$ *satisfies condition* (2.24) *and that* $\|\mathbf{E}\|_2$ *and* $\|\mathbf{f}\|_{-1,p'_\infty}$ *are small enough, then the weak solution* $(\mathbf{E}, \mathbf{v})$ *of* (1.1), (1.2) *ensured by Theorem 2.4 is unique.*

Remark 2.39. It is shown in Theorem 2.3.13 (cf. Picard [101], Werner [128]) that the dimension of the space of harmonic Neumann vector fields $H_N(\Omega)$ is equal to the first Betti number of the domain Ω, i.e. the number of "handles". This number says how many additional conditions are necessary to ensure the uniqueness of $\mathbf{E}$.

PROOF of Proposition 2.38: It is shown in Theorem 2.3.21 that a solution $\mathbf{E}$ of (1.1) which is orthogonal to $H_N(\Omega)$ is uniquely determined. Assume now that $\mathbf{u}, \mathbf{v} \in E_{p,\nu}$ are two weak solutions of (1.2) for given $\mathbf{f}$ and $\mathbf{E}$. Denoting their difference by $\mathbf{w}$, i.e. $\mathbf{w} = \mathbf{u} - \mathbf{v}$, we get by using (2.3), partial integration and div $\mathbf{v} = 0$ that

$$
\int\limits_\Omega \left(\mathbf{S}(\mathbf{D}(\mathbf{u}), \mathbf{E}) - \mathbf{S}(\mathbf{D}(\mathbf{v}), \mathbf{E}) \right) \cdot \mathbf{D}(\mathbf{w}) \, dx = \int\limits_\Omega [\nabla \mathbf{u}] \mathbf{w} \cdot \mathbf{w} \, dx.
\tag{2.40}
$$

The integrand on the left-hand side can be re-written as

$$
\int\limits_0^1 \frac{\partial S_{ij}}{\partial D_{kl}} (\mathbf{D}(s\mathbf{u} + (1-s)\mathbf{v}), \mathbf{E}) \, ds \, D_{kl}(\mathbf{w}) D_{ij}(\mathbf{w})
\tag{2.41}
$$
$$
\geq (C_1 + C_2 |\mathbf{E}|^2) |\mathbf{D}(\mathbf{w})|^2 \int\limits_0^1 (1 + |\mathbf{D}(s\mathbf{u} + (1-s)\mathbf{v})|^2)^{\frac{p(|\mathbf{E}|^2)-2}{2}} \, ds.
$$

Using (2.13) and (2.28) we detect that the last integral is bounded from below by

$$\frac{c}{1 + |\mathbf{D}(\mathbf{u})|^{2-p_\infty} + |\mathbf{D}(\mathbf{v})|^{2-p_\infty}} \qquad \text{if} \quad 1 < p_\infty < 2,$$
$$1 \qquad\qquad\qquad \text{if} \quad 2 \le p_\infty. \tag{2.42}$$

Now assume $9/5 < p_\infty < 2$. From (2.41) and (2.42) we obtain the pointwise inequality

$$|\mathbf{D}(\mathbf{w})|^2 \le c\big(\mathbf{S}(\mathbf{D}(\mathbf{u}), \mathbf{E}) - \mathbf{S}(\mathbf{D}(\mathbf{v}), \mathbf{E})\big) \cdot \mathbf{D}(\mathbf{w})\big(1 + |\mathbf{D}(\mathbf{u})|^{2-p_\infty} + |\mathbf{D}(\mathbf{v})|^{2-p_\infty}\big),$$

which raised to the power $p_\infty/2$, integrated over Ω, together with Hölder's inequality implies that

$$\|\mathbf{D}(\mathbf{w})\|_{p_\infty}^2 \le c\bigg(\int_\Omega \big(\mathbf{S}(\mathbf{D}(\mathbf{u}), \mathbf{E}) - \mathbf{S}(\mathbf{D}(\mathbf{v}), \mathbf{E})\big) \cdot \mathbf{D}(\mathbf{w})\, dx\bigg) \times$$
$$\times \big(1 + \|\mathbf{D}(\mathbf{u})\|_{p_\infty}^{2-p_\infty} + \|\mathbf{D}(\mathbf{v})\|_{p_\infty}^{2-p_\infty}\big). \tag{2.43}$$

The second term in the product on the right-hand side of (2.43) is due to (2.34) estimated by

$$1 + C_6(\mathbf{f}, \mathbf{E}_0)^{\frac{2-p_\infty}{p_\infty}}. \tag{2.44}$$

The right-hand side of (2.40) can be bounded by

$$\|\nabla\mathbf{w}\|_{p_\infty}^2 \|\nabla\mathbf{u}\|_{\frac{3p_\infty}{5p_\infty-6}} \le c\|\mathbf{D}(\mathbf{w})\|_{p_\infty}^2 \|\nabla\mathbf{u}\|_{p_\infty}$$
$$\le c\|\mathbf{D}(\mathbf{w})\|_{p_\infty}^2 C_6(\mathbf{f}, \mathbf{E}_0)^{\frac{1}{p_\infty}}, \tag{2.45}$$

where we used the embeddings $W^{1,p_\infty}(\Omega) \hookrightarrow L^{p_\infty^*}(\Omega)$ and $L^{p_\infty}(\Omega) \hookrightarrow L^{\frac{3p_\infty}{5p_\infty-6}}(\Omega)$, which hold for $p_\infty > 9/5$, also taking into account Korn's inequality and (2.34). From (2.43)–(2.45) we obtain

$$\|\mathbf{D}(\mathbf{w})\|_{p_\infty}^2 \le c\|\mathbf{D}(\mathbf{w})\|_{p_\infty}^2 C_6(\mathbf{f}, \mathbf{E}_0)^{\frac{1}{p_\infty}}\Big(1 + C_6(\mathbf{f}, \mathbf{E}_0)^{\frac{2-p_\infty}{p_\infty}}\Big),$$

which provides the uniqueness if $\mathbf{f}$ and $\mathbf{E}_0$ are sufficiently small, since $C_6(\mathbf{f}, \mathbf{E}_0) \to 0$ as $\|\mathbf{f}\|_{-1,p'_\infty}, \|\mathbf{E}_0\|_{H(\text{div})} \to 0$. For $p_\infty \ge 2$ we estimate the right-hand side of (2.40) by

$$\|\nabla\mathbf{w}\|_2^2 \|\nabla\mathbf{u}\|_{3/2} \le c\|\mathbf{D}(\mathbf{w})\|_2^2 C_6(\mathbf{f}, \mathbf{E}_0)^{\frac{1}{p_\infty}}. \tag{2.46}$$

Therefore we deduce from (2.40)–(2.42) and (2.46) that

$$\|\mathbf{D}(\mathbf{w})\|_2^2 \le c\|\mathbf{D}(\mathbf{w})\|_2^2 C_6(\mathbf{f}, \mathbf{E}_0)^{\frac{1}{p_\infty}}$$

and we conclude the proof for $p_\infty \ge 2$ in the same way as for $p_\infty < 2$.   ∎

3.3 Strong Solutions

Now we shall present another approach to the steady problem, which ensures the existence of solutions to the problem (1.1), (1.2). This technique works mainly with classical Lebesgue and Sobolev spaces and does not use in an essential way the theory of generalized Lebesgue and generalized Sobolev spaces. Since our operator has non-standard growth it will not map the space V_{p_∞}, which is the natural energy space within the context of classical Sobolev spaces, into its dual.

Thus, we cannot apply the theory of monotone operators to ensure that solutions of an appropriate approximation converge to the desired quantities, especially the limiting process in the nonlinear extra stress tensor $\mathbf{S}(\mathbf{D}, \mathbf{E})$ is not clear. However, ideas which have been developed in the context of generalized Newtonian fluids to handle such problems (cf. Málek, Nečas, Růžička [71], [73], Bellout, Bloom Nečas [9], Málek, Nečas, Rokyta, Růžička [70]) will turn out to be useful. This method uses Vitali's theorem for the limiting process in $\mathbf{S}(\mathbf{D}, \mathbf{E})$ and therefore we need almost everywhere convergence of the symmetric velocity gradients $\mathbf{D}(\mathbf{v}^n)$ of some approximate solutions. This convergence is provided from apriori estimates using essentially $-\Delta\mathbf{v}^n$ as a test function. Since $-\Delta\mathbf{v}^n$ is divergence free, but does not vanish on the boundary $\partial\Omega$, i.e. it is not an admissible test function in the weak formulation (2.3), we need a weak formulation including the pressure ϕ. The properties of ϕ are deduced from $(1.2)_1$ interpreted as an operator equation for ϕ.

To clarify the situation let us formally multiply $(1.2)_1$ by $-\Delta\mathbf{v}\xi^2$, where ξ is a usual cut-off function, and consider only the main terms, i.e. we forget the contributions of $\nabla\xi^2$. This yields, using partial integration and $(1.2)_2$

$$
\begin{aligned}
&\int_\Omega \frac{\partial S_{ij}}{\partial D_{kl}}(\mathbf{D}(\mathbf{v}), \mathbf{E}) D_{ij}(\nabla\mathbf{v}) D_{kl}(\nabla\mathbf{v})\, dx \\
&\leq \int_\Omega |\mathbf{v}|\,|\nabla\mathbf{v}|\,|\nabla^2\mathbf{v}|\, dx + \int_\Omega |\phi|\,|\Delta\mathbf{v}|\, dx \\
&\quad + \int_\Omega |\mathbf{f}|\,|\Delta\mathbf{v}|\, dx + \int_\Omega |\mathbf{E}|\,|\nabla\mathbf{E}|\,|\mathbf{D}(\nabla\mathbf{v})|\, dx\,.
\end{aligned}
\tag{3.1}
$$

Assuming that (2.24) holds, the left-hand side is bounded from below by

$$
c\int_\Omega (1 + |\mathbf{E}|^2)(1 + |\mathbf{D}(\mathbf{v})|^2)^{\frac{p_\infty - 2}{2}} |\mathbf{D}(\nabla\mathbf{v})|^2\, dx\,.
\tag{3.2}
$$

Now we should distinguish the cases (i) $p_\infty \geq 2$ and (ii) $p_\infty < 2$. For $p_\infty \geq 2$ we see that (3.2) is bounded from below by

$$
\int_\Omega (1 + |\mathbf{E}|^2)|\mathbf{D}(\nabla\mathbf{v})|^2\, dx\,.
\tag{3.3}
$$

The last two terms on the right-hand side of (3.1) can be treated with the help of

(3.3) under appropriate assumptions on $\mathbf{f}$ and $\nabla\mathbf{E}$. The integral coming from the convective term can be handled e.g. if $\mathbf{v}$ is replaced by its mollification $\mathbf{v}_\epsilon$. However the second integral on the right-hand side of (3.1) causes the main trouble since the operator $-\operatorname{div}\mathbf{S}(.,\mathbf{E})$ maps V_{p_∞} into $(V_{\frac{p_\infty}{p_\infty-p_0+1}})^*$, if $p_0 < p_\infty + 1$, and therefore we see that the pressure ϕ belongs to the space $L^{\frac{p_\infty}{p_0-1}}(\Omega)$ only, since $(\frac{p_\infty}{p_\infty-p_0+1})' = \frac{p_\infty}{p_0-1}$. However $\frac{p_\infty}{p_0-1} \le \frac{p_\infty}{p_\infty-1} \le 2$ and unfortunately we do not have any information about $\Delta\mathbf{v}$ in $L^{\frac{p_\infty}{p_\infty-p_0+1}}(\Omega)$.

To avoid this situation we approximate $\mathbf{S}(\mathbf{D},\mathbf{E})$ by $\mathbf{S}^A(\mathbf{D},\mathbf{E})$, such that $\mathbf{S}^A$ converges locally uniformly to $\mathbf{S}$ for $A \to \infty$ and

$$\mathbf{S}^A(\mathbf{D},\mathbf{E}) \cdot \mathbf{D} \ge c(1+|\mathbf{E}|^2)|\mathbf{D}|^2 ,$$

$$\frac{\partial S^A_{ij}(\mathbf{D},\mathbf{E})}{\partial D_{kl}} B_{ij}B_{kl} \ge c(A)(1+|\mathbf{E}|^2)|\mathbf{B}|^2 , \tag{3.4}$$

$$|\mathbf{S}^A(\mathbf{D},\mathbf{E})| \le c(A)(1+|\mathbf{E}|^2)(1+|\mathbf{D}|^2)^{\frac{1}{2}} .$$

For this approximate operator $\mathbf{S}^A$ one easily checks that $-\operatorname{div}\mathbf{S}^A(.,\mathbf{E})$ maps V_2 into V_2^*, which in turn implies that $\phi^A \in L^2(\Omega)$ and the pressure term can be handled. With this approximation depending on ε and A we have at our disposal well defined approximate solutions $\mathbf{v}^{\varepsilon,A}$. Afterwards we shall derive estimates independent of ε and A which enable us to justify the limiting processes.

The situation for $p_\infty < 2$ is similar. In that case we see that (3.2) is bounded from below by (cf. (5.2.9))

$$\|\mathbf{D}(\nabla\mathbf{v})\|_{3p_\infty}^{p_\infty} . \tag{3.5}$$

Under appropriate assumptions on $\mathbf{f}$ and $\nabla\mathbf{E}$ we can again treat the last two terms in (3.1). The convective term can be handled if $\mathbf{v}$ is replaced by $\mathbf{v}_\varepsilon$, but one must be more careful. The pressure term would be easier for $p_\infty < 2$ if p would be constant, but due to the non-standard growth also this term causes trouble. This can be resolved if we again approximate $\mathbf{S}(\mathbf{D},\mathbf{E})$ by $\mathbf{S}^A(\mathbf{D},\mathbf{E})$ such that $\mathbf{S}^A$ converges locally uniformly to $\mathbf{S}$ and

$$\mathbf{S}^A(\mathbf{D},\mathbf{E}) \cdot \mathbf{D} \ge c(1+|\mathbf{E}|^2)(1+|\mathbf{D}|^2)^{\frac{p_\infty-2}{2}}|\mathbf{D}|^2 ,$$

$$\frac{\partial S^A_{ij}(\mathbf{D},\mathbf{E})}{\partial D_{kl}} B_{ij}B_{kl} \ge c(1+|\mathbf{E}|^2)(1+|\mathbf{D}|^2)^{\frac{p_\infty-2}{2}}|\mathbf{B}|^2 , \tag{3.6}$$

$$|\mathbf{S}^A(\mathbf{D},\mathbf{E})| \le c(A)(1+|\mathbf{E}|^2)(1+|\mathbf{D}|^2)^{\frac{p_\infty-2}{2}}|\mathbf{D}| .$$

Before we conclude this section with the construction of the approximate extra stress tensor $\mathbf{S}^A(\mathbf{D},\mathbf{E})$ and showing precisely its properties (cf. Lemma 3.21, Lemma 3.41, Lemma 3.52 and Lemma 3.65) we formulate our main result as far as strong solutions of the problem (1.1), (1.2) are concerned.

Theorem 3.7. *Let Ω be such that $\partial\Omega \in C^{3,1}$. Assume that $\mathbf{E}_0 \in W^{2,r}(\Omega)$, $r > 3$ and $\mathbf{f} \in L^{p'_\infty}(\Omega)$ are given. Then it holds:*

(i) In the case $2 \le p_\infty \le p_0 < 6$

There exists a strong solution $(\mathbf{E}, \mathbf{v}, \phi)$ of the problem (1.1), (1.2) such that

$$\mathbf{E} - \mathbf{E}_0 \in \overset{\circ}{H}(\mathrm{div}\,) \cap W^{2,r}(\Omega)\,,$$

$$\mathbf{v} \in E_{p(|\mathbf{E}|^2),\nu} \cap W^{2,2}_{\mathrm{loc}}(\Omega) \cap W^{1,\frac{3p_0(p_\infty-1)}{2p_0-3}}_{\mathrm{loc}}(\Omega) \cap V_{p_\infty}\,, \tag{3.8}$$

$$\phi \in L^2_{\mathrm{loc}}(\Omega) \cap L^{p'_0}(\Omega)\,,$$

satisfying (1.1) almost everywhere in Ω and (1.2) in the weak sense, i.e.

$$\int_\Omega \mathbf{S}(\mathbf{D}(\mathbf{v}), \mathbf{E}) \cdot \mathbf{D}(\boldsymbol{\varphi})\,dx + \int_\Omega [\nabla\mathbf{v}]\mathbf{v} \cdot \boldsymbol{\varphi}\,dx - \int_\Omega \phi\,\mathrm{div}\,\boldsymbol{\varphi}\,dx$$
$$= \int_\Omega \mathbf{f} \cdot \boldsymbol{\varphi}\,dx - \chi^E \int_\Omega \mathbf{E} \otimes \mathbf{E} \cdot \mathbf{D}(\boldsymbol{\varphi})\,dx \qquad \forall \boldsymbol{\varphi} \in \mathcal{D}(\Omega)\,. \tag{3.9}$$

(ii) In the case $9/5 < p_\infty \le p_0 < 2$

There exists a strong solution $(\mathbf{E}, \mathbf{v}, \phi)$ of the problem (1.1), (1.2) such that

$$\mathbf{E} - \mathbf{E}_0 \in \overset{\circ}{H}(\mathrm{div}\,) \cap W^{2,r}(\Omega)\,,$$

$$\mathbf{v} \in E_{p(|\mathbf{E}|^2),\nu} \cap W^{2,p_\infty}_{\mathrm{loc}}(\Omega) \cap W^{1,3p_\infty}_{\mathrm{loc}}(\Omega) \cap V_{p_\infty}\,, \tag{3.10}$$

$$\phi \in L^{p'_\infty}_{\mathrm{loc}}(\Omega) \cap L^{p'_0}(\Omega)\,,$$

satisfying (1.1) almost everywhere in Ω and (1.2) in the weak sense (3.9).

(iii) In the case $9/5 < p_\infty < 2 \le p_0 < p^*_\infty$

There exists a strong solution $(\mathbf{E}, \mathbf{v}, \phi)$ of the problem (1.1), (1.2) such that

$$\mathbf{E} - \mathbf{E}_0 \in \overset{\circ}{H}(\mathrm{div}\,) \cap W^{2,r}(\Omega)\,,$$

$$\mathbf{v} \in E_{p(|\mathbf{E}|^2),\nu} \cap W^{2,p_\infty}_{\mathrm{loc}}(\Omega) \cap W^{1,\frac{3p_0(p_\infty-1)}{2p_0-3}}_{\mathrm{loc}}(\Omega) \cap V_{p_\infty}\,, \tag{3.11}$$

$$\phi \in L^{p'_\infty}_{\mathrm{loc}}(\Omega) \cap L^{p'_0}(\Omega)\,,$$

satisfying (1.1) almost everywhere in Ω and (1.2) in the weak sense (3.9).

Remark 3.12. 1) The lower bound

$$p_\infty > 9/5$$

is the same as in Theorem 2.4. This is due to the fact that we have been able to bound the second order derivatives locally, despite of the fact that the pressure is a non-local term. If one would like to prove global estimates of the second order derivatives it is to be expected that there will be a slightly worse lower bound (cf. the treatment of the unsteady problem for generalized Newtonian fluids in Málek, Nečas, Růžička [72], [73]).

2) The upper bound $p_0 < 6$ in the case $p_\infty \geq 2$ is not satisfactory. It is due to the approximation, which is chosen for the treatment of the pressure. However, from the point of view of applications this bound seems to be not too restrictive, since most shear dependent fluids have a shear power p below or near 2.

3) An interesting feature of Theorem 3.7 is, that it provides a method to obtain existence of solutions without using the method of monotone operators. This will be crucial in the treatment of the unsteady problem, since in this situation it is not clear what is the equivalence of the decomposition

$$L^q(I \times \Omega) = L^q(I, L^q(\Omega))$$

if $L^{p(t,x)}(I \times \Omega)$ is considered. This decomposition is crucial for the application of the theory of monotone operators in the parabolic setting.

4) As already discussed in Section 2.1, there are, to the knowledge of the author, only two other results concerning the existence of solutions for an elliptic problem with non-standard growth conditions. In Marcellini [78] the key point is an estimate of the $W^{1,\infty}$-norm via the W^{1,p_∞}-norm. It is very unlikely that this approach can be adapted to the system (1.2) since the elliptic operator $\mathbf{S}$ depends on $|\mathbf{D}(\mathbf{v})|$ and not on $|\nabla \mathbf{v}|$. This additional difficulty is even for $p = $ const. not well understood. Only in the two-dimensional case there are some results in this direction (cf. Ladyzhenskaya, Seregin [59] for the unsteady case and Kaplický, Málek, Stará [50], [51] for the steady case). In Boccardo, Gallouet, Marcellini [12] some special test functions are constructed, which enable the authors to show the almost everywhere convergence of some approximate solutions. For standard growth conditions this method can be adapted to treat generalized Newtonian fluids (cf. Frehse, Málek, Steinhauer [30], Růžička [112]). This however, requires tools which are not available if generalized Sobolev spaces are needed (cf. conjecture in Chapter 2). Most investigations of equations or systems with non-standard growth are concerned with the question of regularity of solutions. The authors usually assume that the solution belongs to the space $W^{1,p_0}(\Omega)$. However, it is in general not clear how such solutions can be constructed.

3.3.1   Approximations

Now we discuss the approximation of the extra stress $\mathbf{S}(\mathbf{D}, \mathbf{E})$. We assume that $\mathbf{S}$ is given by (1.3) and that (1.4)–(1.6) are satisfied. Moreover we require that (1.4.59)–(1.4.63) hold. These conditions ensure that (2.24) holds, i.e. $\mathbf{S}(\mathbf{D}, \mathbf{E})$ is uniformly monotone. We define $\mathbf{S}^A(\mathbf{D}, \mathbf{E})$, for $A \geq 1$, by

$$
\begin{aligned}
S_{ij}^A =& \alpha_{21} V^A(|\mathbf{D}|^2, |\mathbf{E}|^2) E_i E_j + (\alpha_{31} + \alpha_{33} |\mathbf{E}|^2) \partial_{ij} U^A(|\mathbf{D}|^2, |\mathbf{E}|^2) \\
& + \alpha_{51} \big(\partial_{ik} U^A(|\mathbf{D}|^2, |\mathbf{E}|^2) E_i E_k + \partial_{jk} U^A(|\mathbf{D}|^2, |\mathbf{E}|^2) E_i E_k \big) ,
\end{aligned}
\tag{3.13}
$$

where ∂_{ij} denotes the derivative with respect to the components $D_{ij}, i, j = 1, 2, 3$, of the symmetric part of the velocity gradient. This means e.g.

$$\partial_{ij}U^A(|\mathbf{D}|^2, |\mathbf{E}|^2) \equiv \frac{\partial U^A(|\mathbf{D}|^2, |\mathbf{E}|^2)}{\partial D_{ij}}$$

and similar formulas for $V^A, U, V, \mathbf{S}, \mathbf{S}^A$. We will use the letters i, j, k, l as indices for this derivative. The potentials $U^A(|\mathbf{D}|^2, |\mathbf{E}|^2)$ and $V^A(|\mathbf{D}|^2, |\mathbf{E}|^2)$ are defined differently in the cases (i) $p_\infty \geq 2$ and (ii) $p_\infty < 2$.

(i) The case $p_\infty \geq 2$

Since both U^A and V^A depend on $|\mathbf{E}|^2$ only through $p(|\mathbf{E}|^2)$ we will mostly suppress the dependence of U^A and V^A on $|\mathbf{E}|^2$ if there is no danger of confusion. Let us define

$$U^A(|\mathbf{D}|^2, |\mathbf{E}|^2) = \begin{cases} U(|\mathbf{D}|^2, |\mathbf{E}|^2) & |\mathbf{D}| \leq A, \\ a_0 + a_1(1 + |\mathbf{D}|^2)^{\frac{1}{2}} + a_2(1 + |\mathbf{D}|^2) & |\mathbf{D}| \geq A, \end{cases} \tag{3.14}$$

and

$$V^A(|\mathbf{D}|^2, |\mathbf{E}|^2) = \begin{cases} (1 + |\mathbf{D}|^2)^{\frac{p(|\mathbf{E}|^2)-1}{2}} - 1 & |\mathbf{D}| \leq A, \\ b_0 + b_1(1 + |\mathbf{D}|^2)^{\frac{1}{2}} & |\mathbf{D}| \geq A, \end{cases} \tag{3.15}$$

where $a_i = a_i(|\mathbf{E}|^2)$, $i = 0, 1, 2$, $b_i = b_i(|\mathbf{E}|^2)$, $i = 0, 1$, and

$$U(|\mathbf{D}|^2, |\mathbf{E}|^2) = \frac{1}{p(|\mathbf{E}|^2)}\left((1 + |\mathbf{D}|^2)^{\frac{p(|\mathbf{E}|^2)}{2}} - 1\right). \tag{3.16}$$

Note that

$$\partial_{ij}U(|\mathbf{D}|^2, |\mathbf{E}|^2) = (1 + |\mathbf{D}|^2)^{\frac{p(|\mathbf{E}|^2)-2}{2}} D_{ij}, \tag{3.17}$$

$$U(0, |\mathbf{E}|^2) = \partial_{ij}U(0, |\mathbf{E}|^2) = 0. \tag{3.18}$$

From the definition of $\mathbf{S}$ and (3.13)–(3.17) it is clear that

$$\mathbf{S}^A(\mathbf{D}, \mathbf{E}) = \mathbf{S}(\mathbf{D}, \mathbf{E}) \qquad \text{if } |\mathbf{D}| \leq A. \tag{3.19}$$

We require some regularity for U^A and V^A, namely that $U^A(\cdot, |\mathbf{E}|^2) \in C^2(\mathbb{R}^+)$, $U^A(|\mathbf{D}|^2, \cdot) \in C^1(\mathbb{R}^+)$ and $V^A(\cdot, \cdot) \in C^1(\mathbb{R}^+ \times \mathbb{R}^+)$, which allows us to compute the coefficients a_0, a_1, a_2 and b_0, b_1. We obtain

$$a_0 = \left(\frac{1}{p} + \frac{p}{2} - \frac{3}{2}\right)(1 + A^2)^{\frac{p}{2}} - \frac{1}{p},$$

$$a_1 = (2 - p)(1 + A^2)^{\frac{p-1}{2}},$$

$$a_2 = \frac{p-1}{2}(1 + A^2)^{\frac{p-2}{2}}, \tag{3.20}$$

$$b_0 = (2 - p)(1 + A^2)^{\frac{p-1}{2}} - 1,$$

$$b_1 = (p - 1)(1 + A^2)^{\frac{p-2}{2}}.$$

Lemma 3.21. *Let* **S** *be given by* (1.3)*, such that the conditions* (1.4)–(1.6) *and* (1.4.59)–(1.4.63) *are satisfied. Assume that* $\mathbf{S}^A$ *is given by* (3.13)–(3.16) *and* (3.20) *and that* $p_\infty \geq 2$*. Then we have for all* $i, j, k, l = 1, 2, 3$ *all* $\mathbf{B}, \mathbf{D} \in X$*, all* $\mathbf{E} \in \mathbb{R}^3$ *and* A *sufficiently large*

$$|\mathbf{S}^A(\mathbf{D}, \mathbf{E})| \leq c_1(A)(1 + |\mathbf{E}|^2)(1 + |\mathbf{D}|^2)^{\frac{1}{2}}, \tag{3.22}$$

$$|\partial_{ij}\mathbf{S}^A(\mathbf{D}, \mathbf{E})| \leq c_2(A)(1 + |\mathbf{E}|^2), \tag{3.23}$$

$$\mathbf{S}^A(\mathbf{D}, \mathbf{E}) \cdot \mathbf{D} \geq c_3(1 + |\mathbf{E}|^2)|\mathbf{D}|^2 \begin{cases} (1 + |\mathbf{D}|^2)^{\frac{p(|\mathbf{E}|^2)-2}{2}}, \\ (1 + A^2)^{\frac{p(|\mathbf{E}|^2)-2}{2}}, \end{cases} \tag{3.24}$$

$$\partial_{kl}S_{ij}^A(\mathbf{D}, \mathbf{E})B_{ij}B_{kl} \geq c_4(1 + |\mathbf{E}|^2)|\mathbf{B}|^2 \begin{cases} (1 + |\mathbf{D}|^2)^{\frac{p(|\mathbf{E}|^2)-2}{2}}, \\ (1 + A^2)^{\frac{p(|\mathbf{E}|^2)-2}{2}}, \end{cases} \tag{3.25}$$

$$|\mathbf{S}^A| \leq c_5(1 + |\mathbf{E}|^2)(1 + |\mathbf{D}|^2)^{\frac{1}{2}} \begin{cases} (1 + |\mathbf{D}|^2)^{\frac{p(|\mathbf{E}|^2)-2}{2}}, \\ (1 + A^2)^{\frac{p(|\mathbf{E}|^2)-2}{2}}, \end{cases} \tag{3.26}$$

$$|\partial_{ij}\mathbf{S}^A(\mathbf{D}, \mathbf{E})| \leq c_6(1 + |\mathbf{E}|^2) \begin{cases} (1 + |\mathbf{D}|^2)^{\frac{p(|\mathbf{E}|^2)-2}{2}}, \\ (1 + A^2)^{\frac{p(|\mathbf{E}|^2)-2}{2}}, \end{cases} \tag{3.27}$$

$$\partial U^A(|\mathbf{D}|^2, |\mathbf{E}|^2) \cdot \mathbf{D} \leq c_7|\mathbf{D}|^2 \begin{cases} (1 + |\mathbf{D}|^2)^{\frac{p(|\mathbf{E}|^2)-2}{2}}, \\ (1 + A^2)^{\frac{p(|\mathbf{E}|^2)-2}{2}}, \end{cases} \tag{3.28}$$

$$1 + \partial U^A(|\mathbf{D}|^2, |\mathbf{E}|^2) \cdot \mathbf{D} \geq c_8(1 + |\mathbf{D}|^2) \begin{cases} (1 + |\mathbf{D}|^2)^{\frac{p(|\mathbf{E}|^2)-2}{2}}, \\ (1 + A^2)^{\frac{p(|\mathbf{E}|^2)-2}{2}}, \end{cases} \tag{3.29}$$

$$|\partial U^A(|\mathbf{D}|^2, |\mathbf{E}|^2)| \leq c_9|\mathbf{D}| \begin{cases} (1 + |\mathbf{D}|^2)^{\frac{p(|\mathbf{E}|^2)-2}{2}}, \\ (1 + A^2)^{\frac{p(|\mathbf{E}|^2)-2}{2}}, \end{cases} \tag{3.30}$$

where the first line in the above cases holds if $|\mathbf{D}| \leq A$ *while the second one holds for* $|\mathbf{D}| \geq A$*. The constants* c_3–c_9 *depend on* p_0, p_∞*, whereas* c_1, c_2 *depend on* A, p_0 *and* p_∞*.*

PROOF : Since $\mathbf{S}^A(\mathbf{D}, \mathbf{E}) = \mathbf{S}(\mathbf{D}, \mathbf{E})$ if $|\mathbf{D}| \leq A$, we get all assertions for $\mathbf{S}^A$ from the properties of $\mathbf{S}(\mathbf{D}, \mathbf{E})$ and some straightforward computations (cf. (1.9), (1.8) and (2.24)). Inequalities (3.28), (3.29) follow directly from (3.17) and the observation that $(1 + |\mathbf{D}|^2)^{\frac{p-2}{2}}$ is bounded by $2^{\frac{p-2}{2}}$ or $\partial U^A \cdot \mathbf{D}$ if $|\mathbf{D}| \leq 1$ or $1 \leq |\mathbf{D}| \leq A$ respectively. Inequality (3.30) is a consequence of (3.17). Let us therefore assume $|\mathbf{D}| \geq A$ in what follows. One easily computes

$$
\begin{aligned}
S_{ij}^A(\mathbf{D}, \mathbf{E}) = {}& \alpha_{21}\big(b_0 + b_1(1 + |\mathbf{D}|^2)^{1/2}\big)E_i E_j \\
& + \big(\alpha_{31} + \alpha_{33}|\mathbf{E}|^2\big)\Big(\frac{a_1}{(1 + |\mathbf{D}|^2)^{1/2}} + 2a_2\Big)D_{ij} \\
& + \alpha_{51}\Big(\frac{a_1}{(1 + |\mathbf{D}|^2)^{1/2}} + 2a_2\Big)\big(D_{im}E_k E_m + D_{jm}E_m E_i\big)
\end{aligned}
\tag{3.31}
$$

and

$$
\begin{aligned}
\partial_{kl}S_{ij}^A(\mathbf{D}, \mathbf{E}) = {}& \alpha_{21}b_1\frac{D_{kl}}{(1 + |\mathbf{D}|^2)^{1/2}}E_i E_j \\
& + \big(\alpha_{31} + \alpha_{33}|\mathbf{E}|^2\big)\Big(\frac{a_1}{(1 + |\mathbf{D}|^2)^{1/2}} + 2a_2\Big)\delta_{ik}\delta_{jl} \\
& - \big(\alpha_{31} + \alpha_{33}|\mathbf{E}|^2\big)\frac{a_1}{(1 + |\mathbf{D}|^2)^{3/2}}D_{ij}D_{kl} \\
& + \alpha_{51}\Big(\frac{a_1}{(1 + |\mathbf{D}|^2)^{1/2}} + 2a_2\Big)\big(\delta_{ik}E_l E_j + \delta_{jk}E_l E_i\big) \\
& - \alpha_{51}\frac{a_1}{(1 + |\mathbf{D}|^2)^{3/2}}\big(D_{im}E_m D_{kl}E_j + D_{jm}E_m D_{kl}E_i\big),
\end{aligned}
\tag{3.32}
$$

which immediately yields (3.22) and (3.23) for $A \geq 1$. Since $|\mathbf{D}| \geq A$ we obtain

$$
\frac{a_1}{(1 + |\mathbf{D}|^2)^{1/2}} + 2a_2 = (p - 2)(1 + A^2)^{\frac{p-2}{2}}\Big(1 - \Big(\frac{1 + A^2}{1 + |\mathbf{D}|^2}\Big)^{\frac{1}{2}}\Big) + (1 + A^2)^{\frac{p-2}{2}}
\tag{3.33}
$$

$$
\leq (p - 1)(1 + A^2)^{\frac{p-2}{2}}.
\tag{3.34}
$$

Similarly we compute

$$
|b_1| + \frac{|a_1|}{(1 + |\mathbf{D}|^2)^{1/2}} + \frac{|b_0|}{(1 + |\mathbf{D}|^2)^{1/2}} \leq c(1 + A^2)^{\frac{p-2}{2}}
\tag{3.35}
$$

and therefore we easily derive (3.26) and (3.27) from (3.31) and (3.32). From (3.31), (3.20) and (3.33) we compute

$$
\begin{aligned}
\mathbf{S}^A \cdot \mathbf{D} = {}& (p - 2)(1 + A^2)^{\frac{p-2}{2}}\Big(1 - \Big(\frac{1 + A^2}{1 + |\mathbf{D}|^2}\Big)^{\frac{1}{2}}\Big)\Big\{\alpha_{21}(1 + |\mathbf{D}|^2)^{\frac{1}{2}}\mathbf{D}\mathbf{E} \cdot \mathbf{E} \\
& + \big(\alpha_{31} + \alpha_{33}|\mathbf{E}|^2\big)|\mathbf{D}|^2 + 2\alpha_{51}|\mathbf{D}\mathbf{E}|^2\Big\} \\
& + (1 + A^2)^{\frac{p-2}{2}}\Big\{\alpha_{21}\frac{(1 + A^2)^{\frac{p-2}{2}}(1 + |\mathbf{D}|^2)^{\frac{1}{2}} - 1}{(1 + A^2)^{\frac{p-2}{2}}}\mathbf{D}\mathbf{E} \cdot \mathbf{E} \\
& + \big(\alpha_{31} + \alpha_{33}|\mathbf{E}|^2\big)|\mathbf{D}|^2 + 2\alpha_{51}|\mathbf{D}\mathbf{E}|^2\Big\}.
\end{aligned}
\tag{3.36}
$$

Using for $|\mathbf{D}| \geq A \geq \varepsilon^{-1}$

$$(1 + |\mathbf{D}|^2)^{\frac{1}{2}} \leq (1 + \varepsilon^2)|\mathbf{D}| \tag{3.37}$$

the first term with α_{21} can be estimated from below by

$$-(1 + \varepsilon^2)|\alpha_{21}|\,|\mathbf{D}|\,|\mathbf{E}|\,|\mathbf{DE}|\,.$$

Due to the strict inequality in (1.6) and the considerations in Section 1.4 (cf. (1.4.45)) we can choose ε such that the terms in the first squiggly brackets are non-negative. Since also the factor in front of it is so we deduce that the first summand in (3.36) is non-negative. Furthermore one can show that (compare with (1.4.39), (1.4.40))

$$\frac{(1 + A^2)^{\frac{p-2}{2}}(1 + |\mathbf{D}|^2)^{\frac{1}{2}} - 1}{(1 + A^2)^{\frac{p-2}{2}}} \leq k_0(p)|\mathbf{D}| \tag{3.38}$$

for all $|\mathbf{D}| \geq A \geq 1$. Thus, condition (1.6) ensures the validity of (3.24). From (3.32), (3.20) and (3.33) we calculate

$$\begin{aligned}
\partial_{kl}S^A_{ij}(\mathbf{D},\mathbf{E})B_{ij}B_{kl} &= (p-2)(1+A^2)^{\frac{p-2}{2}}\left(1 - \left(\frac{1+A^2}{1+|\mathbf{D}|^2}\right)^{\frac{1}{2}}\right) \times \\
&\quad \times \left\{ \left(\alpha_{31} + \alpha_{33}|\mathbf{E}|^2\right)|\mathbf{B}|^2 + 2\alpha_{51}|\mathbf{BE}|^2 \right\} \\
&\quad + (p-2)(1+A^2)^{\frac{p-1}{2}}\left(\alpha_{31} + \alpha_{33}|\mathbf{E}|^2\right)\frac{(\mathbf{B}\cdot\mathbf{D})^2}{1+|\mathbf{D}|^2} \\
&\quad + (1+A^2)^{\frac{p-2}{2}}\left\{\alpha_{21}(p-1)\frac{(\mathbf{D}\cdot\mathbf{B})(\mathbf{BE}\cdot\mathbf{E})}{(1+|\mathbf{D}|^2)^{1/2}}\right. \\
&\quad + 2(p-2)\alpha_{51}\frac{(\mathbf{BE}\cdot\mathbf{DE})(\mathbf{B}\cdot\mathbf{D})}{1+|\mathbf{D}|^2}\left(\frac{1+A^2}{1+|\mathbf{D}|^2}\right)^{\frac{1}{2}} \\
&\quad \left. + \left(\alpha_{31} + \alpha_{33}|\mathbf{E}|^2\right)|\mathbf{B}|^2 + 2\alpha_{51}|\mathbf{BE}|^2 \right\} \\
&= I_1 + I_2 + I_3\,.
\end{aligned} \tag{3.39}$$

The non-negativity of I_1 and I_2 follows immediately from the assumptions of the lemma. Just using $|\mathbf{D}| \geq A$ the terms in the squiggly brackets in I_3 can be handled in the same way as the right-hand side of (1.4.52). Therefore the conditions (1.4.59)–(1.4.63) together with (1.5), (1.6) ensure the validity of (3.25). Inequality (3.29) follows from (3.36) with $\alpha_{21} = \alpha_{33} = \alpha_{51} = 0$, $\alpha_{31} = 1$ and (3.37) with $\varepsilon = 1$. Finally, from

$$\partial U^A(|\mathbf{D}|^2, |\mathbf{E}|^2) = \left(\frac{a_1}{(1 + |\mathbf{D}|^2)^{\frac{1}{2}}} + 2a_2\right)\mathbf{D}$$

and (3.34) we obtain (3.28) and (3.30). This completes the proof of the lemma. ∎

Remark 3.40. From (3.25) and (3.27) it is clear that for all $\mathbf{B} \in X$ the quantities

$$\partial_{kl} S_{ij}^A(\mathbf{D}, \mathbf{E}) B_{ij} B_{kl}$$

and

$$(1 + |\mathbf{E}|^2) |\mathbf{B}|^2 \begin{cases} (1 + |\mathbf{D}|^2)^{\frac{p(|\mathbf{E}|^2) - 2}{2}} \\ (1 + A^2)^{\frac{p(|\mathbf{E}|^2) - 2}{2}} \end{cases}$$

are equivalent.

Later on we will need also the derivative with respect to the components $E_n, n = 1, 2, 3$, of the electric field. We shall use the notation

$$\partial_n U^A(|\mathbf{D}|^2, |\mathbf{E}|^2) \equiv \frac{\partial U^A(|\mathbf{D}|^2, |\mathbf{E}|^2)}{\partial E_n}$$

and similar formulas for $V^A, \mathbf{S}^A, U, V$ and $\mathbf{S}$. We will use the letters m, n as indices for this derivative.

Lemma 3.41. *Let the assumption of Lemma 3.21 be satisfied. Then there exists constants depending on the function p and α_{ij} such that*

$$|\partial_n \partial_{ij} U^A(|\mathbf{D}|^2, |\mathbf{E}|^2)| \le c |\mathbf{E}| |\mathbf{D}| \begin{cases} (1 + |\mathbf{D}|^2)^{\frac{p(|\mathbf{E}|^2) - 2}{2}} \left(1 + \ln(1 + |\mathbf{D}|^2)\right), \\ (1 + A^2)^{\frac{p(|\mathbf{E}|^2) - 2}{2}} \left(1 + \ln(1 + |A|^2)\right), \end{cases} \tag{3.42}$$

$$|\partial_n V^A(|\mathbf{D}|^2, |\mathbf{E}|^2)| \le c |\mathbf{E}| (1 + |\mathbf{D}|^2)^{\frac{1}{2}} \begin{cases} (1 + |\mathbf{D}|^2)^{\frac{p(|\mathbf{E}|^2) - 2}{2}} \left(1 + \ln(1 + |\mathbf{D}|^2)\right), \\ (1 + A^2)^{\frac{p(|\mathbf{E}|^2) - 2}{2}} \left(1 + \ln(1 + A^2)\right), \end{cases} \tag{3.43}$$

$$|\partial_n \mathbf{S}^A(\mathbf{D}, \mathbf{E})| \le c |\mathbf{E}| (1 + |\mathbf{E}|^2)(1 + |\mathbf{D}|^2)^{\frac{1}{2}} \times$$
$$\times \begin{cases} (1 + |\mathbf{D}|^2)^{\frac{p(|\mathbf{E}|^2) - 2}{2}} \left(1 + \ln(1 + |\mathbf{D}|^2)\right), \\ (1 + A^2)^{\frac{p(|\mathbf{E}|^2) - 2}{2}} \left(1 + \ln(1 + A^2)\right), \end{cases} \tag{3.44}$$

$$|\partial_n U^A(|\mathbf{D}|^2, |\mathbf{E}|^2)| \le c |\mathbf{E}| (1 + |\mathbf{D}|^2) \begin{cases} (1 + |\mathbf{D}|^2)^{\frac{p(|\mathbf{E}|^2) - 2}{2}} \left(1 + \ln(1 + |\mathbf{D}|^2)\right), \\ (1 + A^2)^{\frac{p(|\mathbf{E}|^2) - 2}{2}} \left(1 + \ln(1 + A^2)\right). \end{cases} \tag{3.45}$$

PROOF : The estimates follow immediately from (3.14), (3.15), (3.20) and (3.13). ∎

Remark 3.46. From the inequality

$$1 + \ln(1 + y^2) \leq c(\gamma)(1 + y^2)^{\gamma/2} \qquad \forall \gamma > 0, \tag{3.47}$$

together with (3.45) and (3.29) and a similar argument as in (5.2.14) we deduce for all $s > 1$

$$|\partial_n U^A(|\mathbf{D}|^2, |\mathbf{E}|^2)| \leq c_{10}\left(1 + \partial U^A(|\mathbf{D}|^2, |\mathbf{E}|^2) \cdot \mathbf{D}\right)^s. \tag{3.48}$$

(ii) The case $1 < p_\infty < 2$

In this case the potentials U^A and V^A are chosen differently. We define

$$U^A(|\mathbf{D}|^2, |\mathbf{E}|^2) = \begin{cases} U(|\mathbf{D}|^2, |\mathbf{E}|^2) & |\mathbf{D}| \leq A, \\ a_0 + a_1(1 + |\mathbf{D}|^2)^{\frac{\pi}{2}} + a_2(1 + |\mathbf{D}|^2)^{\frac{p_\infty}{2}} & |\mathbf{D}| \geq A, \end{cases} \tag{3.49}$$

and

$$V^A(|\mathbf{D}|^2, |\mathbf{E}|^2) = \begin{cases} (1 + |\mathbf{D}|^2)^{\frac{p(|\mathbf{E}|^2)-1}{2}} - 1 & |\mathbf{D}| \leq A, \\ b_0 + b_1(1 + |\mathbf{D}|^2)^{\frac{\pi-1}{2}} + b_2(1 + |\mathbf{D}|^2)^{\frac{p_\infty-1}{2}} & |\mathbf{D}| \geq A, \end{cases} \tag{3.50}$$

where $a_i = a_i(|\mathbf{E}|^2)$, $b_i = b_i(|\mathbf{E}|^2)$, $i = 0, 1, 2$. The number $\pi \in (1, p_\infty)$ will be chosen later. If we require the same regularity for U^A and V^A as in the case $p_\infty \geq 2$ we can compute a_0, a_1, a_2 and b_0, b_1, b_2. For b_0, b_1 and b_2 we dispose of a free parameter which will be chosen such that $b_0 = -1$. Therefore we can calculate

$$\begin{aligned} a_0 &= (1 + A^2)^{\frac{p}{2}}\left(\frac{1}{p} - \frac{p_\infty - p}{\pi(p_\infty - \pi)} - \frac{p - \pi}{p_\infty(p_\infty - \pi)}\right) - \frac{1}{p}, \\ a_1 &= \frac{p_\infty - p}{\pi(p_\infty - \pi)}(1 + A^2)^{\frac{p-\pi}{2}}, \\ a_2 &= \frac{p - \pi}{p_\infty(p_\infty - \pi)}(1 + A^2)^{\frac{p-p_\infty}{2}}, \\ b_0 &= -1, \\ b_1 &= \frac{p_\infty - p}{p_\infty - \pi}(1 + A^2)^{\frac{p-\pi}{2}}, \\ b_2 &= \frac{p - \pi}{p_\infty - \pi}(1 + A^2)^{\frac{p-p_\infty}{2}}. \end{aligned} \tag{3.51}$$

We have a similar results as for the case $p_\infty \geq 2$, namely

Lemma 3.52. *Let* $\mathbf{S}$ *be given by* (1.3), *such that the conditions* (1.4)–(1.6) *and* (1.4.59)–(1.4.63) *are satisfied. Assume that* $\mathbf{S}^A$ *is given by* (3.13), (3.49)–(3.51) *and that* $1 < p_\infty < 2$. *Then we have for all* $i, j, k, l = 1, 2, 3$ *and all* $\mathbf{B}, \mathbf{D} \in X$, *all* $\mathbf{E} \in \mathbb{R}^3$ *and* A *sufficiently large*

$$|S^A(\mathbf{D}, \mathbf{E})| \leq c_1(A)(1 + |\mathbf{E}|^2)(1 + |\mathbf{D}|^2)^{\frac{p_\infty-1}{2}}, \tag{3.53}$$

$$|\partial_{ij}\mathbf{S}^A(\mathbf{D},\mathbf{E})| \le c_2(A)(1+|\mathbf{E}|^2)(1+|\mathbf{D}|^2)^{\frac{p_\infty-2}{2}}, \tag{3.54}$$

$$\mathbf{S}^A(\mathbf{D},\mathbf{E})\cdot\mathbf{D} \ge c_3(1+|\mathbf{E}|^2)(1+|\mathbf{D}|^2)^{\frac{p_\infty-2}{2}}|\mathbf{D}|^2 \begin{cases} (1+|\mathbf{D}|^2)^{\frac{p(|\mathbf{E}|^2)-p_\infty}{2}}, \\ (1+A^2)^{\frac{p(|\mathbf{E}|^2)-p_\infty}{2}}, \end{cases} \tag{3.55}$$

$$\partial_{kl}S_{ij}^A(\mathbf{D},\mathbf{E})B_{ij}B_{kl} \ge c_4(1+|\mathbf{E}|^2)(1+|\mathbf{D}|^2)^{\frac{p_\infty-2}{2}}|\mathbf{B}|^2 \begin{cases} (1+|\mathbf{D}|^2)^{\frac{p(|\mathbf{E}|^2)-p_\infty}{2}}, \\ (1+A^2)^{\frac{p(|\mathbf{E}|^2)-p_\infty}{2}}, \end{cases} \tag{3.56}$$

$$|\mathbf{S}^A| \le c_5(1+|\mathbf{E}|^2)(1+|\mathbf{D}|^2)^{\frac{p_\infty-1}{2}} \begin{cases} (1+|\mathbf{D}|^2)^{\frac{p(|\mathbf{E}|^2)-p_\infty}{2}}, \\ (1+A^2)^{\frac{p(|\mathbf{E}|^2)-p_\infty}{2}}, \end{cases} \tag{3.57}$$

$$|\partial_{ij}\mathbf{S}^A(\mathbf{D},\mathbf{E})| \le c_6(1+|\mathbf{E}|^2)(1+|\mathbf{D}|^2)^{\frac{p_\infty-2}{2}} \begin{cases} (1+|\mathbf{D}|^2)^{\frac{p(|\mathbf{E}|^2)-p_\infty}{2}}, \\ (1+A^2)^{\frac{p(|\mathbf{E}|^2)-p_\infty}{2}}, \end{cases} \tag{3.58}$$

$$\partial U^A(|\mathbf{D}|^2,|\mathbf{E}|^2)\cdot\mathbf{D} \le c_7(1+|\mathbf{D}|^2)^{\frac{p_\infty-2}{2}}|\mathbf{D}|^2 \begin{cases} (1+|\mathbf{D}|^2)^{\frac{p(|\mathbf{E}|^2)-p_\infty}{2}}, \\ (1+A^2)^{\frac{p(|\mathbf{E}|^2)-p_\infty}{2}}, \end{cases} \tag{3.59}$$

$$1+\partial U^A(|\mathbf{D}|^2,|\mathbf{E}|^2)\cdot\mathbf{D} \ge c_8(1+|\mathbf{D}|^2)^{\frac{p_\infty}{2}} \begin{cases} (1+|\mathbf{D}|^2)^{\frac{p(|\mathbf{E}|^2)-p_\infty}{2}}, \\ (1+A^2)^{\frac{p(|\mathbf{E}|^2)-p_\infty}{2}}, \end{cases} \tag{3.60}$$

$$|\partial U^A(|\mathbf{D}|^2,|\mathbf{E}|^2)| \le c_9|\mathbf{D}|^{p_\infty-1} \begin{cases} (1+|\mathbf{D}|^2)^{\frac{p(|\mathbf{E}|^2)-p_\infty}{2}}, \\ (1+A^2)^{\frac{p(|\mathbf{E}|^2)-p_\infty}{2}}, \end{cases} \tag{3.61}$$

where the first line in the above cases holds if $|\mathbf{D}| \le A$ while the second one holds for $|\mathbf{D}| \ge A$. The constants c_1, c_2 depend on A, p_0, p_∞ and π, while the constants c_3–c_9 depend only on p_0, p_∞ and π.

PROOF : For $|\mathbf{D}| \le A$ we have $\mathbf{S}^A = \mathbf{S}$, $U^A = U$ and therefore all assertions of the lemma are clear in this case (cf. the proof of Lemma 3.21). Henceforth we assume $|\mathbf{D}| \ge A$. The inequalities (3.53), (3.54) and (3.57), (3.58) follow immediately from the construction of $\mathbf{S}^A$ and some straightforward computations (cf. (3.13), (3.49)–(3.51) and the proof of Lemma 3.21). Moreover, we notice

$$
\mathbf{S}^A \cdot \mathbf{D} = \frac{p - p_\infty}{p_\infty - \pi} \left(1 - \left(\frac{1 + A^2}{1 + |\mathbf{D}|^2} \right)^{\frac{p_\infty - \pi}{2}} \right) (1 + A^2)^{\frac{p - p_\infty}{2}} (1 + |\mathbf{D}|^2)^{\frac{p_\infty - 2}{2}} \times
$$

$$
\times \left\{ \alpha_{21} (1 + |\mathbf{D}|^2)^{\frac{1}{2}} \mathbf{D} \mathbf{E} \cdot \mathbf{E} + (\alpha_{31} + \alpha_{33} |\mathbf{E}|^2) |\mathbf{D}|^2 + 2\alpha_{51} |\mathbf{D} \mathbf{E}|^2 \right\}
$$

$$
+ (1 + A^2)^{\frac{p - p_\infty}{2}} (1 + |\mathbf{D}|^2)^{\frac{p_\infty - 2}{2}} \left\{ \alpha_{21} \frac{(1 + |\mathbf{D}|^2)^{\frac{p_\infty - 1}{2}} - 1}{(1 + |\mathbf{D}|^2)^{\frac{p_\infty - 2}{2}}} \mathbf{D} \mathbf{E} \cdot \mathbf{E} \right.
$$

$$
\left. + (\alpha_{31} + \alpha_{33} |\mathbf{E}|^2) |\mathbf{D}|^2 + 2\alpha_{51} |\mathbf{D} \mathbf{E}|^2 \right\}. \tag{3.62}
$$

Both terms can be handled in the same way as the corresponding terms in (3.36), proving (3.55). Furthermore, we can write

$$
\partial_{kl} S_{ij}^A B_{ij} B_{kl} = (1 + A^2)^{\frac{p - p_\infty}{2}} (1 + |\mathbf{D}|^2)^{\frac{p_\infty - 2}{2}} \times \tag{3.63}
$$

$$
\times \left\{ \frac{p_\infty - p}{p_\infty - \pi} \left(\frac{1 + A^2}{1 + |\mathbf{D}|^2} \right)^{\frac{p_\infty - \pi}{2}} H(\pi) + \frac{p - \pi}{p_\infty - \pi} H(p_\infty) \right\},
$$

where

$$
H(p) = \alpha_{21} (p - 1) \frac{(\mathbf{B} \cdot \mathbf{D})(\mathbf{B} \mathbf{E} \cdot \mathbf{E})}{(1 + |\mathbf{D}|^2)^{1/2}} + (\alpha_{31} + \alpha_{33} |\mathbf{E}|^2) \left(|\mathbf{B}|^2 + (p - 2) \frac{(\mathbf{B} \cdot \mathbf{D})^2}{1 + |\mathbf{D}|^2} \right)
$$

$$
+ 2\alpha_{51} \left(|\mathbf{B} \mathbf{E}|^2 (p - 2) \frac{(\mathbf{B} \mathbf{E} \cdot \mathbf{D} \mathbf{E})(\mathbf{B} \cdot \mathbf{D})}{1 + |\mathbf{D}|^2} \right).
$$

$H(p)$ is exactly the factor which also appears in the discussion for the monotonicity of $\mathbf{S}$ (cf. (1.4.52)). The conditions (1.4.59)–(1.4.63) ensure that $H(p) \geq 0$. But due to the strict inequalities in (1.4.59)–(1.4.63) we can choose $\pi = p_\infty - \varepsilon, \varepsilon$ small enough, such that also $H(\pi) \geq 0$. Therefore we can estimate the terms in the squiggly brackets in (3.63) from below by

$$
\frac{p_\infty - p}{p_\infty - \pi} H(\pi) + \frac{p - \pi}{p_\infty - \pi} H(p_\infty) = H(p).
$$

The computations in Section 1.4 show that for all $p \in [p_\infty, p_0]$

$$
H(p) \geq c(1 + |\mathbf{E}|^2) |\mathbf{B}|^2
$$

which together with (3.63) gives (3.56). Inequality (3.60) follows from (3.62) setting $\alpha_{21} = \alpha_{33} = \alpha_{51} = 0$, $\alpha_{31} = 1$ and (3.37). Finally, from (3.49), (3.51) and $\pi < p_\infty$ one easily deduces

$$
|\partial U^A (|\mathbf{D}|^2, |\mathbf{E}|^2)| \leq c (1 + A^2)^{\frac{p - p_\infty}{2}} (1 + |\mathbf{D}|^2)^{\frac{p_\infty - 2}{2}} |\mathbf{D}|.
$$

Taking into account (3.37) yields (3.61) and (3.59). The proof is complete.   ∎

Remark 3.64. From (3.56) and (3.58) it is clear that for $\mathbf{B} \in X$,

$$\partial_{kl} S_{ij}^A(\mathbf{D}, \mathbf{E}) B_{ij} B_{kl}$$

and

$$(1 + |\mathbf{E}|^2)(1 + |\mathbf{D}|^2)^{\frac{p_\infty - 2}{2}} |\mathbf{B}|^2 \begin{cases} (1 + |\mathbf{D}|^2)^{\frac{p(|\mathbf{E}|^2) - p_\infty}{2}} \\ (1 + A^2)^{\frac{p(|\mathbf{E}|^2) - p_\infty}{2}} \end{cases}$$

are equivalent.

Lemma 3.65. *Let the assumptions of Lemma 3.52 be satisfied. Then there exist constants depending on the function p and on α_{ij} and π such that*

$$|\partial_n \partial_{ij} U^A| \le c \, |\mathbf{E}| (1 + |\mathbf{D}|^2)^{\frac{p_\infty - 1}{2}} \begin{cases} (1 + |\mathbf{D}|^2)^{\frac{p(|\mathbf{E}|^2) - p_\infty}{2}} \left(1 + \ln(1 + |\mathbf{D}|^2)\right), \\ (1 + A^2)^{\frac{p(|\mathbf{E}|^2) - p_\infty}{2}} \left(1 + \ln(1 + A^2)\right), \end{cases} \tag{3.66}$$

$$|\partial_n V^A| \le c \, |\mathbf{E}| (1 + |\mathbf{D}|^2)^{\frac{p_\infty - 1}{2}} \begin{cases} (1 + |\mathbf{D}|^2)^{\frac{p(|\mathbf{E}|^2) - p_\infty}{2}} \left(1 + \ln(1 + |\mathbf{D}|^2)\right), \\ (1 + A^2)^{\frac{p(|\mathbf{E}|^2) - p_\infty}{2}} \left(1 + \ln(1 + A^2)\right), \end{cases} \tag{3.67}$$

$$|\partial_n \mathbf{S}^A(\mathbf{D}, \mathbf{E})| \le c \, |\mathbf{E}| (1 + |\mathbf{E}|^2)(1 + |\mathbf{D}|^2)^{\frac{p_\infty - 1}{2}} \times$$
$$\times \begin{cases} (1 + |\mathbf{D}|^2)^{\frac{p(|\mathbf{E}|^2) - p_\infty}{2}} \left(1 + \ln(1 + |\mathbf{D}|^2)\right), \\ (1 + A^2)^{\frac{p(|\mathbf{E}|^2) - p_\infty}{2}} \left(1 + \ln(1 + A^2)\right), \end{cases} \tag{3.68}$$

$$|\partial_n U^A| \le c \, |\mathbf{E}| (1 + |\mathbf{D}|^2)^{\frac{p_\infty}{2}} \begin{cases} (1 + |\mathbf{D}|^2)^{\frac{p(|\mathbf{E}|^2) - p_\infty}{2}} \left(1 + \ln(1 + |\mathbf{D}|^2)\right), \\ (1 + A^2)^{\frac{p(|\mathbf{E}|^2) - p_\infty}{2}} \left(1 + \ln(1 + A^2)\right). \end{cases} \tag{3.69}$$

PROOF : The estimates can be computed easily from (3.49)–(3.51) and (3.13). ∎

Remark 3.70. From (3.47), (3.60) and (3.69) and a similar argument as in (5.2.14) we deduce for all $s > 1$

$$|\partial_n U^A(|\mathbf{D}|^2, |\mathbf{E}|^2)| \le c_{10} \left(1 + \partial U^A(|\mathbf{D}|^2, |\mathbf{E}|^2) \cdot \mathbf{D}\right)^s, \tag{3.71}$$

with a constant c independent of A.

For the definition of the approximate problem we also need the notion of a mollification. Let $\omega \in \mathcal{D}(\mathbb{R}^3)$ be a usual mollification kernel with support in $B_1(0)$ and $\int_{\mathbb{R}^3} \omega \, dx = 1$. For every $\varepsilon > 0$ we define $\omega_\varepsilon(x) = \varepsilon^{-3} \omega(\frac{x}{\varepsilon})$ and set

$$\mathbf{v}_\varepsilon(x) \equiv \mathbf{v} * \omega_\varepsilon(x) = \frac{1}{\varepsilon^3} \int_{\mathbb{R}^3} \omega(\frac{x - y}{\varepsilon}) \mathbf{v}(y) \, dy.$$

Definition 3.72. *The triple* $(\mathbf{E}, \mathbf{v}, \phi) = (\mathbf{E}, \mathbf{v}^{\varepsilon,A}, \phi^{\varepsilon,A})$, *for* $\varepsilon > 0, A \geq A_0$ *given, is said to be a weak solution of the problem* (1.1) *and*

$$
\begin{aligned}
- \operatorname{div} \mathbf{S}^A(\mathbf{D}(\mathbf{v}), \mathbf{E}) + [\nabla\mathbf{v}]\mathbf{v}_\varepsilon + \nabla\phi &= \mathbf{f} + \chi^E[\nabla\mathbf{E}]\mathbf{E}, \\
\operatorname{div} \mathbf{v} &= 0 \\
\mathbf{v} &= 0
\end{aligned}
\qquad
\begin{aligned}
&\quad in\ \Omega, \\
&\quad on\ \partial\Omega,
\end{aligned}
\qquad (1.2)_{\varepsilon,A}
$$

where $\mathbf{S}^A$ *is given by* (3.13), (3.14), (3.15), (3.20), *if* $p_0 \geq 2$, *or by* (3.49)–(3.51), *if* $1 < p_\infty < 2$, *if and only if*

$$
\begin{aligned}
\mathbf{E} &\in H(\operatorname{div}) \cap H(\operatorname{curl}), \\
\mathbf{v} &\in V_q, \\
\phi &\in L^{q'}(\Omega),
\end{aligned}
\qquad (3.73)
$$

where $q = \min(p_\infty, 2)$; *the system* (1.1)$_{1,2}$ *is satisfied almost everywhere in* Ω, *and* $\mathbf{E} - \mathbf{E}_0 \in \overset{\circ}{H}(\operatorname{div})$. *The weak formulation of* (1.2)$_{\varepsilon,A}$

$$
\int_\Omega \mathbf{S}^A(\mathbf{D}(\mathbf{v}), \mathbf{E}) \cdot \mathbf{D}(\boldsymbol{\varphi})\, dx + \int_\Omega [\nabla\mathbf{v}]\mathbf{v}_\varepsilon \cdot \boldsymbol{\varphi}\, dx - \int_\Omega \phi \operatorname{div} \boldsymbol{\varphi}\, dx
$$
$$
= \langle \mathbf{f}, \boldsymbol{\varphi} \rangle_{1,q} - \chi^E \int_\Omega \mathbf{E} \otimes \mathbf{E} \cdot \mathbf{D}(\boldsymbol{\varphi})\, dx
\qquad (3.74)
$$

is satisfied for all $\boldsymbol{\varphi} \in W_0^{1,q}(\Omega)$.

3.4   Existence of Approximate Solutions

Now, we shall show the existence of weak solutions for the system (1.1), (1.2)$_{\varepsilon,A}$ with the additional property that $\nabla^2\mathbf{v} \in L^q_{\text{loc}}(\Omega)$, where

$$
q \equiv \min(2, p_\infty). \qquad (4.1)
$$

Proposition 4.2. *Let* $\Omega \subseteq \mathbb{R}^3$ *be a bounded domain with* $\partial\Omega \in C^{3,1}$. *Assume that* $\mathbf{E}_0 \in W^{2,r}(\Omega)$, $r > 3$, $\mathbf{f} \in L^{q'}(\Omega)$, $\varepsilon > 0$ *and* $A \geq A_0$ *are given. If* $3/2 < p_\infty$ *then there exists a weak solution* $(\mathbf{E}, \mathbf{v}, \phi) = (\mathbf{E}, \mathbf{v}^{\varepsilon,A}, \phi^{\varepsilon,A})$ *of the problem* (1.1), (1.2)$_{\varepsilon,A}$, *such that*

$$
\|\mathbf{E}\|_{2,r} \leq c\,\|\mathbf{E}_0\|_{2,r}, \qquad (4.3)
$$

$$
\|\mathbf{D}(\mathbf{v})\|_{q,\nu} + \|\nabla\mathbf{v}\|_q \leq c(\mathbf{f}, \mathbf{E}_0), \qquad (4.4)
$$

$$
\|\partial U^A(|\mathbf{D}(\mathbf{v})|^2, |\mathbf{E}|^2) \cdot \mathbf{D}(\mathbf{v})\|_{1,\nu} \leq c(\mathbf{f}, \mathbf{E}_0). \qquad (4.5)
$$

Moreover, we have

$$
\|\phi\|_{p_0'} \leq c(\mathbf{f}, \mathbf{E}_0) + \|[\nabla\mathbf{v}]\mathbf{v}_\varepsilon\|_{(p_0^*)'} \leq c(\mathbf{f}, \mathbf{E}_0, \varepsilon^{-1}), \qquad (4.6)
$$

$$
\|\phi\, \eta^\gamma\|_{q'} \leq c(\mathbf{f}, \mathbf{E}_0) + c\|\mathbf{S}^A\, \eta^\gamma\|_{q'} + \|[\nabla\mathbf{v}]\mathbf{v}_\varepsilon\, \eta^\gamma\|_{(q^*)'} + c\|[\nabla\mathbf{v}]\mathbf{v}_\varepsilon\|_{(p_0^*)'}
$$
$$
\leq c(\mathbf{f}, \mathbf{E}_0, \varepsilon^{-1}, A), \qquad (4.7)
$$

with $\eta \in \mathcal{D}(\Omega), \gamma > 1$, and

$$\|\nabla^2 \mathbf{v}\|_{q,\text{loc}} \leq c(\mathbf{f}, \mathbf{E}_0, \varepsilon^{-1}, A), \tag{4.8}$$

$$\int_\Omega (1 + |\mathbf{D}(\mathbf{v})|^2)^{\frac{q-2}{2}} |\mathbf{D}(\nabla \mathbf{v})|^2 \xi^{2\alpha} \, dx \leq c(\mathbf{f}, \mathbf{E}_0, \varepsilon^{-1}, A), \tag{4.9}$$

where $\xi \in \mathcal{D}(\Omega)$ is a usual cut-off function and $\alpha > 1$. Equation $(1.2)_{\varepsilon,A}$ holds almost everywhere in Ω.

PROOF : The first part of the proposition concerning the existence of a weak solution to the system (1.1), $(1.2)_{\varepsilon,A}$ is standard using apriori estimates, the theory of monotone operators and a compactness argument. However, the remaining part is not so common and we will give a detailed proof.

The existence of $\mathbf{E}$ solving (1.1) and having the properties stated above follows from Theorem 2.3.31. We shall fix one such solution for the following considerations. Let us formally derive an apriori estimate for the system $(1.2)_{\varepsilon,A}$. We use here the Galerkin approach (cf. Lions [68], Section 2.5) and are very brief since the details are similar to the proof of Theorem 2.4. Using the Galerkin approximation $\mathbf{v}^n$ as a test function in the weak formulation of $(1.2)_{\varepsilon,A}$ with divergence free test functions, i.e. the pressure term in (3.74) is absent and only test functions $\varphi \in V$ are allowed, we obtain

$$\int_\Omega \mathbf{S}^A(\mathbf{D}(\mathbf{v}^n), \mathbf{E}) \cdot \mathbf{D}(\mathbf{v}^n) \, dx = \int_\Omega \mathbf{f} \cdot \mathbf{v}^n \, dx - \chi^E \int_\Omega \mathbf{E} \otimes \mathbf{E} \cdot \mathbf{D}(\mathbf{v}^n) \, dx. \tag{4.10}$$

Using (3.24) and (3.55), respectively, the left-hand side is bounded from below by

$$c \int_\Omega (1 + |\mathbf{E}|^2)(1 + |\mathbf{D}(\mathbf{v}^n)|^2)^{\frac{q-2}{2}} |\mathbf{D}(\mathbf{v}^n)|^2 \, dx$$
$$\geq c \int_\Omega (1 + |\mathbf{E}|^2)|\mathbf{D}(\mathbf{v}^n)|^q \, dx + c \int_\Omega |\nabla \mathbf{v}^n|^q \, dx - c \int_\Omega 1 + |\mathbf{E}|^2 \, dx, \tag{4.11}$$

where we employed the pointwise inequality $(1 + x^2)^{\frac{q-2}{2}} x^2 \geq c(x^q - 1)$ and Korn's inequality. The right-hand side of (4.10) is bounded from above by

$$c\|\mathbf{f}\|_{q'} \|\mathbf{v}^n\|_q + c\|\mathbf{E}\|_{2q'}^2 \|\mathbf{D}(\mathbf{v}^n)\|_{q,\nu}. \tag{4.12}$$

Recall that $\|\mathbf{D}(\mathbf{v})\|_{q,\nu}^q = \int_\Omega |\mathbf{D}(\mathbf{v})|^q(1 + |\mathbf{E}|^2) \, dx$. From (4.10)–(4.12) and Young's inequality we immediately obtain (4.4) for $\mathbf{v}^n$. From (3.22), (3.25) and (3.53), (3.56), respectively, and the regularity properties of $\mathbf{E}$ it is clear that the operator $-\operatorname{div} \mathbf{S}^A(\mathbf{D}(\cdot), \mathbf{E}) : V_q \to (V_q)^*$ is uniformly monotone. From (4.4) $[\nabla \mathbf{v}]\mathbf{v}_\varepsilon \in L^{(q^*)'}$ for $p_\infty \geq 3/2$, the estimate (4.4) and the compact embedding $W^{1,q}(\Omega) \hookrightarrow L^{q^*}(\Omega)$, for $q > 3/2$, are enough to justify the limiting process in the Galerkin system by standard arguments (cf. proof of Theorem 2.4 with $p_\infty = p_0 = q$). Thus we obtain the existence of a solution satisfying $(3.73)_2$ and (3.74) for $\varphi \in V_q$. Moreover, for the limiting element we again obtain (4.10) and that the right-hand side is bounded by some constant $c(\mathbf{f}, \mathbf{E}_0)$. Now, we use (3.24), (3.28) and (3.55), (3.59), respectively, on the left-hand side of (4.10) written now for $\mathbf{v}$, which is possible due to (4.4), to arrive at (4.5). Defining $\mathbf{F} \in (W_0^{1,q}(\Omega))^*$ by

$$
\begin{aligned}
\langle \mathbf{F}, \boldsymbol{\varphi} \rangle_{1,q} &\equiv \int_\Omega \mathbf{S}^A(\mathbf{D}(\mathbf{v}), \mathbf{E}) \cdot \mathbf{D}(\boldsymbol{\varphi}) \, dx + \int_\Omega [\nabla \mathbf{v}] \mathbf{v}_\varepsilon \cdot \boldsymbol{\varphi} \, dx \\
&\quad - \int_\Omega \mathbf{f} \cdot \boldsymbol{\varphi} \, dx + \chi^E \int_\Omega \mathbf{E} \otimes \mathbf{E} \cdot \mathbf{D}(\boldsymbol{\varphi}) \, dx \,,
\end{aligned}
\tag{4.13}
$$

we see that

$$
\langle \mathbf{F}, \boldsymbol{\varphi} \rangle_{1,q} = 0 \qquad \forall \boldsymbol{\varphi} \in V_q \,.
$$

This and Theorem 5.1.10 lead to the existence of $\phi \in L^{q'}(\Omega)$, $\int_\Omega \phi \, dx = 0$, such that

$$
\langle \mathbf{F}, \boldsymbol{\varphi} \rangle_{1,q} = -\int_\Omega \phi \operatorname{div} \boldsymbol{\varphi} \, dx \qquad \forall \boldsymbol{\varphi} \in W_0^{1,q}(\Omega) \,.
\tag{4.14}
$$

From (4.13) and (4.14) we obtain that $(\mathbf{v}, \phi)$ is a weak solution of $(1.2)_{\varepsilon,A}$, satisfying $(3.73)_{2,3}$ and (3.74). Using Proposition 5.1.25 we have

$$
\|\phi\|_{L^{q'}(\Omega)} = \sup_{\substack{g \in L_0^q(\Omega) \\ \|g\|_{L_0^q(\Omega)} \le 1}} \left| \int_\Omega \phi g \, dx \right| \le \sup_{\substack{\varphi \in W_0^{1,q}(\Omega) \\ \|\nabla \varphi\|_q \le c}} \left| \int_\Omega \phi \operatorname{div} \boldsymbol{\varphi} \, dx \right|
\tag{4.15}
$$

which together with (4.14), (4.13) implies

$$
\begin{aligned}
\|\phi\|_{q'} &\le c(\mathbf{f}, \mathbf{E}_0) + \|\mathbf{S}^A\|_{q'} + \|[\nabla \mathbf{v}] \mathbf{v}_\varepsilon\|_{(q^*)'} \\
&\le c(\mathbf{f}, \mathbf{E}_0, \varepsilon^{-1}, A) \,,
\end{aligned}
$$

where we used (3.22) and (3.53), respectively, the properties of the mollifier and (4.4). We also obtain an estimate of ϕ independent of A, using that we can derive from (3.26) and (3.57), respectively, and a similar argument as in (5.2.14) the estimate

$$
|\mathbf{S}^A(\mathbf{D}(\mathbf{v}), \mathbf{E})|^{p_0'} \le c(\mathbf{E}_0) \big(1 + \partial U^A(|\mathbf{D}(\mathbf{v})|^2, |\mathbf{E}|^2) \cdot \mathbf{D}(\mathbf{v}) \big) \,,
$$

which together with (4.5) implies

$$
\|\mathbf{S}^A(\mathbf{D}(\mathbf{v}), \mathbf{E})\|_{p_0'} \le c(\mathbf{f}, \mathbf{E}_0) \,.
\tag{4.16}
$$

From (4.15), (4.14), (4.13) and (4.16) we easily deduce (4.6). In order to obtain estimate (4.7) we apply the negative norm theorem (cf. Theorem 5.1.8) for $\eta \in \mathcal{D}(\Omega)$, $\gamma > 1$ to arrive at

$$
\|\phi \eta^\gamma\|_{q'} \le c \|\phi \eta^\gamma\|_{-1,q'} + c \|\nabla(\phi \eta^\gamma)\|_{-1,q'}
\tag{4.17}
$$

$$
\le c \|\phi \eta^\gamma\|_{(q^*)'} + c(\gamma, \nabla \eta) \|\phi \eta^{\gamma-1}\|_{(q^*)'} + c \sup_{\substack{\varphi \in W_0^{1,q}(\Omega) \\ \|\varphi\|_{1,q} \le 1}} \left| \int_\Omega \phi \operatorname{div}(\boldsymbol{\varphi} \eta^\gamma) \, dx \right| \,.
$$

Inserting into the weak formulation (3.74) the test function $\varphi\,\eta^\gamma$ one easily computes that the last term in (4.17) is estimated by

$$c\|\mathbf{S}^A\eta^\gamma\|_{q'} + c(\nabla\xi,\gamma)\|\mathbf{S}^A\eta^{\gamma-1}\|_{(q^*)'} + c\|[\nabla\mathbf{v}]\mathbf{v}_\epsilon\,\eta^\gamma\|_{(q^*)'} + c(\mathbf{f},\mathbf{E}_0)\,. \qquad (4.18)$$

Now, for $p_0 \leq q^*$ we use (4.6) and (4.16) and see

$$\|\phi\,\eta^{\gamma-1}\|_{(q^*)'} + \|\mathbf{S}^A\eta^{\gamma-1}\|_{(q^*)'} \leq c(\mathbf{f},\mathbf{E}_0) + \|[\nabla\mathbf{v}]\mathbf{v}_\epsilon\|_{(p_0^*)'}\,. \qquad (4.19)$$

If $p_0 > q^*$ we use Young's inequality to obtain that

$$\|g\,\eta^{\gamma-1}\|_{(q^*)'} \leq \frac{1}{2}\|g\,\eta^\gamma\|_{q'} + c\|g\|_{p_0'}\,, \qquad (4.20)$$

which is applied to ϕ and $\mathbf{S}^A$. From (4.17)–(4.20) we immediately derive (4.7). It remains to show (4.8) and (4.9), which will be established by the difference quotient method. For $V \subset\subset \Omega$ we consider $\xi \in \mathcal{D}(\Omega)$, $0 \leq \xi \leq 1$, $\xi \equiv 1$ in V and put (cf. (5.1.34), $T = T_{k,h}$)

$$\mathbf{w}(x) \equiv \mathbf{v}(Tx) - \mathbf{v}(x)\,. \qquad (4.21)$$

Using $h^{-2}\mathbf{w}(x)\xi^{2\alpha}(x)$, $\alpha > 1$, as a test function in the weak formulation (3.74) we can derive the following identity:

$$\begin{aligned}
&\frac{1}{h^2}\int_\Omega \Big(\mathbf{S}^A\big(\mathbf{D}(\mathbf{v}(Tx)),\mathbf{E}(Tx)\big) - \mathbf{S}^A\big(\mathbf{D}(\mathbf{v}(x)),\mathbf{E}(x)\big)\Big)\cdot\mathbf{D}\big(\mathbf{w}(x)\xi^{2\alpha}(x)\big)\,dx \\
&+ \frac{1}{h^2}\int_\Omega \big([\nabla\mathbf{v}(Tx)]\mathbf{v}_\epsilon(Tx) - [\nabla\mathbf{v}(x)]\mathbf{v}_\epsilon(x)\big)\cdot\mathbf{w}(x)\xi^{2\alpha}(x)\,dx \\
&- \frac{1}{h^2}\int_\Omega \big(\phi(Tx) - \phi(x)\big)\operatorname{div}\big(\mathbf{w}(x)\xi^{2\alpha}(x)\big)\,dx \\
&- \frac{1}{h^2}\int_\Omega \big(\mathbf{f}(Tx) - \mathbf{f}(x)\big)\cdot\mathbf{w}(x)\xi^{2\alpha}(x)\,dx \\
&+ \frac{\chi^E}{h^2}\int_\Omega \big(\mathbf{E}(Tx)\otimes\mathbf{E}(Tx) - \mathbf{E}(x)\otimes\mathbf{E}(x)\big)\cdot\mathbf{D}\big(\mathbf{w}(x)\xi^{2\alpha}(x)\big)\,dx = 0\,.
\end{aligned} \qquad (4.22)$$

We denote the integrals on the left-hand side by $I_1,\dots,I_5$ and discuss them separately. Denoting $\mathbf{S}^A(y) = \mathbf{S}^A(\mathbf{D}(\mathbf{v}(y)),\mathbf{E}(y))$ we observe

$$\begin{aligned}
h^2 I_1 &= \int_\Omega \big(\mathbf{S}^A(Tx) - \mathbf{S}^A(x)\big)\cdot\mathbf{D}(\mathbf{w}(x))\,\xi^{2\alpha}(x)\,dx \\
&+ 2\alpha\int_\Omega \big(S_{ij}^A(Tx) - S_{ij}^A(x)\big)w_i(x)\,\xi^{2\alpha-1}(x)\frac{\partial\xi(x)}{\partial x_j}\,dx \\
&= h^2(J_1 + J_2)\,.
\end{aligned} \qquad (4.23)$$

The integrals J_1 and J_2 must be treated differently. Let us start with J_1. We use the notation

$$
\begin{aligned}
\mathbf{F}(x) &= \mathbf{E}(Tx) - \mathbf{E}(x)\,, \\
\mathbf{D}_\lambda(x) &= \mathbf{D}(\mathbf{v}(x)) + \lambda\mathbf{D}(\mathbf{w}(x)) = (1 - \lambda)\mathbf{D}(\mathbf{v}(x)) + \lambda\mathbf{D}(\mathbf{v}(Tx))\,, \\
\mathbf{E}_\lambda(x) &= \mathbf{E}(x) + \lambda\mathbf{F}(x)\,,
\end{aligned}
\tag{4.24}
$$

to shorten our formulas. Thus we can re-write $h^2 J_1$ as

$$
\begin{aligned}
h^2 J_1 &= \int_\Omega \int_0^1 \frac{d}{d\lambda} \mathbf{S}^A(\mathbf{D}_\lambda, \mathbf{E}_\lambda)\, d\lambda \cdot \mathbf{D}(\mathbf{w})\xi^{2\alpha}\, dx \\
&= \int_\Omega \int_0^1 \partial_{kl} S_{ij}^A(\mathbf{D}_\lambda, \mathbf{E}_\lambda) D_{ij}(\mathbf{w}) D_{kl}(\mathbf{w})\xi^{2\alpha}\, d\lambda\, dx \\
&\quad + \int_\Omega \int_0^1 \partial_n S_{ij}^A(\mathbf{D}_\lambda, \mathbf{E}_\lambda) F_n D_{ij}(\mathbf{w})\xi^{2\alpha}\, d\lambda\, dx \\
&= h^2(J_{1,1} + J_{1,2})\,.
\end{aligned}
\tag{4.25}
$$

Now using (3.25) and (3.56), respectively, we note that (recall $q = \min(2, p_\infty)$)

$$
\begin{aligned}
J_{1,1} &\geq c_4 \int_\Omega \int_0^1 (1 + |\mathbf{D}_\lambda|^2)^{\frac{q-2}{2}}\, d\lambda\, \left|\frac{\mathbf{D}(\mathbf{w})}{h}\right|^2 \xi^{2\alpha}\, dx\,, \\
&\geq c_{11} \int_\Omega \left(1 + |\mathbf{D}(\mathbf{v}(x))|^2 + |\mathbf{D}(\mathbf{v}(Tx))|^2\right)^{\frac{q-2}{2}} \left|\frac{\mathbf{D}(\mathbf{w}(x))}{h}\right|^2 \xi^{2\alpha}(x)\, dx\,,
\end{aligned}
$$

where we used (5.1.24). If we also employ (5.1.22) and the estimate (4.4) to bound the constant appearing in (5.1.22) we finally obtain that

$$
\begin{aligned}
J_{1,1} &\geq \frac{c_{11}}{2} \int_\Omega \left(1 + |\mathbf{D}(\mathbf{v}(x))|^2 + |\mathbf{D}(\mathbf{v}(Tx))|^2\right)^{\frac{q-2}{2}} \left|\frac{\mathbf{D}(\mathbf{w}(x))}{h}\right|^2 \xi^{2\alpha}\, dx \\
&\quad + c(\mathbf{f}, \mathbf{E}_0)\left(\int_\Omega \left|\frac{\mathbf{D}(\mathbf{w})}{h}\right|^q \xi^{\alpha q}\, dx\right)^{\frac{2}{q}} \\
&\geq \frac{c_{11}}{2} \int_\Omega \left(1 + |\mathbf{D}(\mathbf{v}(x))|^2 + |\mathbf{D}(\mathbf{v}(Tx))|^2\right)^{\frac{q-2}{2}} \left|\frac{\mathbf{D}(\mathbf{w}(x))}{h}\right|^2 \xi^{2\alpha}\, dx \\
&\quad + c_{12}\left(\int_\Omega \left|\frac{\nabla\mathbf{w}}{h}\right|^q \xi^{\alpha q}\, dx\right)^{\frac{2}{q}} - c\left(\int_\Omega \left|\frac{\mathbf{w}}{h}\right|^q |\nabla\xi|^q\, dx\right)^{\frac{2}{q}}\,,
\end{aligned}
\tag{4.26}
$$

where we used in the last line elementary calculations and Korn's inequality. From

(3.44) and (3.68), respectively, we deduce, using Young's inequality and (5.1.24), that

$$
|J_{1,2}| \le c(A) \int_\Omega \int_0^1 |\mathbf{E}_\lambda|(1+|\mathbf{E}_\lambda|^2)(1+|\mathbf{D}_\lambda|^2)^{\frac{q-1}{2}} \left|\frac{\mathbf{F}}{h}\right| \left|\frac{\mathbf{D}(\mathbf{w})}{h}\right| \xi^{2\alpha}\, d\lambda\, dx
$$

$$
\le c(A, \mathbf{E}_0) \int_\Omega \left|\frac{\mathbf{F}}{h}\right|^2 \xi^{2\alpha}(1+|\mathbf{D}(\mathbf{v}(x))|^2 + \mathbf{D}(\mathbf{v}(Tx))|^2)^{\frac{q}{2}}\, dx \tag{4.27}
$$

$$
+ \frac{c_{11}}{8} \int_\Omega (1+|\mathbf{D}(\mathbf{v}(x)|^2 + |\mathbf{D}(\mathbf{v}(Tx))|^2)^{\frac{q-2}{2}} \left|\frac{\mathbf{D}(\mathbf{w})}{h}\right|^2 \xi^{2\alpha}\, dx\,.
$$

Now we turn our attention to the integral J_2. We notice that (cf. (5.1.31), (5.1.29))

$$
J_2 = 2\alpha \int_\Omega \frac{1}{h} \int_0^1 \frac{d}{d\lambda} S_{ij}^A(T_\lambda x)d\lambda\, \frac{w_i(x)}{h} \xi^{2\alpha-1}(x) \frac{\partial \xi(x)}{\partial x_j}\, dx
$$

$$
= 2\alpha \int_\Omega \frac{\partial}{\partial x_k} \int_0^1 S_{ij}^A(T_\lambda x)\, d\lambda\, \frac{w_i(x)}{h} \xi^{2\alpha-1}(x) \frac{\partial \xi(x)}{\partial x_j}\, dx
$$

$$
= -2\alpha \int_\Omega \int_0^1 S_{ij}^A(T_\lambda x)\, d\lambda\, \frac{\partial}{\partial x_k}\Big(\frac{w_i(x)}{h}\Big) \xi^{2\alpha-1}(x) \frac{\partial \xi(x)}{\partial x_j}\, dx \tag{4.28}
$$

$$
- 2\alpha \int_\Omega \int_0^1 S_{ij}^A(T_\lambda x)\, d\lambda\, \frac{w_i(x)}{h} \Big(\frac{\partial \xi^{2\alpha-1}(x)}{\partial x_k} \frac{\partial \xi(x)}{\partial x_j} + \xi^{2\alpha-1}(x) \frac{\partial^2 \xi(x)}{\partial x_k \partial x_j}\Big)\, dx
$$

$$
= J_{2,1} + J_{2,2}\,.
$$

From (3.22) and (3.53), respectively, Hölder's and Young's inequalities we obtain

$$
|J_{2,1}| \le c(A, \mathbf{E}_0) \int_\Omega \int_0^1 (1+|\mathbf{D}(\mathbf{v}(T_\lambda x))|^2)^{\frac{q-1}{2}}\, d\lambda \left|\frac{\nabla \mathbf{w}}{h}\right| \xi^{2\alpha-1}\, |\nabla \xi|\, dx
$$

$$
\le \frac{c_{12}}{4}\Big(\int_\Omega \left|\frac{\nabla \mathbf{w}}{h}\right|^q \xi^{\alpha q}\, dx\Big)^{\frac{2}{q}} \tag{4.29}
$$

$$
+ c(A, \mathbf{E}_0)\Big(\int_\Omega \Big(\int_0^1 (1+|\mathbf{D}(\mathbf{v}(T_\lambda x))|^2)^{\frac{q-1}{2}}\, d\lambda\Big)^{\frac{q}{q-1}} |\nabla \xi|^{q'} \xi^{(\alpha-1)q'}\, dx\Big)^{\frac{2(q-1)}{q}}
$$

and

$$
|J_{2,2}| \le c(A, \mathbf{E}_0, \nabla^2 \xi) \int_\Omega \Big(\int_0^1 (1+|\mathbf{D}(\mathbf{v}(T_\lambda x))|^2)^{\frac{q-1}{2}}\, d\lambda\Big)^{\frac{q}{q-1}}\, dx
$$

$$
+ c(A, \mathbf{E}_0, \nabla^2 \xi) \int_\Omega \left|\frac{\mathbf{w}}{h}\right|^q\, dx\,. \tag{4.30}
$$

Note, that the last term in (4.29) and both terms in (4.30) are bounded by

$$c(A, \mathbf{E}_0, \nabla^2 \xi)(1 + \|\nabla \mathbf{v}\|_q^2),$$

which is finite due to (4.4). Furthermore, the convective term gives

$$h^2 I_2 = \int_\Omega v_{\varepsilon k}(x) \frac{\partial w_i(x)}{\partial x_k} w_i(x) \xi^{2\alpha}(x)\, dx + \int_\Omega w_{\varepsilon k}(x) \frac{\partial v_i(Tx)}{\partial x_k} w_i(x) \xi^{2\alpha}(x)\, dx$$

$$= \int_\Omega v_{\varepsilon k}(x) \frac{\partial w_i(x)}{\partial x_k} w_i(x) \xi^{2\alpha}(x)\, dx - \int_\Omega w_{\varepsilon k}(x) v_i(Tx) \frac{\partial w_i(x)}{\partial x_k} \xi^{2\alpha}(x)\, dx$$

$$- 2\alpha \int_\Omega w_{\varepsilon k}(x) v_i(Tx) w_i(x) \xi^{2\alpha-1}(x) \frac{\partial \xi(x)}{\partial x_k}\, dx \qquad (4.31)$$

$$= h^2 (J_3 + J_4 + J_5),$$

where we took into account div $\mathbf{v}_\varepsilon(Tx) = 0$. Using the regularity of the mollified function, the properties of the difference quotient, the embedding $W^{1,q}(\Omega) \hookrightarrow L^{\frac{3q}{3-q}}(\Omega)$ and the interpolation of $L^{q'}$ between L^q and $L^{\frac{3q}{3-q}}$ we arrive at

$$|J_3| \le \frac{c_{12}}{16} \left(\int_\Omega \left| \frac{\nabla \mathbf{w}}{h} \right|^q \xi^{\alpha q}\, dx \right)^{\frac{2}{q}} + c(\nabla \xi)\|\mathbf{v}_\varepsilon\|_\infty^2 \left\| \frac{\mathbf{w}}{h} \xi^{\alpha-1} \right\|_q^2,$$

$$|J_4| \le \frac{c_{12}}{16} \left(\int_\Omega \left| \frac{\nabla \mathbf{w}}{h} \right|^q \xi^{\alpha q}\, dx \right)^{\frac{2}{q}} + c\|\nabla \mathbf{v}_\varepsilon\|_\infty^2 \|\mathbf{v}\|_{q'}^2, \qquad (4.32)$$

$$|J_5| \le c(\nabla \xi)\|\nabla \mathbf{v}_\varepsilon\|_\infty \left\| \frac{\mathbf{w}}{h} \xi^\alpha \right\|_q \|\mathbf{v}\|_{q'}.$$

Since $q^* \ge q'$ for $q \ge 3/2$ we obtain from (4.31), (4.32) and (4.4) that

$$|I_2| \le \frac{c_{12}}{8} \left(\int_\Omega \left| \frac{\nabla \mathbf{w}}{h} \right|^q \xi^{\alpha q}\, dx \right)^{\frac{2}{q}} + c(\mathbf{f}, \mathbf{E}_0, \nabla \xi, \varepsilon^{-1}, \alpha). \qquad (4.33)$$

The integral $|I_3|$ can be re-written as

$$h^2 I_3 = \int_\Omega \phi(x) \Big(\text{div} \left(\mathbf{w}(x)\xi^{2\alpha}(x)\right) - \text{div} \left(\mathbf{w}(T^{-1}x)\xi^{2\alpha}(T^{-1}x)\right) \Big) \, dx \qquad (4.34)$$

$$= \int_\Omega \phi(x) \Big(\text{div} \, \mathbf{w}(x) \, \xi^{2\alpha}(x) - \text{div} \, \mathbf{w}(T^{-1}x) \, \xi^{2\alpha}(T^{-1}x) \Big) \, dx$$

$$+ 2\alpha \int_\Omega \phi(x) \left(w_i(x)\xi^{2\alpha-1}(x)\frac{\partial \xi(x)}{\partial x_i} - w_i(T^{-1}x)\xi^{2\alpha-1}(T^{-1}x)\frac{\partial \xi(T^{-1}x)}{\partial x_i} \right) dx$$

$$= 2\alpha \int_\Omega \phi(x) \Big(v_i(Tx) - 2v_i(x) + v_i(T^{-1}x) \Big) \xi^{2\alpha-1}(x)\frac{\partial \xi(x)}{\partial x_i} \, dx$$

$$+ 2\alpha \int_\Omega \phi(x) w_i(T^{-1}x) \left(\xi^{2\alpha-1}(x)\frac{\partial \xi(x)}{\partial x_i} - \xi^{2\alpha-1}(T^{-1}x)\frac{\partial \xi(T^{-1}x)}{\partial x_i} \right) dx$$

$$= h^2 (J_6 + J_7) \, .$$

Denoting

$$\mathbf{g}(x) \equiv \frac{\mathbf{w}(x)}{h} = \frac{\mathbf{v}(Tx) - \mathbf{v}(x)}{h} \qquad (4.35)$$

we can re-write J_6 as follows:

$$J_6 = 2\alpha \int_\Omega \phi(x)\frac{g_i(x) - g_i(T^{-1}x)}{h}\xi^{2\alpha-1}(x)\frac{\partial \xi(x)}{\partial x_i} \, dx$$

$$= 2\alpha \int_\Omega \phi(x)\xi^{\alpha-1}(x)\frac{1}{h}\Big(g_i(x)\xi^\alpha(x) - g_i(T^{-1}x)\xi^\alpha(T^{-1}x) \Big)\frac{\partial \xi(x)}{\partial x_i} \, dx \qquad (4.36)$$

$$+ 2\alpha \int_\Omega \phi(x)\xi^{\alpha-1}(x)\frac{1}{h}\big(\xi^\alpha(T^{-1}x) - \xi^\alpha(x) \big) g_i(T^{-1}x)\frac{\partial \xi(x)}{\partial x_i} \, dx \, .$$

Since $\mathbf{g}\xi^\alpha \in W_0^{1,q}(\Omega)$ for h fixed, we can use (5.1.30) and conclude that

$$|J_6| \leq \frac{c_{12}}{8} \left(\int_\Omega \left|\frac{\nabla \mathbf{w}}{h}\right|^q \xi^{\alpha q} \, dx \right)^{\frac{2}{q}} + c(\nabla \xi) \big(\|\phi \xi^{\alpha-1}\|_{q'}^2 + \|\nabla \mathbf{v}\|_q^2 \big) \, . \qquad (4.37)$$

For the term J_7 we have

$$h^2 J_7 = 2\alpha \int_\Omega \phi(x)\xi^{\alpha-1}(x)\left(\xi^\alpha(x)\frac{\partial \xi(x)}{\partial x_i} - \xi^\alpha(T^{-1}x)\frac{\partial \xi(T^{-1}x)}{\partial x_i} \right) w_i(T^{-1}x) \, dx$$

$$+ 2\alpha \int_\Omega \phi(x)\xi^{\alpha-1}(T^{-1}x)\big(\xi^\alpha(x) - \xi^\alpha(T^{-1}x) \big)\frac{\partial \xi(T^{-1}x)}{\partial x_i} w_i(T^{-1}x) \, dx$$

and therefore

$$|J_7| \leq c(\nabla^2 \xi) \|\nabla \mathbf{v}\|_q^2 + c \|\phi \xi^{\alpha-1}\|_{q'}^2 + c \|\phi(x) \xi^{\alpha-1}(T^{-1}x)\|_{q'}^2 \, . \tag{4.38}$$

Thus, we obtain

$$|I_3| \leq \frac{c_{12}}{8} \left(\int\limits_\Omega \left|\frac{\nabla \mathbf{w}}{h}\right|^q \xi^{\alpha q} \, dx \right)^{\frac{2}{q}} + c(\mathbf{f}, \mathbf{E}_0, \nabla \xi, \alpha)$$
$$+ c \|\phi \xi^{\alpha-1}\|_{q'}^2 + c \|\phi(x) \xi^{\alpha-1}(T^{-1}x)\|_{q'}^2 \, . \tag{4.39}$$

The term I_4 in (4.22) can be written as

$$h^2 I_4 = \int\limits_\Omega \mathbf{f}(x) \cdot \big(\mathbf{v}(Tx) - 2\mathbf{v}(x) + \mathbf{v}(T^{-1}x)\big) \xi^{2\alpha}(x) \, dx$$
$$+ \int\limits_\Omega \mathbf{f}(x) \cdot \big(\mathbf{v}(x) - \mathbf{v}(T^{-1}x)\big) \big(\xi^{2\alpha}(x) - \xi^{2\alpha}(T^{-1}x)\big) \, dx \, . \tag{4.40}$$

The first integral in (4.40) may be treated as J_6 and we obtain

$$|I_4| \leq \frac{c_{12}}{8} \left(\int\limits_\Omega \left|\frac{\nabla \mathbf{w}}{h}\right|^q \xi^{\alpha q} \, dx \right)^{\frac{2}{q}} + c(\nabla \xi) \big(\|\mathbf{f}\|_{q'}^2 + \|\nabla \mathbf{v}\|_q^2 \big) \, . \tag{4.41}$$

Finally, we write I_5 as

$$h^2 I_5 = \chi^E \int\limits_\Omega (\mathbf{E}(Tx) \otimes \mathbf{E}(Tx) - \mathbf{E}(x) \otimes \mathbf{E}(x)) \mathbf{D}(\mathbf{w}(x)) \xi^{2\alpha}(x) \, dx \tag{4.42}$$
$$+ 2\chi^E \int\limits_\Omega (\mathbf{E}(Tx) \otimes \mathbf{E}(Tx) - \mathbf{E}(x) \otimes \mathbf{E}(x)) \, \mathbf{w}(x) \otimes \nabla \xi(x) \xi^{2\alpha-1}(x) \, dx \, ,$$

which provides

$$|I_5| \leq \frac{c_{12}}{8} \left(\int\limits_\Omega \left|\frac{\nabla \mathbf{w}}{h}\right|^q \xi^{\alpha q} \, dx \right)^{\frac{2}{q}} + c(\mathbf{E}_0, \nabla \xi) \big(\|\nabla \mathbf{E}\|_{q'}^2 + \left\|\frac{\mathbf{w}}{h} \xi^\alpha\right\|_q \|\nabla \mathbf{E}\|_{q'} \big) \, . \tag{4.43}$$

Putting all calculations between (4.22) and (4.43) together, also using (4.4) and (4.7) we arrive at (with $\gamma = \alpha - 1$, $\eta = \xi$ and $\eta = \xi(T^{-1}x)$, respectively)

$$c_{11} \int\limits_\Omega \big(1 + |\mathbf{D}(\mathbf{v}(x))|^2 + |\mathbf{D}(\mathbf{v}(Tx))|^2\big)^{\frac{q-2}{2}} \left|\frac{\mathbf{D}(\mathbf{w}(x))}{h}\right|^2 \xi^{2\alpha} \, dx$$
$$+ c_{12} \left(\int\limits_\Omega \left|\frac{\nabla \mathbf{w}}{h}\right|^q \xi^{\alpha q} \, dx \right)^{\frac{2}{q}} \leq c(\mathbf{E}_0, \mathbf{f}, \varepsilon^{-1}, A, \nabla^2 \xi, \alpha) \, . \tag{4.44}$$

From the properties of the difference quotient we conclude that

$$\sum_{i,j=1}^{3} \int_{\Omega} \left| \frac{\partial^2 \mathbf{v}(x)}{\partial x_i \partial x_j} \right|^q \xi^{\alpha q}(x)\, dx \leq c(\mathbf{E}_0, \mathbf{f}, \varepsilon^{-1}, A),\tag{4.45}$$

which is exactly (4.8) and moreover that (cf. (5.1.35))

$$\frac{\nabla \mathbf{w}}{h} \to \nabla^2 \mathbf{v} \qquad \text{strongly in } L^q_{\text{loc}}(\Omega).$$

Hence, we have established that

$$\mathbf{D}(\mathbf{v}(T_h x)) \to \mathbf{D}(\mathbf{v}(x)) \qquad \text{a.e. in } \Omega$$

and thus Fatou's lemma and (4.44) imply

$$\int_{\Omega} \left(1 + |\mathbf{D}(\mathbf{v})|^2\right)^{\frac{q-2}{2}} |\mathbf{D}(\nabla \mathbf{v})|^2 \xi^{2\alpha}\, dx \leq c,\tag{4.46}$$

which is exactly (4.9). Therefore the estimates (4.8) and (4.9) are proved and the system $(1.2)_{\varepsilon,A}$ holds almost everywhere in Ω. The proof is complete. ∎

3.5 Limiting Process $A \to \infty$

In this section we will derive estimates independent of A, which enable the limiting process $A \to \infty$. Thus we will come to an approximation of the problem (1.1), (1.2) where only the convective term is mollified. In preparation for the limiting process $\varepsilon \to 0$ we will indicate the dependence of the estimates on ε. Let us introduce some notation. Let χ_A be the characteristic function of the set

$$\Omega_A = \{x \in \Omega, |\mathbf{D}(\mathbf{v}(x))| \geq A\}.\tag{5.1}$$

Furthermore let $\mathbf{u}$ be an integrable function, then we denote

$$I_{p,A}(\mathbf{u}, \xi) = \int_{\Omega} \left(1 + |\mathbf{E}|^2\right)\left(1 + |\mathbf{D}(\mathbf{v})|^2\right)^{\frac{q-2}{2}} |\mathbf{u}|^2 \xi^2 \times$$
$$\times \left\{ \left(1 + |\mathbf{D}(\mathbf{v})|^2\right)^{\frac{p-q}{2}} \chi_A + (1 + A^2)^{\frac{p-q}{2}} (1 - \chi_A) \right\} dx,\tag{5.2}$$

where $\mathbf{u}$ will be usually some second order derivative of $\mathbf{v}$, and ξ will be either identical 1 or a cut-off function ξ. Note, that it follows from Remark 3.40 and Remark 3.64, respectively, that there are constants c_{13}, c_{14} such that

$$c_{13}\, I_{p,A}(\mathbf{D}(\nabla \mathbf{v}), \xi) \leq \int_{\Omega} \partial_{kl} S_{ij}^A(\mathbf{D}(\mathbf{v}), \mathbf{E}) D_{ij}(\nabla \mathbf{v}) D_{kl}(\nabla \mathbf{v}) \xi^2\, dx$$
$$\leq c_{14}\, I_{p,A}(\mathbf{D}(\nabla \mathbf{v}), \xi).\tag{5.3}$$

Now we define weak solutions to the problem (1.1), $(1.2)_\varepsilon$.

Definition 5.4. *The triple* $(\mathbf{E}, \mathbf{v}, \phi) = (\mathbf{E}, \mathbf{v}^\varepsilon, \phi^\varepsilon)$, *for* $\varepsilon > 0$ *given, is said to be a weak solution of the problem* (1.1) *and*

$$
\begin{aligned}
- \operatorname{div} \mathbf{S}(\mathbf{D}(\mathbf{v}), \mathbf{E}) + [\nabla \mathbf{v}] \mathbf{v}_\varepsilon + \nabla \phi &= \mathbf{f} + \chi^E [\nabla \mathbf{E}] \mathbf{E} \\
\operatorname{div} \mathbf{v} &= 0 \qquad\qquad\qquad \begin{aligned} &\text{in } \Omega, \\ \end{aligned} \\
\mathbf{v} &= 0 \qquad\qquad\qquad \text{on } \partial\Omega,
\end{aligned} \tag{1.2}_\varepsilon
$$

where $\mathbf{S}$ *is given by* (1.3), *if and only if*

$$
\begin{aligned}
\mathbf{E} &\in H(\operatorname{div}) \cap H(\operatorname{curl}), \\
\mathbf{v} &\in E_{p(|\mathbf{E}|^2)}, \\
\phi &\in L^{p_0'}(\Omega),
\end{aligned} \tag{5.5}
$$

and the system $(1.1)_{1,2}$ *is satisfied almost everywhere in* Ω, $\mathbf{E} - \mathbf{E}_0 \in \overset{\circ}{H}(\operatorname{div})$ *and the weak formulation of* $(1.2)_\varepsilon$

$$
\begin{aligned}
\int_\Omega \mathbf{S}(\mathbf{D}(\mathbf{v}), \mathbf{E}) \cdot \mathbf{D}(\boldsymbol{\varphi}) \, dx &+ \int [\nabla \mathbf{v}] \mathbf{v}_\varepsilon \cdot \boldsymbol{\varphi} \, dx - \int \phi \operatorname{div} \boldsymbol{\varphi} \, dx \\
&= \langle \mathbf{f}, \boldsymbol{\varphi} \rangle_{1,q} - \chi^E \int_\Omega \mathbf{E} \otimes \mathbf{E} \cdot \mathbf{D}(\boldsymbol{\varphi}) \, dx
\end{aligned} \tag{5.6}
$$

is satisfied for all $\boldsymbol{\varphi} \in \mathcal{D}(\Omega)$.

Proposition 5.7. *Let* $p_\infty > 9/5$, *then for* $\varepsilon > 0$ *and* $A \geq A_0$ *given, the weak solutions* $(\mathbf{E}, \mathbf{v}^{\varepsilon,A}, \phi^{\varepsilon,A})$ *of the problem* (1.1), $(1.2)_{\varepsilon,A}$ *satisfy the following estimates independent of* A:

$$
\|\mathbf{E}\|_{2,r} \leq c \, \|\mathbf{E}_0\|_{2,r}, \tag{5.8}
$$

$$
\|\mathbf{D}(\mathbf{v}^{\varepsilon,A})\|_{q,\nu} + \|\nabla \mathbf{v}^{\varepsilon,A}\|_q \leq c(\mathbf{f}, \mathbf{E}_0), \tag{5.9}
$$

$$
\|\partial U^A(|\mathbf{D}(\mathbf{v}^{\varepsilon,A})|^2, |\mathbf{E}|^2) \cdot \mathbf{D}(\mathbf{v}^{\varepsilon,A})\|_{1,\nu} \leq c(\mathbf{f}, \mathbf{E}_0), \tag{5.10}
$$

$$
\|\mathbf{S}^A(\mathbf{D}(\mathbf{v}^{\varepsilon,A}), \mathbf{E})\|_{p_0'} + \|\phi^{\varepsilon,A}\|_{p_0'} \leq c(\mathbf{f}, \mathbf{E}_0). \tag{5.11}
$$

Moreover, we have for $p_0 < q^*$ *and* $\alpha \geq \alpha_0, \gamma > 1$

$$
\begin{aligned}
\|\phi^{\varepsilon,A} \, \xi^\gamma\|_{q'} &\leq c(\mathbf{f}, \mathbf{E}_0) \left(1 + \|[\nabla \mathbf{v}^{\varepsilon,A}] \mathbf{v}_\varepsilon^{\varepsilon,A} \, \xi^\gamma\|_{q'}\right) \\
&\leq c(\mathbf{f}, \mathbf{E}_0, \varepsilon^{-1}),
\end{aligned} \tag{5.12}
$$

$$
\begin{aligned}
\|\nabla \mathbf{S}^A(\mathbf{D}(\mathbf{v}^{\varepsilon,A}), \mathbf{E}) \, \xi^\alpha\|_{s'}^2 &+ \|\nabla \mathbf{v}^{\varepsilon,A} \, \xi^{\frac{2\alpha}{q}}\|_{3q}^q + \|\nabla^2 \mathbf{v}^{\varepsilon,A} \, \xi^\alpha\|_q^2 \\
&\leq c(\mathbf{f}, \mathbf{E}_0) \left(1 + \|[\nabla \mathbf{v}^{\varepsilon,A}] \mathbf{v}_\varepsilon^{\varepsilon,A} \, \xi^\alpha\|_{q'}^2\right) \leq c(\mathbf{f}, \mathbf{E}_0, \varepsilon^{-1}),
\end{aligned} \tag{5.13}
$$

where $q = \min(2, p_\infty)$ *and* $s = \max(2, p_0)$.

PROOF : The estimates (5.8)–(5.10) have already been proved in Proposition 4.2. For $q > 9/5$ we observe

$$\|[\nabla \mathbf{v}^{\varepsilon,A}]\mathbf{v}_\varepsilon^{\varepsilon,A}\|_{(q^*)'} \le c\,\|\nabla \mathbf{v}^{\varepsilon,A}\|_q^{\frac{3q}{4q-3}}\|\mathbf{v}^{\varepsilon,A}\|_{\frac{3q}{4q-6}}^{\frac{3q}{4q-3}} \tag{5.14}$$
$$\le c\,\|\nabla \mathbf{v}^{\varepsilon,A}\|_q^{\frac{6q}{4q-3}},$$

since $\frac{3q}{4q-6} < \frac{3q}{3-q} = q^*$. Because $(p_0^*)' \le (q^*)'$ we obtain estimate (5.11) from (4.6), (4.16); (5.14) and (5.9). Moreover from (4.7) and (5.14) we derive

$$\|\phi^{\varepsilon,A}\,\xi^\gamma\|_{q'} \le c(\mathbf{f}, \mathbf{E}_0) + \|\mathbf{S}^A(\mathbf{D}(\mathbf{v}^{\varepsilon,A}), \mathbf{E})\,\xi^\gamma\|_{q'}. \tag{5.15}$$

From now on we will drop the indices ε, A and write $\mathbf{v} = \mathbf{v}^{\varepsilon,A}$ and $\phi = \phi^{\varepsilon,A}$. Proposition 4.2 ensures that $-\Delta \mathbf{v}\,\xi^{2\alpha}$, $\alpha \ge 1$ is an admissible test function in the weak formulation (3.74). Recall that ξ is the cut-off function for the interior regularity. Let us denote by $I_1, \ldots, I_5$ the integrals in (3.74) for this test function. We will treat the resulting terms separately. We see that

$$I_1 = \int_\Omega \frac{\partial}{\partial x_k} S_{ij}^A(\mathbf{D}(\mathbf{v}), \mathbf{E}) D_{ij}\Big(\frac{\partial \mathbf{v}}{\partial x_k}\Big)\xi^{2\alpha}\,dx$$
$$+ 2\alpha \int_\Omega S_{ij}^A(\mathbf{D}(\mathbf{v}), \mathbf{E}) D_{ij}\Big(\frac{\partial \mathbf{v}}{\partial x_k}\Big)\xi^{2\alpha-1}\frac{\partial \xi}{\partial x_k}\,dx \tag{5.16}$$
$$- \alpha \int_\Omega S_{ij}^A(\mathbf{D}(\mathbf{v}), \mathbf{E})\Big(\Delta v_i\,\xi^{2\alpha-1}\frac{\partial \xi}{\partial x_j} + \Delta v_j\,\xi^{2\alpha-1}\frac{\partial \xi}{\partial x_i}\Big)\,dx$$
$$= J_1 + J_2 + J_3.$$

The term J_1 can be written as

$$J_1 = \int_\Omega \partial_{lm}S_{ij}^A(\mathbf{D}(\mathbf{v}), \mathbf{E}) D_{lm}\Big(\frac{\partial \mathbf{v}}{\partial x_k}\Big) D_{ij}\Big(\frac{\partial \mathbf{v}}{\partial x_k}\Big)\xi^{2\alpha}\,dx$$
$$+ \int_\Omega \partial_n S_{ij}^A(\mathbf{D}(\mathbf{v}), \mathbf{E})\frac{\partial E_n}{\partial x_k} D_{ij}\Big(\frac{\partial \mathbf{v}}{\partial x_k}\Big)\xi^{2\alpha}\,dx \tag{5.17}$$
$$= J_{1,1} + J_{1,2}.$$

Using (3.25) and (3.56), respectively, and (5.3), together with (5.1.22) and (5.9) we obtain

$$J_{1,1} \ge \frac{c_{13}}{2} I_{p,A}(\mathbf{D}(\nabla \mathbf{v}), \xi^\alpha) + c(\mathbf{f}, \mathbf{E}_0)\Big(\int_\Omega |\mathbf{D}(\nabla \mathbf{v})|^q \xi^{\alpha q}\,dx\Big)^{\frac{2}{q}}$$
$$\ge \frac{c_{13}}{2} I_{p,A}(\mathbf{D}(\nabla \mathbf{v}), \xi^\alpha) + c_{15}\Big(\int_\Omega |\nabla^2 \mathbf{v}|^q \xi^{\alpha q}\,dx\Big)^{\frac{2}{q}} \tag{5.18}$$
$$- c(\nabla \xi)\Big(\int_\Omega |\nabla \mathbf{v}|^q\,dx\Big)^{\frac{2}{q}},$$

where we also used Korn's inequality and elementary calculations. From (3.44) and (3.68), respectively, we obtain, for $s > 1$,

$$|J_{1,2}| \leq c_{10} \int_\Omega |\mathbf{E}|(1 + |\mathbf{E}|^2)(1 + |\mathbf{D}(\mathbf{v})|^2)^{\frac{q-1}{2}} |\mathbf{D}(\nabla\mathbf{v})| \, |\nabla\mathbf{E}| \, \xi^{2\alpha} \times$$

$$\times \left\{ \chi_A (1 + |\mathbf{D}|^2)^{\frac{p-q}{2}} \left(1 + \ln(1 + |\mathbf{D}|^2)\right) \right.$$

$$\left. + (1 - \chi_A)(1 + A^2)^{\frac{p-q}{2}} \left(1 + \ln(1 + A^2)\right) \right\} dx$$

$$\leq c(\mathbf{E}_0) \int_\Omega \left\{ (1 + |\mathbf{E}|^2)(1 + |\mathbf{D}|^2)^{\frac{q-2}{2}} (1 + A^2)^{\frac{p-q}{2}} |\mathbf{D}(\nabla\mathbf{v})|^2 \, \xi^{2\alpha} \right\}^{\frac{1}{2}} \times \qquad (5.19)$$

$$\times \left\{ (1 + |\mathbf{D}|^2)^{\frac{q}{2}} (1 + A^2)^{\frac{p-q+(s-1)p}{2}} \xi^{2\alpha} \right\}^{\frac{1}{2}} \chi_A \, dx$$

$$+ c(\mathbf{E}_0) \int_\Omega \left\{ (1 + |\mathbf{E}|^2)(1 + |\mathbf{D}|^2)^{\frac{q-2}{2}} (1 + |\mathbf{D}|^2)^{\frac{p-q}{2}} |\mathbf{D}(\nabla\mathbf{v})|^2 \, \xi^{2\alpha} \right\}^{\frac{1}{2}} \times$$

$$\times \left\{ (1 + |\mathbf{D}|^2)^{\frac{q}{2}} (1 + |\mathbf{D}|^2)^{\frac{p-q+(s-1)p}{2}} \xi^{2\alpha} \right\}^{\frac{1}{2}} (1 - \chi_A) \, dx$$

$$\leq \frac{c_{13}}{4} I_{p,A}(\mathbf{D}(\nabla\mathbf{v}), \xi^\alpha) + c(\mathbf{E}_0) \left(\int_\Omega \left(1 + \partial U^A(|\mathbf{D}(\mathbf{v})|^2, |\mathbf{E}|^2) \cdot \mathbf{D}(\mathbf{v})\right)^s \xi^{2\alpha} \, dx \right),$$

where we also used (3.48) and (3.71), respectively. The term in the last line can be further estimated by (cf. (3.24), (3.28), (3.55), (3.59))

$$c(\mathbf{E}_0) \left(1 + \|\nabla\mathbf{v}\|_q^s \, \|\mathbf{S}^A \xi^\alpha\|_{\frac{sq}{q-s}}^s \right). \qquad (5.20)$$

Next, we notice that

$$|J_2 + J_3| \leq \frac{c_{15}}{8} \left(\int_\Omega |\nabla^2\mathbf{v}|^q \, \xi^{\alpha q} \, dx \right)^{\frac{2}{q}} + c\|\mathbf{S}^A \xi^{\alpha-1}\|_{q'}^2. \qquad (5.21)$$

For the convective term we have

$$|I_2| \leq \frac{c_{15}}{8} \|\nabla^2\mathbf{v}\, \xi^\alpha\|_q^2 + c \left(\int_\Omega |\nabla\mathbf{v}|^{q'} |\mathbf{v}_\varepsilon|^{q'} \xi^{\alpha q'} \, dx \right)^{\frac{2}{q'}} \qquad (5.22)$$

and for the pressure term we obtain

$$|I_3| = |2\alpha \int_\Omega \phi \Delta v_i \, \xi^{2\alpha-1} \frac{\partial\xi}{\partial x_i} \, dx| \leq \frac{c_{15}}{8} \|\nabla^2\mathbf{v}\, \xi^\alpha\|_q^2 + c\|\phi\xi^{\alpha-1}\|_{q'}^2. \qquad (5.23)$$

The remaining two terms are estimated by

$$|I_4| \leq \frac{c_{15}}{8} \|\nabla^2\mathbf{v}\, \xi^\alpha\|_q^2 + c\|\mathbf{f}\|_{q'}^2, \qquad (5.24)$$

and

$$|I_5| = 2\chi^E \int_\Omega E_i \frac{\partial E_j}{\partial x_k} D_{ij}\Big(\frac{\partial \mathbf{v}}{\partial x_k}\Big) \xi^{2\alpha}\, dx$$

$$- \alpha\chi^E \int_\Omega E_i E_j \Big(\Delta v_i \frac{\partial \xi}{\partial x_j} + \Delta v_j \frac{\partial \xi}{\partial x_i} - 2D_{ij}\Big(\frac{\partial \mathbf{v}}{\partial x_k}\Big)\frac{\partial \xi}{\partial x_k} \Big) \xi^{2\alpha-1}\, dx \qquad (5.25)$$

$$\leq \frac{c_{15}}{8}\|\nabla^2 \mathbf{v}\xi^\alpha\|_q^2 + c(\mathbf{E}_0)\,.$$

Putting all computations between (5.16) and (5.25) together we arrive at

$$\frac{c_{13}}{4} I_{p,A}(\mathbf{D}(\nabla \mathbf{v}^{\varepsilon,A}), \xi^\alpha) + \frac{c_{15}}{2}\|\nabla^2 \mathbf{v}^{\varepsilon,A}\, \xi^\alpha\|_q^2$$

$$\leq c(\mathbf{f}, \mathbf{E}_0, \nabla\xi)\Big(1 + \|\mathbf{S}^A(\mathbf{D}(\mathbf{v}^{\varepsilon,A}), \mathbf{E})\,\xi^\alpha\|_{\frac{sq}{q-s}}^s + \|\mathbf{S}^A(\mathbf{D}(\mathbf{v}^{\varepsilon,A}), \mathbf{E})\,\xi^{\alpha-1}\|_{q'}^2$$

$$+ \|[\nabla \mathbf{v}^{\varepsilon,A}]\mathbf{v}_\varepsilon^{\varepsilon,A}\,\xi^\alpha\|_{q'}^2 + \|\phi^{\varepsilon,A}\,\xi^{\alpha-1}\|_{q'}^2\Big) \qquad (5.26)$$

$$\leq c(\mathbf{f}, \mathbf{E}_0, \nabla\xi)\Big(1 + \|\mathbf{S}^A(\mathbf{D}(\mathbf{v}^{\varepsilon,A}), \mathbf{E})\,\xi^{\alpha-1}\|_{sq'}^2 + \|[\nabla \mathbf{v}^{\varepsilon,A}]\mathbf{v}_\varepsilon^{\varepsilon,A}\,\xi^{\alpha-1}\|_{q'}^2\Big)\,,$$

where $s > 1$ is chosen appropriately. Moreover, (5.9) and (5.15), with $\gamma = \alpha - 1$, have been taken into account. Based on estimate (5.26) and Proposition 5.2.23 we must distinguish the cases

(i) $q = 2 \leq p_\infty \leq p_0$,

(ii) $9/5 < q = p_\infty \leq p_0 < 2$,

(iii) $9/5 < q = p_\infty < 2 \leq p_0$.

Let us start with

(i) The case $2 \leq p_\infty \leq p_0$

From (5.26), (5.2.25) and (5.2) it follows

$$\|\nabla\mathbf{S}^A(\mathbf{D}(\mathbf{v}), \mathbf{E})\,\xi^\alpha\|_{p_0'}^2 + \|\nabla^2\mathbf{v}\,\xi^\alpha\|_2^2$$

$$\leq c(\mathbf{f}, \mathbf{E}_0)\Big(1 + \|\mathbf{S}^A(\mathbf{D}(\mathbf{v}), \mathbf{E})\,\xi^{\alpha-1}\|_{2s}^2 + \|[\nabla\mathbf{v}]\mathbf{v}_\varepsilon\,\xi^\alpha\|_2^2\Big)\,, \qquad (5.27)$$

where we employed (5.9), (5.10), $s < p_0'$ and $\xi \leq 1$. We will use the global estimate (5.11) and the first term on the left-hand side of (5.27) to bound the term with $\mathbf{S}^A$ on the right-hand side of (5.27). From the embedding $W^{1,p_0'}(\Omega) \hookrightarrow L^{\frac{3p_0}{2p_0-3}}(\Omega)$ we obtain the restriction that $p_0 < 6$, since we must require that $2s < \frac{3p_0}{2p_0-3}$. Motivated by the interpolation of L^{2s} between $L^{p_0'}$ and $L^{\frac{3p_0}{2p_0-3}}$ we have for $\lambda = \frac{3(2s(p_0-1)-p_0)}{2sp_0}$,

$$
\begin{aligned}
\int_\Omega |\mathbf{S}^A|^{2s}\xi^{(\alpha-1)2s}\,dx &= \int_\Omega \left(|\mathbf{S}^A|\xi^\alpha\right)^{2s\lambda}|\mathbf{S}^A|^{2s(1-\lambda)}\xi^{2s(\alpha(1-\lambda)-1)}\,dx \\
&\le \|\mathbf{S}^A\xi^\alpha\|_{\frac{3p_0}{2p_0-3}}^{2s\lambda}\|\mathbf{S}^A\|_{p_0'}^{2s(1-\lambda)} \\
&\le c(\mathbf{f},\mathbf{E}_0) + \frac{1}{2}\|\nabla\mathbf{S}^A\xi^\alpha\|_{p_0'}^2,
\end{aligned}
\tag{5.28}
$$

as long as $\alpha \ge (1-\lambda)^{-1}$. Inequalities (5.27), (5.28) and the embedding $W^{1,2}(\Omega) \hookrightarrow L^6(\Omega)$ therefore lead to

$$
\begin{aligned}
&\|\nabla\mathbf{S}^A(\mathbf{D}(\mathbf{v}),\mathbf{E})\,\xi^\alpha\|_{p_0'}^2 + \|\nabla^2\mathbf{v}\,\xi^\alpha\|_2^2 + \|\nabla\mathbf{v}\,\xi^\alpha\|_6^2 \\
&\quad \le c(\mathbf{f},\mathbf{E}_0)\left(1 + \|[\nabla\mathbf{v}]\mathbf{v}_\varepsilon\,\xi^\alpha\|_2^2\right),
\end{aligned}
\tag{5.29}
$$

as long as $2 \le p_\infty \le p_0 < 6$ and $\alpha \ge (1-\lambda)^{-1}$. From (5.9) and the properties of the mollifier one gets that the right-hand side of (5.29) is bounded by $c(\mathbf{f},\mathbf{E}_0,\varepsilon^{-1})$.

(ii) The case $9/5 < p_\infty \le p_0 < 2$

From (5.26), (5.2.24) and (5.2.9) we conclude

$$
\begin{aligned}
&\|\nabla\mathbf{S}^A(\mathbf{D}(\mathbf{v}),\mathbf{E})\,\xi^\alpha\|_2^2 + \|\nabla\mathbf{v}\,\xi^{\frac{2\alpha}{p_\infty}}\|_{3p_\infty}^{p_\infty} + \|\nabla^2\mathbf{v}\,\xi^\alpha\|_{p_\infty}^2 \\
&\quad \le c(\mathbf{f},\mathbf{E}_0)\left(1 + \|\mathbf{S}^A(\mathbf{D}(\mathbf{v}),\mathbf{E})\,\xi^{\alpha-1}\|_{sp_\infty'}^2 + \|[\nabla\mathbf{v}]\mathbf{v}_\varepsilon\,\xi^\alpha\|_{p_\infty'}^2\right),
\end{aligned}
\tag{5.30}
$$

where we used (5.9) and (5.10). We proceed analogously to the case (i). We do not obtain any additional restriction on p_∞ and p_0. As in (5.28) we can derive that

$$
\|\mathbf{S}^A\,\xi^{\alpha-1}\|_{sp_\infty'}^2 \le c(\mathbf{f},\mathbf{E}_0) + \frac{1}{2}\|\nabla\mathbf{S}^A\,\xi^\alpha\|_2^2,
\tag{5.31}
$$

as long as $\alpha < (1-\lambda)^{-1}$, where $\lambda = \frac{6((p_0-1)sp_\infty - p_0(p_\infty-1))}{(5p_0-6)sp_\infty}$, which comes from the interpolation of L^{sp_∞} between $L^{p_0'}$ and L^6. Thus we have arrived at

$$
\begin{aligned}
&\|\nabla\mathbf{S}^A(\mathbf{D}(\mathbf{v}),\mathbf{E})\,\xi^\alpha\|_2^2 + \|\nabla\mathbf{v}\,\xi^{\frac{2\alpha}{p_\infty}}\|_{3p_\infty}^{p_\infty} + \|\nabla^2\mathbf{v}\,\xi^\alpha\|_{p_\infty}^2 \\
&\quad \le c(\mathbf{f},\mathbf{E}_0)\left(1 + \|[\nabla\mathbf{v}]\mathbf{v}_\varepsilon\,\xi^\alpha\|_{p_\infty'}^2\right),
\end{aligned}
\tag{5.32}
$$

as long as $9/5 < p_\infty \le p_0 < 2$ and $\alpha \ge (1-\lambda)^{-1}$. With the help of the interpolation of $L^{p_\infty'}$ between L^{p_∞} and $L^{p_\infty^*}$ we obtain

$$
\begin{aligned}
\|[\nabla\mathbf{v}]\mathbf{v}_\varepsilon\,\xi^\alpha\|_{p_\infty'} &\le c(\varepsilon^{-1})\|\nabla\mathbf{v}\,\xi^\alpha\|_{p_\infty'} \\
&\le \frac{1}{2}\|\nabla^2\mathbf{v}\,\xi^\alpha\|_{p_\infty} + c(\varepsilon^{-1})\|\nabla\mathbf{v}\,\xi^\alpha\|_{p_\infty}.
\end{aligned}
$$

This inserted into (5.32) together with (5.9) and the properties of the mollifier implies that the right-hand side of (5.32) is bounded by $c(\mathbf{f}, \mathbf{E}_0, \varepsilon^{-1})$. Finally, we come to

(iii) The case $9/5 < p_\infty < 2 \le p_0$

From (5.26), (5.2.25) and (5.2.9) we obtain

$$
\begin{aligned}
&\|\nabla \mathbf{S}^A(\mathbf{D}(\mathbf{v}), \mathbf{E})\, \xi^\alpha\|_{p_0'}^2 + \|\nabla \mathbf{v}\, \xi^{\frac{2\alpha}{p_\infty}}\|_{3p_\infty}^{p_\infty} + \|\nabla^2 \mathbf{v}\, \xi^\alpha\|_{p_\infty}^2 \\
&\le c(\mathbf{f}, \mathbf{E}_0)\left(1 + \|\mathbf{S}^A(\mathbf{D}(\mathbf{v}), \mathbf{E})\, \xi^{\alpha-1}\|_{sp_\infty'}^2 + \|[\nabla \mathbf{v}]\mathbf{v}_\varepsilon\, \xi^\alpha\|_{p_\infty'}^2\right)
\end{aligned}
\tag{5.33}
$$

where we applied (5.9), (5.10), $s < p_0'$ and $\xi \le 1$. We will use the interpolation of $L^{sp_\infty'}$ between $L^{p_0'}$ and $L^{\frac{3p_0}{2p_0-3}}$, which is possible for $p_0 < p_\infty^*$. As in (5.31) we conclude

$$
\|\mathbf{S}^A\, \xi^{\alpha-1}\|_{sp_\infty'}^2 \le \frac{1}{2}\|\nabla \mathbf{S}^A\, \xi^\alpha\|_{p_0'}^2 + c(\mathbf{f}, \mathbf{E}_0)
$$

as long as $\alpha \ge (1-\lambda)^{-1}$, with $\lambda = \frac{3(sp_\infty(p_0-1)-p_0(p_\infty-1))}{sp_\infty p_0}$. Thus we note that

$$
\begin{aligned}
&\|\nabla \mathbf{S}^A(\mathbf{D}(\mathbf{v}), \mathbf{E})\, \xi^\alpha\|_{p_0'}^2 + \|\nabla \mathbf{v}\, \xi^{\frac{2\alpha}{p_\infty}}\|_{3p_\infty}^{p_\infty} + \|\nabla^2 \mathbf{v}\, \xi^\alpha\|_{p_\infty}^2 \\
&\le c(\mathbf{f}, \mathbf{E}_0)\left(1 + \|[\nabla \mathbf{v}]\mathbf{v}_\varepsilon\, \xi^\alpha\|_{p_\infty'}^2\right),
\end{aligned}
\tag{5.34}
$$

as long as $9/5 < p_\infty < 2 \le p_0 \le p_\infty^*$ and $\alpha \ge (1-\lambda)^{-1}$. As in the case (ii) we obtain that the right-hand side of (5.34) is bounded by some constant $c(\mathbf{f}, \mathbf{E}_0, \varepsilon^{-1})$.

The estimates (5.29), (5.32) and (5.34) together imply (5.13). In all cases (i)–(iii) we have shown that

$$
\|\mathbf{S}^A\, \xi^\alpha\|_{q'} \le c(\mathbf{f}, \mathbf{E}_0) + c\|\nabla \mathbf{S}^A\, \xi^\alpha\|_{s'},
$$

where $s = \max(2, p_0)$. Therefore we deduce from (5.15) and (5.13) the estimate (5.12). The proof is complete. ∎

In Proposition 5.7 we have prepared everything for the limiting process $A \to \infty$. From (5.9), (5.11)–(5.13) we obtain the existence of $\mathbf{v}^\varepsilon, \phi^\varepsilon$ such that

$$
\begin{aligned}
\nabla \mathbf{v}^{\varepsilon,A} &\to \nabla \mathbf{v}^\varepsilon &&\text{strongly in } L^\alpha_{loc}(\Omega), \quad \alpha < \tfrac{3q}{3-q}, \\
\mathbf{v}^{\varepsilon,A} &\rightharpoonup \mathbf{v}^\varepsilon &&\text{weakly in } W^{2,q}_{loc}(\Omega) \cap W^{1,3q}_{loc}(\Omega) \cap W^{1,q}_0(\Omega), \\
\phi^{\varepsilon,A} &\rightharpoonup \phi^\varepsilon &&\text{weakly in } L^{q'}_{loc}(\Omega) \cap L^{p_0'}(\Omega),
\end{aligned}
\tag{5.35}
$$

as $A \to \infty$ at least for a subsequence. From $(5.35)_1$ it follows that

$$
\nabla \mathbf{v}^{\varepsilon,A} \to \nabla \mathbf{v}^\varepsilon \qquad \text{a.e. in } \Omega,
\tag{5.36}
$$

and since $\mathbf{S}^A$ converges locally uniformly to $\mathbf{S}$, and thus also ∂U^A converges locally uniformly to ∂U we conclude

$$
\begin{aligned}
\mathbf{S}^A(\mathbf{D}(\mathbf{v}^{\varepsilon,A}), \mathbf{E}) &\to \mathbf{S}(\mathbf{D}(\mathbf{v}^\varepsilon), \mathbf{E}) &&\text{a.e. in } \Omega, \\
\partial U^A(|\mathbf{D}(\mathbf{v}^{\varepsilon,A})|^2, |\mathbf{E}|^2) &\to \partial U(|\mathbf{D}(\mathbf{v}^\varepsilon)|^2, |\mathbf{E}|^2) &&\text{a.e. in } \Omega.
\end{aligned}
\tag{5.37}
$$

Therefore we obtain from (5.13) and (5.11) that

$$\mathbf{S}^A(\mathbf{D}(\mathbf{v}^{\varepsilon,A}), \mathbf{E}) \rightharpoonup \mathbf{S}(\mathbf{D}(\mathbf{v}^{\varepsilon}), \mathbf{E}) \qquad \text{weakly in } W^{1,s'}_{\text{loc}}(\Omega) \cap L^{p'_0}(\Omega), \qquad (5.38)$$

since weak and almost everywhere limits coincide. Moreover from (5.10), (5.37)$_2$, (5.36) and Fatou's lemma we deduce

$$\|\partial U(|\mathbf{D}(\mathbf{v}^{\varepsilon})|^2, |\mathbf{E}|^2) \cdot \mathbf{D}(\mathbf{v}^{\varepsilon})\|_{1,\nu} \le c(\mathbf{f}, \mathbf{E}_0), \qquad (5.39)$$

which in turn implies[4]

$$\|\nabla \mathbf{v}^{\varepsilon}\|_{p_\infty} + \|\mathbf{D}(\mathbf{v}^{\varepsilon})\|_{p(|\mathbf{E}|^2),\nu} \le c(\mathbf{f}, \mathbf{E}_0), \qquad (5.40)$$

where we used (2.2.49) and Korn's inequality. We also observe that

$$\lim_{A \to \infty} \int_\Omega |[\nabla \mathbf{v}^{\varepsilon,A}] \mathbf{v}^{\varepsilon,A}_\varepsilon|^{q'} \xi^{\alpha q'} \, dx = \int_\Omega |[\nabla \mathbf{v}^{\varepsilon}] \mathbf{v}^{\varepsilon}_\varepsilon|^{q'} \xi^{\alpha q'} \, dx, \qquad (5.41)$$

which follows from the almost everywhere convergence of the integrand and Vitali's theorem, since $q' < q^*$ (cf. (5.35)$_1$). From (5.41) and the weak lower semicontinuity of the norm we deduce that the estimates (5.9)–(5.13) remain valid with $\mathbf{v}^{\varepsilon,A}$ and $\phi^{\varepsilon,A}$ replaced by $\mathbf{v}^{\varepsilon}$ and ϕ^{ε}, respectively.

It remains to show that $\mathbf{v}^{\varepsilon}$ satisfies the weak formulation (5.6). The limiting process $A \to \infty$ in the weak formulation (3.74) is clear in all terms except the term with $\mathbf{S}^A$ (cf. the proof of Theorem 2.4). For this term we obtain for $\varphi \in \mathcal{D}(\Omega)$

$$\int_\Omega \mathbf{S}^A(\mathbf{D}(\mathbf{v}^{\varepsilon,A}), \mathbf{E}) \cdot \mathbf{D}(\varphi) \, dx \to \int_\Omega \mathbf{S}(\mathbf{D}(\mathbf{v}^{\varepsilon}), \mathbf{E}) \cdot \mathbf{D}(\varphi) \, dx,$$

using Vitali's theorem. This is possible due to (5.37)$_1$ and the estimate (5.11), which gives the uniform integrability of $\mathbf{S}^A(\mathbf{D}(\mathbf{v}^{\varepsilon,A}), \mathbf{E}) \cdot \mathbf{D}(\varphi)$. Therefore we have proved

Proposition 5.42. *Let* $9/5 < p_\infty \le p_0 < q^*$, *where* $q = \min(2, p_\infty)$. *Assume that* $\varepsilon > 0$, $\mathbf{E}_0 \in W^{2,r}(\Omega)$, $r > 3$ *and* $\mathbf{f} \in L^{p'_\infty}(\Omega)$ *are given. Then there exists a weak solution* $(\mathbf{E}, \mathbf{v}^{\varepsilon}, \phi^{\varepsilon})$ *of the problem* (1.1), (1.2)$_\varepsilon$ *which satisfies the estimates*

$$\|\mathbf{E}\|_{2,r} \le c \|\mathbf{E}_0\|_{2,r}, \qquad (5.43)$$

$$\|\nabla \mathbf{v}^{\varepsilon}\|_{p_\infty} + \|\mathbf{D}(\mathbf{v}^{\varepsilon})\|_{p,\nu} \le c(\mathbf{f}, \mathbf{E}_0), \qquad (5.44)$$

$$\|\partial U(|\mathbf{D}(\mathbf{v}^{\varepsilon})|^2, |\mathbf{E}|^2) \cdot \mathbf{D}(\mathbf{v}^{\varepsilon})\|_{1,\nu} \le c(\mathbf{f}, \mathbf{E}_0), \qquad (5.45)$$

$$\|\mathbf{S}(\mathbf{D}(\mathbf{v}^{\varepsilon}), \mathbf{E})\|_{p'_0} + \|\phi^{\varepsilon}\|_{p'_0} \le c(\mathbf{f}, \mathbf{E}_0), \qquad (5.46)$$

$$\|\phi^{\varepsilon} \xi^{\gamma}\|_{q'} \le c(\mathbf{f}, \mathbf{E}_0)\big(1 + \|[\nabla \mathbf{v}^{\varepsilon}] \mathbf{v}^{\varepsilon}_\varepsilon \xi^{\gamma}\|_{q'}\big), \qquad (5.47)$$

$$\|\nabla \mathbf{S}(\mathbf{D}(\mathbf{v}^{\varepsilon}), \mathbf{E}) \xi^{\alpha}\|^2_{s'} + \|\nabla \mathbf{v}^{\varepsilon} \xi^{\frac{2\alpha}{q}}\|^q_{3q} + \|\nabla^2 \mathbf{v}^{\varepsilon} \xi^{\alpha}\|^2_{q}$$
$$\le c(\mathbf{f}, \mathbf{E}_0)\big(1 + \|[\nabla \mathbf{v}^{\varepsilon}] \mathbf{v}^{\varepsilon}_\varepsilon \xi^{\alpha}\|^2_{q'}\big), \qquad (5.48)$$

where $q = \min(2, p_\infty)$ *and* $s = \max(2, p_0)$.

---

[4]Note, that in the case $p_\infty \ge 2$ the estimate (5.40) is better than (5.9).

3.6 Limiting Process $\varepsilon \to 0$

Because of the results in the preceding section the situation is now considerably easier. We just need to check if the convective term in the $L^{q'}_{loc}$-norm can be handled by the left-hand sides in the estimates established in Proposition 5.42. As in the previous section we distinguish the cases (i)–(iii).

(i) The case $2 \le p_\infty \le p_0 < 6$

We first consider the situation when $3 \le p_\infty$. We have, using (5.44) that

$$\|[\nabla \mathbf{v}^\varepsilon]\mathbf{v}^\varepsilon_\varepsilon \, \xi^\alpha\|_2 \le c\|\nabla \mathbf{v}^\varepsilon\|_3\|\nabla \mathbf{v}^\varepsilon\|_2$$
$$\le c(\mathbf{f}, E_0)\,. \tag{6.1}$$

If $p_\infty \in [2, 3)$ we obtain

$$\|[\nabla \mathbf{v}^\varepsilon]\mathbf{v}^\varepsilon_\varepsilon \, \xi^\alpha\|_2 \le \|\mathbf{v}^\varepsilon\|_4\|\nabla \mathbf{v}^\varepsilon \, \xi^\alpha\|_4$$
$$\le \|\nabla \mathbf{v}^\varepsilon\|_2^{5/4}\|\nabla \mathbf{v}^\varepsilon \, \xi^\alpha\|_6^{3/4} \tag{6.2}$$
$$\le c(\mathbf{f}, E_0) + \frac{1}{2}\|\nabla \mathbf{v}^\varepsilon \, \xi^\alpha\|_6\,,$$

where we used the interpolation of L^4 between L^2 and L^6, the embedding $W^{1,2}(\Omega) \hookrightarrow L^6(\Omega)$, Young's inequality and (5.44). Moreover, Lemma 5.2.32 yields

$$I_{p,p'_0}(\mathbf{D}(\nabla \mathbf{v}^\varepsilon),\, \xi^\alpha) \le \|\nabla \mathbf{S}(\mathbf{D}(\mathbf{v}^\varepsilon), E)\, \xi^\alpha\|_{p'_0}^{p'_0} + \|\partial_n \mathbf{S}(\mathbf{D}(\mathbf{v}^\varepsilon), E)\nabla E_n \, \xi^\alpha\|_{p'_0}^{p'_0}$$
$$\le c(\mathbf{f}, E_0)\big(1 + \|\nabla \mathbf{S}(\mathbf{D}(\mathbf{v}^\varepsilon), E)\, \xi^\alpha\|_{p'_0}^{p'_0}\big)\,, \tag{6.3}$$

where we used (5.44), (5.2.39) and a similar argument as in (5.28) to handle the right-hand side in (5.2.39). Therefore we deduce from (5.2.41), (6.3), (5.48) and (6.2) and (6.1), respectively,

$$\|\nabla^2 \mathbf{v}^\varepsilon \, \xi^\alpha\|_2^2 + \|\nabla \mathbf{v}^\varepsilon \, \xi^\alpha\|_6^2 + \|\nabla \mathbf{v}^\varepsilon \, \xi^\alpha\|_{\frac{3p_0}{2p_0-3}(p_\infty-1)}^{p_\infty-1} \le c(\mathbf{f}, E_0)\,. \tag{6.4}$$

Note, that $\frac{3p_0}{2p_0-3}(p_\infty - 1) > 6$ if $p_0 < \frac{6}{5-p_\infty}$. Since we have shown that $\|[\nabla \mathbf{v}^\varepsilon]\mathbf{v}^\varepsilon_\varepsilon \, \xi^\alpha\|_2$ is finite, we see from (5.47) that

$$\|\phi^\varepsilon \, \xi^\alpha\|_2 \le c(\mathbf{f}, E_0)\,. \tag{6.5}$$

(ii) The case $9/5 < p_\infty \le p_0 < 2$

In this case we observe for the convective term

$$\|[\nabla \mathbf{v}^\varepsilon]\mathbf{v}^\varepsilon_\varepsilon \, \xi^\alpha\|_{p'_\infty}^2 \le c\|\nabla \mathbf{v}^\varepsilon\|_{p_\infty}^2\|\nabla \mathbf{v}^\varepsilon \, \xi^\alpha\|_{\frac{3p_\infty}{4p_\infty-6}}^2\,. \tag{6.6}$$

Furthermore we obtain, for $\lambda = \frac{9-4p_\infty}{2}$

$$\|\nabla \mathbf{v}^\varepsilon \, \xi^\alpha\|^2_{\frac{3p_\infty}{4p_\infty-6}} = \left(\int\limits_\Omega \left(|\nabla \mathbf{v}^\varepsilon| \xi^{\frac{2\alpha}{p_\infty}} \right)^{\frac{3p_\infty}{4p_\infty-6}\lambda} |\nabla \mathbf{v}^\varepsilon|^{\frac{3p_\infty}{4p_\infty-6}(1-\lambda)} \xi^{\alpha \frac{3p_\infty}{4p_\infty-6}(1-\lambda\frac{2}{p_\infty})} \, dx \right)^{4\frac{2p_\infty-3}{3p_\infty}}$$

$$\leq \|\nabla \mathbf{v}^\varepsilon \, \xi^{\frac{2\alpha}{p_\infty}}\|^{2\lambda}_{3p_\infty} \|\nabla \mathbf{v}^\varepsilon\|^{2(1-\lambda)}_{p_\infty}$$

as long as $2\lambda < p_\infty$, which is equivalent to $p_\infty > 9/5$. From this and (6.6) we therefore get

$$\|[\nabla \mathbf{v}^\varepsilon] \mathbf{v}^\varepsilon_\varepsilon \, \xi^\alpha\|^2_{p'_\infty} \leq \|\nabla \mathbf{v}^\varepsilon\|^{2(2-\lambda)}_{p_\infty} \|\nabla \mathbf{v}^\varepsilon \, \xi^{\frac{2\alpha}{p_\infty}}\|^{2\lambda}_{3p_\infty}$$

$$\leq c(\mathbf{f}, \mathbf{E}_0) + \frac{1}{2}\|\nabla \mathbf{v}^\varepsilon \, \xi^{\frac{2\alpha}{p_\infty}}\|^{p_\infty}_{3p_\infty} \, , \tag{6.7}$$

again for $2\lambda < p_\infty$ or equivalently $p_\infty > 9/5$. Thus (5.48) and (6.7) yields

$$\|\nabla \mathbf{S}(\mathbf{D}(\mathbf{v}^\varepsilon), \mathbf{E}) \, \xi^\alpha\|^2_2 + \|\nabla \mathbf{v}^\varepsilon \, \xi^{\frac{2\alpha}{p_\infty}}\|^{p_\infty}_{3p_\infty} + \|\nabla^2 \mathbf{v} \, \xi^\alpha\|^2_{p_\infty} \leq c(\mathbf{f}, \mathbf{E}_0) \, .$$

With regard to the first term in the above estimate we use Lemma 5.2.32, (5.2.39) and (5.2.41) to derive

$$\|\nabla \mathbf{v}^\varepsilon \, \xi^\alpha\|^{2(p_\infty-1)}_{6(p_\infty-1)} \leq c(\mathbf{f}, \mathbf{E}_0)\big(1 + I_{p,2}(\mathbf{D}(\nabla \mathbf{v}^\varepsilon), \, \xi^\alpha)\big)$$

$$\leq c(\mathbf{f}, \mathbf{E}_0)\big(1 + \|\nabla \mathbf{S}(\mathbf{D}(\mathbf{v}^\varepsilon), \mathbf{E}) \, \xi^\alpha\|^2_2\big)$$

$$\leq c(\mathbf{f}, \mathbf{E}_0) \, .$$

However $3p_\infty > 6(p_\infty - 1)$ for $p_\infty < 2$ and thus we do not gain a better information. Hence we have proved

$$\|\nabla \mathbf{v}^\varepsilon \, \xi^{\frac{2\alpha}{p_\infty}}\|^{p_\infty}_{3p_\infty} + \|\nabla^2 \mathbf{v} \, \xi^\alpha\|^2_{p_\infty} \leq c(\mathbf{f}, \mathbf{E}_0) \, . \tag{6.8}$$

Since we have shown that $\|[\nabla \mathbf{v}^\varepsilon] \mathbf{v}^\varepsilon_\varepsilon \, \xi^\alpha\|^2_{p'_\infty}$ is finite, we obtain from (5.47) that

$$\|\phi^\varepsilon \, \xi^\alpha\|_{p'_\infty} \leq c(\mathbf{f}, \mathbf{E}_0) \, . \tag{6.9}$$

(iii) **The case** $9/5 < p_\infty < 2 \leq p_0 \leq p^*_\infty$

We treat the convective term as in the case (ii). Thus we obtain from (5.48) and (6.7)

$$\|\nabla \mathbf{S}(\mathbf{D}(\mathbf{v}^\varepsilon), \mathbf{E}) \, \xi^\alpha\|^2_{p'_0} + \|\nabla \mathbf{v}^\varepsilon \, \xi^{\frac{2\alpha}{p_\infty}}\|^{p_\infty}_{3p_\infty} + \|\nabla^2 \mathbf{v} \, \xi^\alpha\|^2_{p_\infty} \leq c(\mathbf{f}, \mathbf{E}_0) \, .$$

With regard to the first term in the above estimate we proceed as in the case (i) and deduce

$$\|\nabla \mathbf{v}^\varepsilon \, \xi^\alpha\|^{p_\infty-1}_{\frac{3p_0}{2p_0-3}(p_\infty-1)} + \|\nabla^2 \mathbf{v} \, \xi^\alpha\|^2_{p_\infty} \leq c(\mathbf{f}, \mathbf{E}_0) \tag{6.10}$$

since for $p_0 < p_\infty^*$ we have that $\frac{3p_0}{2p_0-3}(p_\infty - 1) > 3p_\infty$. As in the previous cases we deduce from (5.47)

$$\|\phi^\varepsilon \, \xi^\alpha\|_{p'_\infty} \leq c(\mathbf{f}, \mathbf{E}_0) \, . \tag{6.11}$$

In all cases we have derived estimates which are sufficient for the limiting process $\varepsilon \to 0$. In fact, from (6.4), (6.5) and (6.8)–(6.11) we deduce the existence of $\mathbf{v}, \phi$ such that

$$\begin{aligned}
\nabla \mathbf{v}^\varepsilon \to \nabla \mathbf{v} \qquad & \text{strongly in } L^\alpha_{\text{loc}}(\Omega) \, , \quad \alpha < \tfrac{3q}{3-q} \, , \\
\mathbf{v}^\varepsilon \rightharpoonup \mathbf{v} \qquad & \text{weakly in } W^{2,q}_{\text{loc}}(\Omega) \cap W^{1,\beta}_{\text{loc}}(\Omega) \cap W^{1,p_\infty}_0(\Omega) \, , \\
\phi^\varepsilon \rightharpoonup \phi \qquad & \text{weakly in } L^{q'}_{\text{loc}}(\Omega) \cap L^{p'_0}(\Omega) \, ,
\end{aligned}$$

where $q = \min(2, p_\infty)$ and $\beta = \max(6, \frac{3p_0}{2p_0-3}(p_\infty - 1)), \beta = 3p_\infty, \beta = \frac{3p_0}{2p_0-3}(p_\infty - 1)$ in the cases (i), (ii), (iii). Moreover, the estimates (6.4), (6.5) and (6.8)–(6.11) remain valid for the limiting elements $\mathbf{v}, \phi$.

In the same way as in the proof of Proposition 5.42 we deduce that $\mathbf{v}, \phi$ satisfy the weak formulation (2.3). Thus we have proved all assertions of Theorem 3.7. $\blacksquare$

4 Electrorheological Fluids with Shear Dependent Viscosities: Unsteady Flows

4.1 Setting of the Problem and Main Results

In this chapter we investigate unsteady flows of a special incompressible electrorheological fluid in a bounded domain $\Omega \subseteq \mathbb{R}^3$. This motion is governed by the initial-boundary value problem[1]

$$
\begin{aligned}
\operatorname{div} \mathbf{E} &= 0 \\
\operatorname{curl} \mathbf{E} &= 0 \qquad & \text{in } Q_T, \\
\mathbf{E} \cdot \mathbf{n} &= \mathbf{E}_0 \cdot \mathbf{n} \qquad & \text{on } I \times \partial\Omega,
\end{aligned}
\tag{1.1}
$$

$$
\begin{aligned}
\frac{\partial \mathbf{v}}{\partial t} - \operatorname{div} \mathbf{S}(\mathbf{D}(\mathbf{v}), \mathbf{E}) + [\nabla \mathbf{v}]\mathbf{v} + \nabla\phi &= \mathbf{f} + \chi^E[\nabla \mathbf{E}]\mathbf{E}, \qquad & \text{in } Q_T, \\
\operatorname{div} \mathbf{v} &= 0 \\
\mathbf{v} &= 0 \qquad & \text{on } I \times \partial\Omega, \\
\mathbf{v}(0) &= \mathbf{v}_0 \qquad & \text{in } \Omega,
\end{aligned}
\tag{1.2}
$$

where $Q_T = I \times \Omega = (0, T) \times \Omega$. The data $T > 0, \mathbf{f}, \mathbf{E}_0$ and $\mathbf{v}_0$ are given. We will investigate (1.1), (1.2) for two different models, which will be described now.

In the first model, to which the main part of this chapter is devoted, we assume that the extra stress tensor $\mathbf{S}$ has the form

$$
\mathbf{S} = \alpha_{31}(1 + |\mathbf{E}|^2)(1 + |\mathbf{D}(\mathbf{v})|^2)^{\frac{p-2}{2}} \mathbf{D}(\mathbf{v}),
\tag{1.3}
$$

where $p = p(|\mathbf{E}|^2)$ is a C^1-function such that

$$
1 < p_\infty \le p(|\mathbf{E}|^2) \le p_0,
\tag{1.4}
$$

and where

$$
\alpha_{31} > 0.
\tag{1.5}
$$

---

[1]Similarly as in (3.1.2), we have divided (2.1.2) by the constant density ρ_0 and adapted the notation appropriately.

Obviously, we have for $\mathbf{D}, \mathbf{B} \in X = \{\mathbf{D} \in \mathbb{R}^{3 \times 3}_{\text{sym}}, \operatorname{tr} \mathbf{D} = 0\}$

$$\mathbf{S}(\mathbf{D}, \mathbf{E}) \cdot \mathbf{D} \geq \alpha_{31}(1 + |\mathbf{E}|^2)(1 + |\mathbf{D}|^2)^{\frac{p(|\mathbf{E}|^2)-2}{2}} |\mathbf{D}|^2, \tag{1.6}$$

$$\partial_{ij} S_{kl}(\mathbf{D}, \mathbf{E}) B_{ij} B_{kl} \geq \alpha_{31}(1 + |\mathbf{E}|^2)(1 + |\mathbf{D}|^2)^{\frac{p(|\mathbf{E}|^2)-2}{2}} |\mathbf{B}|^2. \tag{1.7}$$

The above system (1.1)–(1.5) describes the unsteady motion of an incompressible electrorheological fluid in a bounded domain Ω over the time interval $I = (0, T)$. In contrast to the steady problem we have restricted ourselves to the case when $\mathbf{S}$ has the pseudo–potential form (1.3) and discuss only the case $p_\infty \geq 2$. These are not very realistic assumptions, since we have lost the dependence of the material response function on the direction of the electric field (cf. Section 1.4) and since most electrorheological fluids possess a shear thinning viscosity, i.e. $p_\infty < 2$. While the restriction $p_\infty \geq 2$ can surely be weakened as in the steady case (cf. Chapter 3), it seems that it is hard to relax the pseudo–potential form (1.3) of $\mathbf{S}$ in the case of Dirichlet boundary conditions. However, the results presented here are the first ones for the unsteady motion of an electrorheological fluid with Dirichlet boundary conditions and together with Růžička [113] [2] and [114] [3] they are the first results for unsteady systems with non-standard growth conditions and a nonlinear right-hand side.

Let us make some comments about the solvability of the unsteady system (1.2)–(1.5). As already pointed out in the previous chapter, this system can not be handled by monotonicity methods. This arises from the fact, that in the parabolic setting it is fundamental to treat the space and time variable differently and to have the equivalence of the spaces

$$L^q(Q_T) \qquad \text{and} \qquad L^q(I, L^q(\Omega)).$$

In the case when we deal with the space $L^{p(t,x)}(Q_T)$ it is not clear how to achieve this equivalence. However, we can adapt the second method presented in Chapter 3 to obtain solutions also in the time dependent case. The presented approach uses and generalizes techniques and ideas developed for the treatment of unsteady problems arising from the theory of generalized Newtonian fluids (cf. Málek, Nečas, Růžička [71], [73], Bellout, Bloom Nečas [9], Málek, Nečas, Rokyta, Růžička [70]).

Similar considerations as in the beginning of Section 3.3 imply that we must use again an approximation of the elliptic operator and that we have to mollify the convective term. Moreover, let us point out that in the time dependent case the properties of both the time derivative $\frac{\partial \mathbf{v}}{\partial t}$ and the pressure ϕ are governed and computed from the system (1.2). This prevents us from working with local estimates as in Chapter 3, because we would run into a circle argument. Thus we are forced to derive global in space-time estimates of the second order spatial derivative $\nabla^2 \mathbf{v}$. Let us now define weak and strong solutions to the system (1.1)–(1.5) and state the main result of this

---

[2]Here the full system (1.1), (1.2) with Dirichlet boundary conditions and a stress tensor of the form $\mathbf{S} = \alpha_{31}(1 + |\mathbf{E}|^2)(1 + |\nabla \mathbf{v}|^2)^{\frac{p(|\mathbf{E}|^2)-2}{2}} \nabla \mathbf{v}$ is discussed.

[3]Here the system (1.2) is treated with the general stress tensor $\mathbf{S}$ given by (3.1.3) but under space periodic boundary conditions.

chapter. Recall the following definition of function spaces of divergence free functions
(cf. Section 2.2)

$$\mathcal{V} \equiv \{\mathbf{u} \in \mathcal{D}(\Omega), \, \text{div } \mathbf{u} = 0\},$$
$$V_q \equiv \text{closure of } \mathcal{V} \text{ in } \|\nabla . \|_q\text{-norm},$$
$$V_{p(x)} \equiv \text{closure of } \mathcal{V} \text{ in } \| . \|_{1,p(x)}\text{-norm},$$
$$E_{p(x)} \equiv \text{closure of } \mathcal{V} \text{ in } \|\mathbf{D}(.)\|_{p(x)}\text{-norm}.$$

Definition 1.8. *The couple* $(\mathbf{E}, \mathbf{v})$ *is said to be a weak solution of the problem* (1.1)–
(1.5) *if and only if*

$$\mathbf{E} \in C^1(\bar{Q}_T),$$
$$\mathbf{v} \in L^\infty(I, H) \cap L^{p_\infty}(I, V_{p_\infty}), \tag{1.9}$$
$$\mathbf{D}(\mathbf{v}) \in L^{p(|\mathbf{E}|^2)}(Q_T),$$

the system (1.1) *is satisfied in the classical sense and the weak formulation of* (1.2)

$$\left\langle \frac{\partial \mathbf{v}(t)}{\partial t}, \boldsymbol{\varphi} \right\rangle_{V_q} + \int_{\Omega_t} \mathbf{S}(\mathbf{D}(\mathbf{v}), \mathbf{E}) \cdot \mathbf{D}(\boldsymbol{\varphi}) + [\nabla \mathbf{v}]\mathbf{v}_\varepsilon \cdot \boldsymbol{\varphi} \, dx$$
$$= \int_{\Omega_t} \mathbf{f} \cdot \boldsymbol{\varphi} - \chi^E \mathbf{E} \otimes \mathbf{E} \cdot \mathbf{D}(\boldsymbol{\varphi}) \, dx \tag{1.10}$$

is satisfied for all $\boldsymbol{\varphi} \in \mathcal{V}$ *and almost all* $t \in I$, *with* $q = \max(p_0, \frac{5p_\infty}{5p_\infty - 6})$.

Definition 1.11. *The triple* $(\mathbf{E}, \mathbf{v}, \phi)$ *is said to be a strong solution of the problem*
(1.1)–(1.5) *if and only if*

$$\mathbf{E} \in C^1(\bar{Q}_T),$$
$$\mathbf{v} \in L^{2\frac{p_\infty - 1}{p_0 - 1}}(I, V_{6\frac{p_\infty - 1}{p_0 - 1}}) \cap L^{\frac{2}{p_0 - 1}}(I, W^{2, \frac{6}{p_0 + 1}}(\Omega)),$$
$$\mathbf{D}(\mathbf{v}) \in L^\infty(I, L^{p(|\mathbf{E}|^2)}(\Omega)) \cap L^{\frac{p_0 + 3}{2p_0} p(|\mathbf{E}|^2)}(Q_T), \tag{1.12}$$
$$\frac{\partial \mathbf{v}}{\partial t}, \phi \in L^2(Q_T),$$

the system (1.1) *is satisfied in the classical sense and the weak formulation of* (1.2)

$$\int_{\Omega_t} \frac{\partial \mathbf{v}}{\partial t} \cdot \boldsymbol{\varphi} \, dx + \int_{\Omega_t} \mathbf{S}(\mathbf{D}(\mathbf{v}), \mathbf{E}) \cdot \mathbf{D}(\boldsymbol{\varphi}) \, dx + \int_{\Omega_t} [\nabla \mathbf{v}]\mathbf{v}_\varepsilon \cdot \boldsymbol{\varphi} \, dx - \int_{\Omega_t} \phi \, \text{div } \boldsymbol{\varphi} \, dx$$
$$= \int_{\Omega_t} \mathbf{f} \cdot \boldsymbol{\varphi} \, dx - \chi^E \int_{\Omega_t} \mathbf{E} \otimes \mathbf{E} \cdot \mathbf{D}(\boldsymbol{\varphi}) \, dx \tag{1.13}$$

is satisfied for all $\boldsymbol{\varphi} \in \mathcal{D}(\Omega)$ *and almost all* $t \in I$.

Theorem 1.14. *Let $\Omega \subset \mathbb{R}^3$ be a bounded domain with $\partial\Omega \in C^{3,1}$ and assume that $T > 0$, $\mathbf{E}_0 \in C^1(\bar{I}, W^{2,r}(\Omega))$, $r > 3$, $\mathbf{f} \in L^2(Q_T)$ and $\mathbf{v}_0 \in E_{p(|\mathbf{E}|^2)}$ are given. Then there exists a weak solution $(\mathbf{E}, \mathbf{v})$ of the problem (1.1)–(1.5) whenever*

$$2 \le p_\infty \le p_0 < \min\left(\frac{8}{3}, \frac{3 + p_\infty}{2(3 - p_\infty)}\right). \tag{1.15}$$

Weak solutions are strong if we additionally know that

$$\frac{9}{4} \le p_\infty \le p_0 \le \frac{3(3 - p_\infty)}{2(5 - 2p_\infty)} \tag{1.16}$$

and unique if we require that $\mathbf{E}$ is orthogonal to $H_N(\Omega)$. Moreover, strong solutions satisfy for all $\Omega_0 \subset\subset \Omega$ the estimate

$$\int_0^T \int_{\Omega_0} (1 + |\mathbf{E}|^2)(1 + |\mathbf{D}(\mathbf{v})|^2)^{\frac{p(|\mathbf{E}|^2)-2}{2}} |\mathbf{D}(\nabla\mathbf{v})|^2 \, dx \, dt \le c(\mathbf{f}, \mathbf{v}_0, \mathbf{E}),$$

and belong in particular to $L^2(I, W^{2,2}_{\text{loc}}(\Omega))$. For the tangential derivatives holds

$$\int_0^T \int_\Omega \sum_{s=1}^2 \sum_{n=1}^N (1 + |\mathbf{E}|^2)(1 + |\mathbf{D}(\mathbf{v})|^2)^{\frac{p(|\mathbf{E}|^2)-2}{2}} |\mathbf{D}(\frac{\partial\mathbf{v}}{\partial\tau^s})|^2 \, \xi_1^n \, dx \, dt \le c(\mathbf{f}, \mathbf{v}_0, \mathbf{E}),$$

and in particular $\nabla\frac{\partial\mathbf{v}}{\partial\tau^s}$, $s = 1, 2$, belong to the space $L^2(Q_T)$.

Remark 1.17. 1) The above theorem shows the strength of the applied method, since the lower bound for the existence of weak solutions is 2. Even if the method of monotone operators could be applied, the lower bound would be at best $11/5$.

2) The lower bound for the existence of strong solutions coincides with the bound established by Málek, Nečas, Růžička [73] in the case $p = \text{const.}$, $\mathbf{E} = \mathbf{0}$. However the upper bound here is more restrictive because of the additional term $\|\mathbf{S}\|_{2s}^{2s}$ on the right-hand side of the estimate for the time derivative (cf. (3.13)).

3) The possible range for weak and strong solutions is

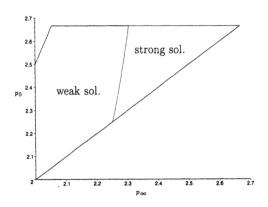

Let us now come to the second model for the extra stress tensor $\mathbf{S}$ in (1.2). Here we assume that $\mathbf{S}$ is linear in $\mathbf{D}$ and thus the extra stress $\mathbf{S}$ is given by (cf. Lemma 1.3.34)

$$\mathbf{S} = (\alpha_{31} + \alpha_{33}|\mathbf{E}|^2)\mathbf{D} + \alpha_{51}(\mathbf{DE} \otimes \mathbf{E} + \mathbf{E} \otimes \mathbf{DE}), \tag{1.18}$$

where

$$\alpha_{31} > 0, \quad \alpha_{33} > 0, \quad \alpha_{33} + \tfrac{4}{3}\alpha_{51} > 0. \tag{1.19}$$

In this model we have completely lost the shear dependency of the viscosity. However, we have a dependence of the stress tensor on the direction of the electric field, which is also important. In this situation we can prove

Theorem 1.20. *Let Ω be a bounded domain with $\partial\Omega \in C^{2,1}$ and assume that $\mathbf{v}_0 \in H$, $\mathbf{f} \in L^2(I, W^{-1,2}(\Omega))$, and $\mathbf{E}_0 \in L^\infty(I, W^{1,2}(\Omega))$ are given. Then there exists a weak solution $(\mathbf{v}, \mathbf{E})$ of the system (1.1), (1.2),(1.18), (1.19) such that*

$$\begin{aligned}
\mathbf{v} &\in L^\infty(I, H) \cap L^2(I, V_2), \\
\frac{\partial\mathbf{v}}{\partial t} &\in L^2(I, (V^{3/2})^*), \\
\mathbf{E} &\in L^\infty(I, W^{1,2}(\Omega)), \\
\mathbf{E} - \mathbf{E}_0 &\in L^\infty(I, \overset{\circ}{H}(\mathrm{div})),
\end{aligned} \tag{1.21}$$

which satisfies (1.1) almost everywhere in Q_T and (1.2) in the weak sense, i.e.

$$\begin{aligned}
&\left\langle \frac{\partial\mathbf{v}}{\partial t}(t), \boldsymbol{\varphi} \right\rangle_{V^{3/2}} + \int\limits_{\Omega_t} \mathbf{S}(\mathbf{D}(\mathbf{v}), \mathbf{E}) \cdot \mathbf{D}(\boldsymbol{\varphi})\,dx \\
&+ \int\limits_{\Omega_t} [\nabla\mathbf{v}]\mathbf{v} \cdot \boldsymbol{\varphi}\,dx = \int\limits_{\Omega_t} \mathbf{f} \cdot \boldsymbol{\varphi}\,dx - \chi^E \int\limits_{\Omega_t} \mathbf{E} \otimes \mathbf{E} \cdot \mathbf{D}(\boldsymbol{\varphi})\,dx
\end{aligned} \tag{1.22}$$

is satisfied for all $\boldsymbol{\varphi} \in V$ and almost every $t \in I$.

The existence of the electric field $\mathbf{E}$ with the above properties follows directly from Theorem 2.3.31. Having the electric field at our disposal, the proof of the existence of the velocity field follows essentially along the same lines as the proof of the corresponding statement in the theory of the unsteady Navier-Stokes equations (see e.g. Ladyzhenskaya [58], Lions [68] or Temam [120]) and therefore we will skip it here. The essential tools are the apriori estimates

$$\sup_{t\in I} \|\mathbf{v}(t)\|_2^2 + \int\limits_0^T \|\mathbf{D}(\mathbf{v})\|_{2,\nu}^2 + \|\nabla\mathbf{v}\|_2^2\,dt \le c(\mathbf{f}, \mathbf{E}_0, \mathbf{v}_0),$$

$$\left\|\frac{\partial\mathbf{v}}{\partial t}\right\|_{L^2(I,(V^{3/2})^*)} \le c(\mathbf{f}, \mathbf{E}_0, \mathbf{v}_0), \tag{1.23}$$

$$\|\mathbf{E}\|_{L^\infty(I,W^{1,2}(\Omega))} \le c\,\|\mathbf{E}_0\|_{L^\infty(I,W^{1,2}(\Omega))},$$

together with the Aubin-Lions lemma, which enables the limiting process in the convective term. The limiting process for the other terms in (1.22), which are linear in $\mathbf{v}$, is no problem due to the estimates (1.23) and the fact that the electric field $\mathbf{E}$ is known apriori.

The remaining part of this chapter is devoted to the study of problem (1.1)–(1.5).

4.1.1 Approximations

We will use the same type of approximation $\mathbf{S}^A(\mathbf{D}(\mathbf{v}), \mathbf{E})$ for the elliptic operator $\mathbf{S}(\mathbf{D}(\mathbf{v}), \mathbf{E})$ (cf. Section 3.3.1). For the convenience of the reader we recall the relevant definitions and properties of $\mathbf{S}^A(\mathbf{D}(\mathbf{v}), \mathbf{E})$ in the special case treated in this chapter. We define, for $A \geq 1$,

$$S_{ij}^A = \alpha_{31}(1 + |\mathbf{E}|^2)\partial_{ij}U^A(|\mathbf{D}(\mathbf{v})|^2, |\mathbf{E}|^2), \tag{1.24}$$

where ∂_{ij} denotes the derivative with respect to the components $D_{ij}(\mathbf{v})$, $i, j = 1, 2, 3$, of the symmetric part of the velocity gradient, i.e.

$$\partial_{ij}U^A(|\mathbf{D}(\mathbf{v})|^2, |\mathbf{E}|^2) = \frac{\partial U^A(|\mathbf{D}(\mathbf{v})|^2, |\mathbf{E}|^2)}{\partial D_{ij}}.$$

We define for all $\mathbf{D} \in X$

$$U^A(|\mathbf{D}|^2, |\mathbf{E}|^2) = \begin{cases} U(|\mathbf{D}|^2, |\mathbf{E}|^2) & |\mathbf{D}| \leq A, \\ a_0 + a_1(1 + |\mathbf{D}|^2)^{\frac{1}{2}} + a_2(1 + |\mathbf{D}|^2) & |\mathbf{D}| \geq A, \end{cases} \tag{1.25}$$

where $a_i = a_i(|\mathbf{E}|^2)$, $i = 0, 1, 2$, and

$$U(|\mathbf{D}|^2, |\mathbf{E}|^2) = \frac{1}{p(|\mathbf{E}|^2)}\left((1 + |\mathbf{D}|^2)^{\frac{p(|\mathbf{E}|^2)}{2}} - 1\right). \tag{1.26}$$

Note, that

$$\partial_{ij}U(|\mathbf{D}|^2, |\mathbf{E}|^2) = (1 + |\mathbf{D}|^2)^{\frac{p(|\mathbf{E}|^2)-2}{2}} D_{ij}, \tag{1.27}$$

$$U(0, |\mathbf{E}|^2) = \partial_{ij}U(0, |\mathbf{E}|^2) = 0. \tag{1.28}$$

From the definition of $\mathbf{S}$ and (1.24)–(1.27) it is clear that

$$\mathbf{S}^A(\mathbf{D}, \mathbf{E}) = \mathbf{S}(\mathbf{D}, \mathbf{E}) \qquad \text{if } |\mathbf{D}| \leq A. \tag{1.29}$$

We require that $U^A(\cdot, |\mathbf{E}|^2) \in C^2(\mathbb{R}^+)$ and $U^A(|\mathbf{D}|^2, \cdot) \in C^1(\mathbb{R}^+)$, which allows us to compute the coefficients a_0, a_1, a_2 as

$$a_0 = \left(\frac{1}{p} + \frac{p}{2} - \frac{3}{2}\right)\left(1 + A^2\right)^{\frac{p}{2}} - \frac{1}{p},$$

$$a_1 = (2 - p)\left(1 + A^2\right)^{\frac{p-1}{2}}, \tag{1.30}$$

$$a_2 = \frac{p-1}{2}\left(1 + A^2\right)^{\frac{p-2}{2}}.$$

Lemma 1.31. *Let* $\mathbf{S}$ *be given by* (1.3), *such that the conditions* (1.5) *and* $2 \leq p_\infty \leq p_0$ *are satisfied. Let* $\mathbf{S}^A$ *be given by* (1.24)–(1.26) *and* (1.30). *Then we have for all* $i, j, k, l = 1, 2, 3$ *all* $\mathbf{D}, \mathbf{B} \in X$, *all* $\mathbf{E} \in \mathbb{R}^3$ *and* A *sufficiently large*

$$|\mathbf{S}^A(\mathbf{D}, \mathbf{E})| \leq c_1(A)(1 + |\mathbf{E}|^2)(1 + |\mathbf{D}|^2)^{\frac{1}{2}}, \tag{1.32}$$

$$|\partial_{ij}\mathbf{S}^A(\mathbf{D}, \mathbf{E})| \leq c_2(A)(1 + |\mathbf{E}|^2), \tag{1.33}$$

$$\mathbf{S}^A(\mathbf{D}, \mathbf{E}) \cdot \mathbf{D} \geq c_3(1 + |\mathbf{E}|^2)|\mathbf{D}|^2 \begin{cases} (1 + |\mathbf{D}|^2)^{\frac{p(|\mathbf{E}|^2)-2}{2}}, \\ (1 + A^2)^{\frac{p(|\mathbf{E}|^2)-2}{2}}, \end{cases} \tag{1.34}$$

$$\partial_{kl}S_{ij}^A(\mathbf{D}, \mathbf{E})B_{ij}B_{kl} \geq c_4(1 + |\mathbf{E}|^2)|\mathbf{B}|^2 \begin{cases} (1 + |\mathbf{D}|^2)^{\frac{p(|\mathbf{E}|^2)-2}{2}}, \\ (1 + A^2)^{\frac{p(|\mathbf{E}|^2)-2}{2}}, \end{cases} \tag{1.35}$$

$$|\mathbf{S}^A(\mathbf{D}, \mathbf{E})| \leq c_5(1 + |\mathbf{E}|^2)|\mathbf{D}| \begin{cases} (1 + |\mathbf{D}|^2)^{\frac{p(|\mathbf{E}|^2)-2}{2}}, \\ (1 + A^2)^{\frac{p(|\mathbf{E}|^2)-2}{2}}, \end{cases} \tag{1.36}$$

$$|\partial_{ij}\mathbf{S}^A(\mathbf{D}, \mathbf{E})| \leq c_6(1 + |\mathbf{E}|^2) \begin{cases} (1 + |\mathbf{D}|^2)^{\frac{p(|\mathbf{E}|^2)-2}{2}}, \\ (1 + A^2)^{\frac{p(|\mathbf{E}|^2)-2}{2}}, \end{cases} \tag{1.37}$$

$$c_7|\mathbf{D}|^2 \begin{cases} (1 + |\mathbf{D}|^2)^{\frac{p(|\mathbf{E}|^2)-2}{2}} \\ (1 + A^2)^{\frac{p(|\mathbf{E}|^2)-2}{2}} \end{cases} \leq U^A(|\mathbf{D}|^2, |\mathbf{E}|^2) \leq c_8|\mathbf{D}|^2 \begin{cases} (1 + |\mathbf{D}|^2)^{\frac{p(|\mathbf{E}|^2)-2}{2}}, \\ (1 + A^2)^{\frac{p(|\mathbf{E}|^2)-2}{2}}, \end{cases} \tag{1.38}$$

$$|\mathbf{S}^A(\mathbf{D}, \mathbf{E})| \leq c_9\left(1 + U^A(|\mathbf{D}|^2, |\mathbf{E}|^2)\right)^{\frac{p(|\mathbf{E}|^2)-1}{p(|\mathbf{E}|^2)}}, \tag{1.39}$$

$$|\partial U^A(|\mathbf{D}|^2, |\mathbf{E}|^2)| \leq c_{10}|\mathbf{D}| \begin{cases} (1 + |\mathbf{D}|^2)^{\frac{p(|\mathbf{E}|^2)-2}{2}}, \\ (1 + A^2)^{\frac{p(|\mathbf{E}|^2)-2}{2}}, \end{cases} \tag{1.40}$$

$$1 + |\mathbf{S}^A(\mathbf{D}, \mathbf{E})| \geq c_{11} \begin{cases} (1 + |\mathbf{D}|^2)^{\frac{p(|\mathbf{E}|^2)-1}{2}}, \\ (1 + A^2)^{\frac{p(|\mathbf{E}|^2)-1}{2}}, \end{cases} \tag{1.41}$$

where the first line in the above cases holds if $|\mathbf{D}| \leq A$ *while the second one holds for* $|\mathbf{D}| \geq A$. *The constants* c_3–c_{11} *depend on* p_0, p_∞, *whereas* c_1, c_2 *depend on* A, p_0 *and* p_∞.

PROOF :   The statements (1.32)–(1.37) and (1.40) have been already proved in
Lemma 3.3.21. Inequalities (1.38), (1.39) and (1.41) are clear for $|\mathbf{D}| \leq A$. For
the case $|\mathbf{D}| \geq A$ we write, using (1.25) and (1.30)

$$
\begin{aligned}
U^A(|\mathbf{D}|^2) &= \frac{p-2}{2}(1+A^2)^{\frac{p-2}{2}}\left((1+|\mathbf{D}|^2)^{\frac{1}{2}} - (1+A^2)^{\frac{1}{2}}\right)^2 \\
&\quad + \frac{1}{2}(1+A^2)^{\frac{p-2}{2}}(|\mathbf{D}|^2 - A^2) + \frac{1}{p}\left((1+A^2)^{\frac{p-2}{2}} - 1\right) + \frac{1}{p}(1+A^2)^{\frac{p-2}{2}}A^2 \\
&\geq \frac{1}{p}(1+A^2)^{\frac{p-2}{2}}(|\mathbf{D}|^2 - A^2) + \frac{1}{p}\left((1+A^2)^{\frac{p-2}{2}} - 1\right) + \frac{1}{p}(1+A^2)^{\frac{p-2}{2}}A^2 \\
&\geq \frac{1}{p}(1+A^2)^{\frac{p-2}{2}}|\mathbf{D}|^2 ,
\end{aligned}
$$

where we employed $p_\infty \geq 2$ and $|\mathbf{D}| \geq A \geq 1$. This gives the lower bound in (1.38).
The upper bound follows from the above formula for U^A, inequality (3.3.37) and

$$
c(p)(1+x^2)^{\frac{p-2}{2}}x^2 \leq (1+x^2)^{\frac{p}{2}} - 1 \leq \tilde{c}(p)(1+x^2)^{\frac{p-2}{2}}x^2 , \tag{1.42}
$$

which holds for $x \in [0, \infty)$ and $p > 1$. Inequality (1.39) is a consequence of (1.36) and
a argument similar to (5.2.14). In order to prove (1.41) we recall that (cf. (3.3.31))

$$
S_{ij}^A(\mathbf{D}, \mathbf{E}) = \alpha_{31}(1+|\mathbf{E}|^2)\left(\frac{a_1}{(1+|\mathbf{D}|^2)^{1/2}} + 2a_2\right)D_{ij} .
$$

Due to (3.3.33), $p_\infty \geq 2$ and $|\mathbf{D}| \geq A \geq 1$ we thus obtain

$$
|\mathbf{S}^A(\mathbf{D}, \mathbf{E})| \geq \alpha_{31}(1+A^2)^{\frac{p-2}{2}}|\mathbf{D}| \geq c_{11}(1+A^2)^{\frac{p-1}{2}} .
$$

This completes the proof of the lemma.   ∎

Remark 1.43. From (1.35) and (1.37) it is clear that

$$
\partial_{kl} S_{ij}^A(\mathbf{D}, \mathbf{E}) B_{ij} B_{kl}
$$

and

$$
(1+|\mathbf{E}|^2)|\mathbf{B}|^2 \begin{cases} (1+|\mathbf{D}|^2)^{\frac{p(|\mathbf{E}|^2)-2}{2}} \\ (1+A^2)^{\frac{p(|\mathbf{E}|^2)-2}{2}} \end{cases}
$$

are equivalent quantities. Moreover, from (1.34), (1.36) and (1.38) it follows that

$$
\mathbf{S}^A(\mathbf{D}, \mathbf{E}) \cdot \mathbf{D} \qquad \text{and} \qquad U^A(|\mathbf{D}|^2, |\mathbf{E}|^2)
$$

are equivalent quantities.

Again we also need the derivative with respect to the components $E_n, n = 1, 2, 3$,
of the electric field. We use the notation

$$
\partial_n U^A(|\mathbf{D}|^2, |\mathbf{E}|^2) \equiv \frac{\partial U^A(|\mathbf{D}|^2, |\mathbf{E}|^2)}{\partial E_n} \tag{1.44}
$$

and similar formulas for $\mathbf{S}^A, U$ and $\mathbf{S}$. We shall use the letters m, n as indices for this
derivative.

Lemma 1.45. *Let the assumption of Lemma 1.31 be satisfied. Then there exist constants depending on the function p and α_{ij} such that*

$$|\partial_n \partial_{ij} U^A(|\mathbf{D}|^2, |\mathbf{E}|^2)| \leq c|\mathbf{E}|\,|\mathbf{D}| \begin{cases} (1+|\mathbf{D}|^2)^{\frac{p(|\mathbf{E}|^2)-2}{2}} \left(1 + \ln(1+|\mathbf{D}|^2)\right), \\ (1+A^2)^{\frac{p(|\mathbf{E}|^2)-2}{2}} \left(1 + \ln(1+|A|^2)\right), \end{cases} \tag{1.46}$$

$$|\partial_n \mathbf{S}^A(\mathbf{D}, \mathbf{E})| \leq c|\mathbf{E}|(1+|\mathbf{E}|^2)\,|\mathbf{D}| \begin{cases} (1+|\mathbf{D}|^2)^{\frac{p(|\mathbf{E}|^2)-2}{2}} \left(1 + \ln(1+|\mathbf{D}|^2)\right), \\ (1+A^2)^{\frac{p(|\mathbf{E}|^2)-2}{2}} \left(1 + \ln(1+A^2)\right), \end{cases} \tag{1.47}$$

$$|\partial_n U^A(|\mathbf{D}|^2, |\mathbf{E}|^2)| \leq c|\mathbf{E}|(1+|\mathbf{D}|^2) \begin{cases} (1+|\mathbf{D}|^2)^{\frac{p(|\mathbf{E}|^2)-2}{2}} \left(1 + \ln(1+|\mathbf{D}|^2)\right), \\ (1+A^2)^{\frac{p(|\mathbf{E}|^2)-2}{2}} \left(1 + \ln(1+A^2)\right). \end{cases} \tag{1.48}$$

PROOF : These estimates follow immediately from (1.24), (1.25) and (1.30). ∎

Remark 1.49. From (3.3.47), (1.48) and (1.38) and a similar argument as in (5.2.14) we deduce for all $s > 1$

$$|\partial_n U^A(|\mathbf{D}|^2, |\mathbf{E}|^2)| \leq c\left(1 + U^A(|\mathbf{D}|^2, |\mathbf{E}|^2)\right)^s. \tag{1.50}$$

Let us finish this section by defining the approximate problem (1.1), $(1.2)_{\varepsilon,A}$.

Definition 1.51. *The triple* $(\mathbf{E}, \mathbf{v}, \phi) = (\mathbf{E}, \mathbf{v}^{\varepsilon,A}, \phi^{\varepsilon,A})$, *for* $\varepsilon > 0$, $A \geq A_0$ *given, is said to be a strong solution of the problem (1.1) and*

$$\frac{\partial \mathbf{v}}{\partial t} - \operatorname{div} \mathbf{S}^A(\mathbf{D}(\mathbf{v}), \mathbf{E}) + [\nabla \mathbf{v}]\mathbf{v}_\varepsilon + \nabla\phi = \mathbf{f} + \chi^E[\nabla \mathbf{E}]\mathbf{E}, \qquad \text{in } Q_T,$$

$$\operatorname{div} \mathbf{v} = 0$$
$$\mathbf{v} = 0 \qquad \text{on } I \times \partial\Omega, \tag{1.2}_{\varepsilon,A}$$
$$\mathbf{v}(0) = \mathbf{v}_0 \qquad \text{in } \Omega,$$

where $\mathbf{S}^A$ is given by (1.24), (1.25), (1.30), if and only if

$$\mathbf{E} \in C^1(\bar{Q}_T),$$
$$\mathbf{v} \in C(\bar{I}, V_2) \cap L^2\big(I, W^{2,2}(\Omega)\big),$$
$$\frac{\partial \mathbf{v}}{\partial t}, \ \phi \in L^2(Q_T); \tag{1.52}$$

the system $(1.1)_{1,2}$ *is satisfied almost everywhere in* Q_T, *for almost all* $t \in I$ *we have* $\mathbf{E}(t) - \mathbf{E}_0(t) \in \overset{\circ}{H}(\mathrm{div})$ *and the weak formulation of* $(1.2)_{\varepsilon,A}$

$$\int_{\Omega_t} \frac{\partial \mathbf{v}}{\partial t} \cdot \boldsymbol{\varphi} \, dx + \int_{\Omega_t} \mathbf{S}^A(\mathbf{D}(\mathbf{v}), \mathbf{E}) \cdot \mathbf{D}(\boldsymbol{\varphi}) \, dx + \int_{\Omega_t} [\nabla \mathbf{v}] \mathbf{v}_\varepsilon \cdot \boldsymbol{\varphi} \, dx - \int_{\Omega_t} \phi \, \mathrm{div} \, \boldsymbol{\varphi} \, dx$$

$$= \int_{\Omega_t} \mathbf{f} \cdot \boldsymbol{\varphi} \, dx - \chi^E \int_{\Omega_t} \mathbf{E} \otimes \mathbf{E} \cdot \mathbf{D}(\boldsymbol{\varphi}) \, dx \tag{1.53}$$

is satisfied for all $\boldsymbol{\varphi} \in \mathcal{D}(\Omega)$ *and almost every* $t \in I$.

4.2   Existence of Approximate Solutions

Now we shall show the existence of strong solutions for the system (1.1), $(1.2)_{\varepsilon,A}$.

Proposition 2.1. *Let* $\Omega \subseteq \mathbb{R}^3$ *be a bounded domain with* $\partial\Omega \in \mathcal{C}^{3,1}$ *and assume that* $T > 0$, $\mathbf{E}_0 \in C^1(\bar{I}, W^{2,r}(\Omega))$, $r > 3$, $\mathbf{f} \in L^2(Q_T)$, $\mathbf{v}_0 \in E_{p(|\mathbf{E}|^2)}$, $\varepsilon > 0$ *and* $A \geq A_0$ *are given. For* $p_\infty \geq 2$ *there exists a strong solution* $(\mathbf{E}, \mathbf{v}, \phi) = (\mathbf{E}, \mathbf{v}^{\varepsilon,A}, \phi^{\varepsilon,A})$ *of the problem* (1.1), $(1.2)_{\varepsilon,A}$ *satisfying the estimates*

$$\|\mathbf{E}\|_{C^1(\bar{I}, W^{2,r}(\Omega))} \leq c \, \|\mathbf{E}_0\|_{C^1(\bar{I}, W^{2,r}(\Omega))}, \tag{2.2}$$

$$\|\mathbf{v}\|^2_{L^\infty(I,H)} + \int_0^T \|\nabla \mathbf{v}\|^2_{2,\nu} \, dt \leq c(\mathbf{f}, \mathbf{v}_0, \mathbf{E}_0), \tag{2.3}$$

$$\int_0^T \|U^A(|\mathbf{D}(\mathbf{v})|^2, |\mathbf{E}|^2)\|_{1,\nu} \, dt \leq c(\mathbf{f}, \mathbf{v}_0, \mathbf{E}_0), \tag{2.4}$$

$$\int_0^T \left\| \frac{\partial \mathbf{v}}{\partial t} \right\|^2_2 dt + \|\nabla \mathbf{v}\|^2_{L^\infty(I, L^2(\Omega, \nu))} \leq c(\mathbf{f}, \mathbf{v}_0, \mathbf{E}_0, \varepsilon^{-1}, A), \tag{2.5}$$

$$\int_0^T \|\phi\|^2_2 \, dt \leq c(\mathbf{f}, \mathbf{v}_0, \mathbf{E}_0) \left(1 + \int_0^T \|\mathbf{S}^A\|^2_2 + \|[\nabla \mathbf{v}]\mathbf{v}_\varepsilon\|^2_2 + \|U^A\|^s_s \, dt \right)$$

$$\leq c(\mathbf{f}, \mathbf{v}_0, \mathbf{E}_0, \varepsilon^{-1}, A), \tag{2.6}$$

where $s > 1$. *Moreover, the estimate*

$$\|\mathbf{v}\|_{L^2(I, W^{2,2}(\Omega))} \leq c(\mathbf{f}, \mathbf{v}_0, \mathbf{E}_0, \varepsilon^{-1}, A) \tag{2.7}$$

holds and $(1.2)_{\varepsilon,A}$ *is satisfied almost everywhere in* Q_T.

PROOF :   The first part of this proposition concerning the existence of a solution to the system (1.1), $(1.2)_{\varepsilon,A}$ is standard using apriori estimates and the theory of monotone operators and thus we will be brief in this part of the proof. However, the part concerning the $W^{2,2}$-estimate of $\mathbf{v}$ is not so common and we will give a detailed proof. We will drop the indices ε, A and write $\mathbf{v} = \mathbf{v}^{\varepsilon,A}$ and $\phi = \phi^{\varepsilon,A}$.

The existence of $\mathbf{E}$ solving (1.1) and having the properties stated above follows from Proposition 2.3.35. We shall fix one such solution for the following considerations. The existence of a solution $\mathbf{v}$ to $(1.2)_{\varepsilon,A}$ satisfying the estimates (2.3)–(2.5) uses the Galerkin approach (cf. Lions [68], Section 2.5). Let $\boldsymbol{\omega}^j$, $j = 1, 2, \ldots$ be a smooth basis of V_2 and let us denote the linear hull of $\boldsymbol{\omega}^1, \cdots, \boldsymbol{\omega}^n$ by X_n. We define the Galerkin approximation $\mathbf{v}^n$ by

$$\mathbf{v}^n(t, x) = \sum_{j=1}^{n} a_j(t) \boldsymbol{\omega}^j(x),$$

which solves the Galerkin system

$$\frac{d}{dt} \int_\Omega \mathbf{v}^n \cdot \boldsymbol{\varphi} \, dx + \int_\Omega \mathbf{S}(\mathbf{D}(\mathbf{v}^n), \mathbf{E}) \cdot \mathbf{D}(\boldsymbol{\varphi}) \, dx + \int_\Omega [\nabla \mathbf{v}^n] \mathbf{v}^n \cdot \boldsymbol{\varphi} \, dx$$

$$= \int_\Omega \mathbf{f} \cdot \boldsymbol{\varphi} \, dx - \chi^E \int_\Omega \mathbf{E} \otimes \mathbf{E} \cdot \mathbf{D}(\boldsymbol{\varphi}) \, dx \qquad \forall \boldsymbol{\varphi} \in X_n, \tag{2.8}$$

$$\int_\Omega \mathbf{v}^n(0) \cdot \boldsymbol{\varphi} \, dx = \int_\Omega \mathbf{v}_0 \cdot \boldsymbol{\varphi} \, dx, \qquad \forall \boldsymbol{\varphi} \in X_n.$$

This ordinary differential equation for $a_j(t)$ is at least locally in time solvable. The global solvability follows from the apriori estimates (2.3)–(2.5), which will be proved next.

Using $\mathbf{v}^n$ as a test function in (2.8) we obtain after integration over $(0, t)$ and using (1.34) and (1.38)

$$\|\mathbf{v}^n(t)\|_{2,\Omega_t}^2 + \int_0^t \|\nabla \mathbf{v}^n\|_{2,\nu,\Omega_\tau}^2 \, d\tau + \int_0^t \|U^A(|\mathbf{D}(\mathbf{v}^n)|^2, |\mathbf{E}|^2)\|_{1,\nu,\Omega_\tau} \, d\tau$$

$$\leq c\|\mathbf{v}_0\|_2^2 + c \int_0^t \|\mathbf{f}\|_2^2 \, d\tau + c \int_0^t \|\mathbf{E}\|_4^4 \, d\tau \leq c(\mathbf{f}, \mathbf{v}_0, \mathbf{E}_0). \tag{2.9}$$

Choosing $\frac{\partial \mathbf{v}^n}{\partial t}$ as a test function in (2.8) we obtain after some simple manipulations, for all $t \in I$

$$\left\| \frac{\partial \mathbf{v}^n}{\partial t} \right\|_{2,\Omega_t}^2 + \int_{\Omega_t} \mathbf{S}^A(\mathbf{D}(\mathbf{v}^n), \mathbf{E}) \cdot \mathbf{D}\left(\frac{\partial \mathbf{v}^n}{\partial t}\right) dx$$

$$\leq c(\mathbf{E}_0) + c\left(\|\mathbf{f}\|_{2,\Omega_t}^2 + \|[\nabla \mathbf{v}^n]\mathbf{v}_\varepsilon^n\|_{2,\Omega_t}^2\right). \tag{2.10}$$

The second term on the left-hand side can be re-written as (recall the notation (1.44))

$$
\alpha_{31} \frac{d}{dt} \int_\Omega U^A(|\mathbf{D}(\mathbf{v}^n)|^2, |\mathbf{E}|^2)(1 + |\mathbf{E}|^2) \, dx
$$

$$
- 2\alpha_{31} \int_\Omega U^A(|\mathbf{D}(\mathbf{v}^n)|^2, |\mathbf{E}|^2) \, \mathbf{E} \cdot \frac{\partial \mathbf{E}}{\partial t} \, dx \tag{2.11}
$$

$$
- \alpha_{31} \int_\Omega \partial_n U^A(|\mathbf{D}(\mathbf{v}^n)|^2, |\mathbf{E}|^2) \frac{\partial E_n}{\partial t} \, (1 + |\mathbf{E}|^2) \, dx \, .
$$

From (2.10), (2.11), the regularity properties of $\mathbf{E}$ and (1.38), (1.48) we deduce for every $t \in I$

$$
\int_0^t \left\| \frac{\partial \mathbf{v}^n}{\partial t} \right\|_2^2 d\tau + \|\nabla \mathbf{v}^n\|_{2,\nu,\Omega_t}^2
$$

$$
\leq c(\mathbf{f}, \mathbf{E}_0, A, \varepsilon^{-1}) \left(1 + \int_0^T \|\nabla \mathbf{v}^n\|_2^2 \, dt \right) \leq c(\mathbf{f}, \mathbf{v}_0, \mathbf{E}_0, \varepsilon^{-1}, A) \, , \tag{2.12}
$$

where we also used Korn's inequality, the properties of the mollifier and (2.9). We easily see from (2.9) and the properties of the mollifier that

$$
[\nabla \mathbf{v}^n] \mathbf{v}_\varepsilon^n \in L^2(Q_T) \, .
$$

From the inequality (1.34) and the regularity of $\mathbf{E}$ it is clear that the operator $- \operatorname{div} \mathbf{S}^A(\mathbf{D}(\cdot), \mathbf{E}) : L^2(I, V_2) \to L^2(I, (V_2)^*)$ is uniformly monotone. Therefore we may prove the existence of a solution $\mathbf{v}$ satisfying for almost all $t \in I$ and all $\boldsymbol{\varphi} \in V_2$ the weak formulation

$$
\int_{\Omega_t} \frac{\partial \mathbf{v}}{\partial t} \cdot \boldsymbol{\varphi} \, dx + \int_{\Omega_t} \mathbf{S}^A(\mathbf{D}(\mathbf{v}), \mathbf{E}) \cdot \mathbf{D}(\boldsymbol{\varphi}) \, dx + \int_{\Omega_t} [\nabla \mathbf{v}] \mathbf{v}_\varepsilon \cdot \boldsymbol{\varphi} \, dx
$$

$$
= \int_{\Omega_t} \mathbf{f} \cdot \boldsymbol{\varphi} \, dx - \chi^E \int_{\Omega_t} \mathbf{E} \otimes \mathbf{E} \cdot \mathbf{D}(\boldsymbol{\varphi}) \, dx \, , \tag{2.13}
$$

and the estimates (2.3), (2.4) and (2.5) by standard arguments using the estimates (2.9), (2.12), the lower semicontinuity of norms and of $\|U^A(|\mathbf{D}(\boldsymbol{\varphi})|^2, |\mathbf{E}|^2)\|_{1,\nu}$ (cf. (1.34)), the Aubin-Lions lemma and the theory of monotone operators (cf. Lions [68] for more details).

Defining $\mathbf{F} \in L^2(I, (W_0^{1,2}(\Omega))^*)$ by

$$
\langle \mathbf{F}(t), \boldsymbol{\varphi} \rangle_{1,2} \equiv \int_{\Omega_t} \frac{\partial \mathbf{v}}{\partial t} \cdot \boldsymbol{\varphi} \, dx + \int_{\Omega_t} \mathbf{S}^A(\mathbf{D}(\mathbf{v}), \mathbf{E}) \cdot \mathbf{D}(\boldsymbol{\varphi}) \, dx + \int_{\Omega_t} [\nabla \mathbf{v}] \mathbf{v}_\varepsilon \cdot \boldsymbol{\varphi} \, dx
$$

$$
- \int_{\Omega_t} \mathbf{f} \cdot \boldsymbol{\varphi} \, dx + \chi^E \int_{\Omega_t} \mathbf{E} \otimes \mathbf{E} \cdot \mathbf{D}(\boldsymbol{\varphi}) \, dx \, , \tag{2.14}
$$

we see that for almost all $t \in I$ and all $\varphi \in V_2$

$$\langle \mathbf{F}(t), \varphi \rangle_{1,2} = 0.$$

From this and Theorem 5.1.10 we deduce the existence of $\phi \in L^2(Q_T)$, $\int_\Omega \phi(t)\,dx = 0$ such that for almost all $t \in I$

$$\langle \mathbf{F}(t), \varphi \rangle_{1,2} = -\int_{\Omega_t} \phi(t)\,\mathrm{div}\,\varphi\,dx \qquad \forall \varphi \in W_0^{1,2}(\Omega), \qquad (2.15)$$

which is the weak formulation (1.53). From (2.14) and (2.15), the dual definition of the norm in $L^2(\Omega)$, and Proposition 5.1.25 we obtain, for almost all $t \in I$, (cf. (3.4.15))

$$\|\phi\|_{2,\Omega_t}^2 \le c(\mathbf{E}_0)\left(1 + \|\mathbf{f}\|_{2,\Omega_t}^2 + \|\mathbf{S}^A\|_{2,\Omega_t}^2 + \|[\nabla\mathbf{v}]\mathbf{v}_\epsilon\|_{2,\Omega_t}^2 + \left\|\frac{\partial\mathbf{v}}{\partial t}\right\|_{2,\Omega_t}^2\right), \qquad (2.16)$$

which, after integration with respect to time, together with (1.32), the properties of the mollifier and (2.5) gives (2.6). Now one clearly sees the difference to the steady case treated in Chapter 3, where we have been able to work with local estimates. If one tries to derive a local estimate for the pressure ϕ then also a local norm of the time derivative of $\mathbf{v}$ would appear on the right–hand side. However, a local estimate for $\frac{\partial\mathbf{v}}{\partial t}$ (cf. (1.53)) would involve the pressure ϕ on the right–hand side since the appropriate test function would not be divergence free and thus we would run into circles. To avoid this situation we must derive global estimates in the unsteady case.

It remains to show (2.7). For the proof of it we have to establish estimates in the interior and estimates near the boundary. We will proceed similar as in the proof of Proposition 3.4.2. Due to the notation and the properties of the mapping T introduced in the Appendix (cf. (5.1.26)–(5.1.34)) we can treat interior regularity and regularity in tangential directions analogously. Since the former case was already accomplished in Section 3.4 and since in the latter one arise some additional difficulties we will present the proof for the regularity in tangential directions here. Let Ω_0^n and U^n, $n = 1, \cdots, N$, be two coverings of the boundary $\partial\Omega$ (cf. Definition 5.1.1) such that $U^n \subset\subset \Omega_0^n \subseteq V^n$, $n = 1, \cdots, N$. For simplicity we drop the index n and denote $\Omega_1 = \Omega \cap \Omega_0$ and consider $\xi_1 \in \mathcal{D}(\Omega_0)$, $0 \le \xi_1 \le 1, \xi_1 \equiv 1$ in U and put (cf. (5.1.26), $T = T_{r,h}$)

$$\mathbf{w}(x) \equiv \mathbf{v}(Tx) - \mathbf{v}(x). \qquad (2.17)$$

Due to the definition of the mapping T we see that $h^{-2}\mathbf{w}(x)\,\xi_1^2(x) \in W_0^{1,2}(\Omega)$ is an admissible test function in the weak formulation (1.53) we can derive the following identity:

$$\frac{1}{h^2} \int\limits_{\Omega_1} \left(\frac{\partial \mathbf{v}}{\partial t}(Tx) - \frac{\partial \mathbf{v}}{\partial t}(x)\right) \cdot \mathbf{w}(x)\,\xi^2(x)\,dx$$

$$+ \frac{1}{h^2} \int\limits_{\Omega_1} \left(\mathbf{S}^A\big(\mathbf{D}(\mathbf{v}(Tx)), \mathbf{E}(Tx)\big) - \mathbf{S}^A\big(\mathbf{D}(\mathbf{v}(x)), \mathbf{E}(x)\big)\right) \cdot \mathbf{D}\big(\mathbf{w}(x)\,\xi^2(x)\big)\,dx$$

$$+ \frac{1}{h^2} \int\limits_{\Omega_1} \big([\nabla\mathbf{v}(Tx)]\mathbf{v}_\varepsilon(Tx) - [\nabla\mathbf{v}(x)]\mathbf{v}_\varepsilon(x)\big) \cdot \mathbf{w}(x)\,\xi^2(x)\,dx$$

$$- \frac{1}{h^2} \int\limits_{\Omega_1} \big(\phi(Tx) - \phi(x)\big)\,\mathrm{div}\,\big(\mathbf{w}(x)\,\xi^2(x)\big)\,dx \qquad (2.18)$$

$$- \frac{1}{h^2} \int\limits_{\Omega_1} \big(\mathbf{f}(Tx) - \mathbf{f}(x)\big) \cdot \mathbf{w}(x)\,\xi^2(x)\,dx$$

$$+ \frac{\chi^E}{h^2} \int\limits_{\Omega_1} \big(\mathbf{E}(Tx) \otimes \mathbf{E}(Tx) - \mathbf{E}(x) \otimes \mathbf{E}(x)\big) \cdot \mathbf{D}\big(\mathbf{w}(x)\,\xi^2(x)\big)\,dx = 0\,,$$

where we dropped the index 1 from the cut-off function for brevity. We denote the integrals on the left-hand side by $I_0, \ldots, I_5$ and discuss them separately. We see that

$$I_0 = \frac{1}{2}\frac{d}{dt} \int\limits_{\Omega_1} \left|\frac{\mathbf{w}}{h}\right|^2 \xi^2\,dx\,. \qquad (2.19)$$

The terms I_1, I_2, I_4 and I_5 can be treated, with some small modifications, in the same way as in the proof of Proposition 3.4.2 (put $q = 2$ and $\alpha = 1$). Thus, we obtain as a contribution of I_1 on the left-hand side (cf. (3.4.26)) [4]

$$c_{12} \int\limits_{\Omega_1} \left|\frac{\nabla\mathbf{w}}{h}\right|^2 \xi^2\,dx \qquad (2.20)$$

and contributions of I_1, I_2, I_4 and I_5 on the right-hand side (cf. (3.4.27), (3.4.29), (3.4.30), (3.4.32), (3.4.33), (3.4.41), (3.4.43))

$$\frac{c_{11}}{8} \int\limits_{\Omega_1} \left|\frac{\nabla\mathbf{w}}{h}\right|^2 \xi^2\,dx + c\int\limits_{\Omega_1} \left|\frac{\mathbf{E}(Tx) - \mathbf{E}(x)}{h}\right|^2 \big(|\mathbf{D}(\mathbf{v}(Tx))|^2 + |\mathbf{D}(\mathbf{v}(x))|^2\big)\,\xi^2\,dx$$

$$+ c\left(\|\nabla\mathbf{v}\|_2^2 + \left\|\frac{\mathbf{w}}{h}\right\|_{2,\Omega_1}^2 + \|\mathbf{f}\|_2^2 + \|\nabla\mathbf{E}\|_2^2 + \|\mathbf{v}\|_2^4 + \|\mathbf{v}\|_2^2 \left\|\frac{\mathbf{w}}{h}\xi\right\|_{2,\Omega_t}^2\right), \qquad (2.21)$$

where $c = c(A, \varepsilon^{-1}, \mathbf{E}_0, a, \nabla^2\xi)$ and where we also used

$$\|\nabla\mathbf{v}_\varepsilon\|_\infty \le \varepsilon^{-3}\|\mathbf{v}\|_2\,. \qquad (2.22)$$

---

[4] Since $h^{-2}\mathbf{w}(x)\,\xi_1^2(x)$ belongs to the space $W_0^{1,2}(\Omega)$ we can as in (3.4.26) use Korn's inequality.

With the pressure term I_3 we must be more carefully. It can be re-written as

$$
\begin{aligned}
h^2 I_3 &= \int_{\Omega_1} \phi(x)\Big(\operatorname{div}\big(\mathbf{w}(x)\xi^2(x)\big) - \operatorname{div}\big(\mathbf{w}(T^{-1}x)\xi^2(T^{-1}x)\big)\Big)\,dx \\
&= \int_{\Omega_1} \phi(x)\Big(\operatorname{div}\mathbf{w}(x)\,\xi^2(x) - \operatorname{div}\mathbf{w}(T^{-1}x)\,\xi^2(T^{-1}x)\Big)\,dx \\
&\quad + 2\int_{\Omega_1} \phi(x)\Big(w_i(x)\xi(x)\frac{\partial\xi(x)}{\partial x_i} - w_i(T^{-1}x)\xi(T^{-1}x)\frac{\partial\xi(T^{-1}x)}{\partial x_i}\Big)\,dx \\
&= \int_{\Omega_1} \phi(x)\operatorname{div}\big(\mathbf{v}(Tx) - 2\mathbf{v}(x) + \mathbf{v}(T^{-1}x)\big)\xi^2(x)\,dx \\
&\quad + \int_{\Omega_1} \phi(x)\operatorname{div}\mathbf{w}(T^{-1}x)\big(\xi^2(x) - \xi^2(T^{-1}x)\big)\,dx \\
&\quad + 2\int_{\Omega_1} \phi(x)\big(v_i(Tx) - 2v_i(x) + v_i(T^{-1}x)\big)\xi(x)\frac{\partial\xi(x)}{\partial x_i}\,dx \\
&\quad + 2\int_{\Omega_1} \phi(x)w_i(T^{-1}x)\Big(\xi(x)\frac{\partial\xi(x)}{\partial x_i} - \xi(T^{-1}x)\frac{\partial\xi(T^{-1}x)}{\partial x_i}\Big)\,dx \\
&= h^2\big(J_6 + J_7 + J_8 + J_9\big).
\end{aligned}
\tag{2.23}
$$

Using (5.1.27), (5.1.28) and $\operatorname{div}\mathbf{v} = 0$ we can re-write J_6 as follows

$$
\begin{aligned}
h^2 J_6 &= \int_{\Omega_1} \sum_{s=1}^{2} \phi(x)\Big(\frac{\partial v_s(Tx)}{\partial y_3}\frac{\partial\Delta^+ a(x')}{\partial x_s} + \frac{\partial v_s(T^{-1}x)}{\partial y_3}\frac{\partial\Delta^- a(x')}{\partial x_s}\Big)\xi^2(x)\,dx \\
&= \int_{\Omega_1} \sum_{s=1}^{2} \phi(x)\Big(\frac{\partial v_s(Tx)}{\partial y_3} - \frac{\partial v_s(x)}{\partial y_3}\Big)\frac{\partial\Delta^+ a(x')}{\partial x_s}\xi^2(x)\,dx \\
&\quad + \int_{\Omega_1} \sum_{s=1}^{2} \phi(x)\Big(\frac{\partial v_s(x)}{\partial y_3} - \frac{\partial v_s(T^{-1}x)}{\partial y_3}\Big)\frac{\partial\Delta^- a(x')}{\partial x_s}\xi^2(x)\,dx \\
&\quad + \int_{\Omega_1} \sum_{s=1}^{2} \phi(x)\frac{\partial v_s(x)}{\partial y_3}\frac{\partial}{\partial x_s}\Big(a(x' + h\hat{e}^r) - 2a(x') + a(x' - h\hat{e}^r)\Big)\xi^2(x)\,dx \\
&= h^2\big(J_{6,1} + J_{6,2} + J_{6,3}\big).
\end{aligned}
\tag{2.24}
$$

Moreover, note that

$$J_{6,2} = \int_{\Omega_1} \sum_{s=1}^{2} \phi(Tx) \left(\frac{\partial v_s(Tx)}{\partial y_3} - \frac{\partial v_s(x)}{\partial y_3} \right) \frac{\partial \Delta^+ a(x')}{\partial x_s} \xi^2(x) \, dx$$

$$+ \int_{\Omega_1} \sum_{s=1}^{2} \phi(Tx) \left(\frac{\partial v_s(Tx)}{\partial y_3} - \frac{\partial v_s(x)}{\partial y_3} \right) \frac{\partial \Delta^+ a(x')}{\partial x_s} \left(\xi^2(Tx) - \xi^2(x) \right) dx$$

and therefore

$$|J_{6,1} + J_{6,2}| \le c(\nabla^2 a)\|\phi\|_2^2 + \frac{c_{11}}{8} \int_{\Omega_1} \left| \frac{\nabla \mathbf{w}}{h} \right|^2 \xi^2 \, dx + c(\nabla \xi)\|\nabla \mathbf{v}\|_2^2 . \tag{2.25}$$

The term $J_{6,3}$ can be handled similarly as J_6 in Section 3.4 and we obtain that

$$|J_{6,3}| \le \frac{c_{11}}{8} \int_{\Omega_1} \left| \frac{\nabla \mathbf{w}}{h} \right|^q \xi^2 \, dx + c(\nabla \xi) \left(\|\phi\|_2^2 + \|\nabla \mathbf{v}\|_2^2 \right) . \tag{2.26}$$

Since div $\mathbf{w}(x) = \sum_{s=1}^{2} \frac{\partial v_s(Tx)}{\partial y_3} \frac{\partial \Delta^+ a(x')}{\partial x_s}$, we see that

$$|J_7| \le \int_{\Omega_1} |\phi(Tx)| \, |\nabla \mathbf{v}(Tx)| \left| \frac{\nabla \Delta^+ a(x')}{h} \right| \left| \frac{\xi^2(Tx) - \xi^2(x)}{h} \right| dx$$

$$\le c(\nabla^2 a, \nabla \xi)\|\phi\|_2 \|\nabla \mathbf{v}\|_2 . \tag{2.27}$$

The terms J_8 and J_9 can be treated exactly as the terms J_6 and J_7 in Section 3.4. Thus we deduce

$$|I_3| \le c(\nabla^2 a, \nabla \xi)\|\phi\|_2^2 + c(\nabla \xi)\|\nabla \mathbf{v}\|_2^2 + \frac{c_{11}}{4} \int_{\Omega_1} \left| \frac{\nabla \mathbf{w}}{h} \right|^2 \xi^2 \, dx . \tag{2.28}$$

Putting all estimates between (2.18) and (2.28) together we arrive, after integration over $(0, T)$, at (now we use again the notation ξ_1 for the cut-off function near the boundary)

$$\int_0^T \int_{\Omega_1} \left| \frac{\nabla \mathbf{w}}{h} \right|^2 \xi_1^2 \, dx \, dt \le c(\mathbf{f}, \mathbf{E}_0, \mathbf{v}_0) + c \int_0^T \|\nabla \mathbf{v}\|_2^2 + \|\phi\|_2^2 \, dt , \tag{2.29}$$

where both constants may depend on $\varepsilon^{-1}, A, \nabla^2 a$ and $\nabla^2 \xi$. From (5.1.32), (5.1.33) and (2.29) we therefore obtain that

$$\sum_{s=1}^{2} \sum_{i=1}^{3} \int_0^T \int_{\Omega_1} \left| \frac{\partial^2 \mathbf{v}(x)}{\partial x_i \partial \tau^s} \right|^2 \xi_1^2 \, dx \le c . \tag{2.30}$$

In the case of interior regularity, i.e. T is now given by (5.1.34) instead of (5.1.26), we proceed in an analogous way with some simplifications due to the simpler structure of the mapping T. Denoting now the cut-off function in the interior case by ξ_0, i.e. $\xi_0 \in \mathcal{D}(\Omega)$, we obtain

$$\sum_{i,j=1}^{3} \int_0^T \int_\Omega \left| \frac{\partial^2 \mathbf{v}(x)}{\partial x_i \partial x_j} \right|^2 \xi_0^2(x)\, dx \leq c. \tag{2.31}$$

This estimate implies that (2.7) holds for all $\Omega' \subset\subset \Omega$ and therefore the first equation in $(1.2)_{\varepsilon,A}$ holds almost everywhere in Q_T. In order to get (2.7) globally we must also estimate the normal derivatives near the boundary. This will be accomplished by applying the curl-operator to the first equation in $(1.2)_{\varepsilon,A}$. We obtain three equations in $W^{-1,2}(\Omega)$, however, only the first two are useful for us. The first equation reads

$$\frac{\partial}{\partial x_2}\frac{\partial v_3}{\partial t} - \frac{\partial}{\partial x_3}\frac{\partial v_2}{\partial t} + \frac{\partial}{\partial x_3}[E_k \frac{\partial E_2}{\partial x_k}] - \frac{\partial}{\partial x_2}[E_k \frac{\partial E_3}{\partial x_k}]$$

$$+ \frac{\partial}{\partial x_2}[v_{\varepsilon k}\frac{\partial v_3}{\partial x_k}] - \frac{\partial}{\partial x_3}[v_{\varepsilon k}\frac{\partial v_2}{\partial x_k}] - \frac{\partial f_3}{\partial x_2} + \frac{\partial f_2}{\partial x_3} \tag{2.32}$$

$$- \frac{\partial^2 S_{31}^A}{\partial x_1 \partial x_2} - \frac{\partial^2 S_{32}^A}{\partial x_2 \partial x_2} - \frac{\partial^2 S_{33}^A}{\partial x_2 \partial x_3} + \frac{\partial^2 S_{21}^A}{\partial x_3 \partial x_1} + \frac{\partial^2 S_{22}^A}{\partial x_3 \partial x_2} + \frac{\partial^2 S_{23}^A}{\partial x_3 \partial x_3} = 0,$$

while the second equation has the form

$$\frac{\partial}{\partial x_3}\frac{\partial v_1}{\partial t} - \frac{\partial}{\partial x_1}\frac{\partial v_3}{\partial t} + \frac{\partial}{\partial x_1}[E_k \frac{\partial E_3}{\partial x_k}] - \frac{\partial}{\partial x_3}[E_k \frac{\partial E_1}{\partial x_k}]$$

$$+ \frac{\partial}{\partial x_3}[v_{\varepsilon k}\frac{\partial v_1}{\partial x_k}] - \frac{\partial}{\partial x_1}[v_{\varepsilon k}\frac{\partial v_3}{\partial x_k}] - \frac{\partial f_1}{\partial x_3} + \frac{\partial f_3}{\partial x_1} \tag{2.33}$$

$$+ \frac{\partial^2 S_{31}^A}{\partial x_1 \partial x_1} + \frac{\partial^2 S_{32}^A}{\partial x_1 \partial x_2} + \frac{\partial^2 S_{33}^A}{\partial x_1 \partial x_3} - \frac{\partial^2 S_{11}^A}{\partial x_1 \partial x_3} - \frac{\partial^2 S_{12}^A}{\partial x_2 \partial x_3} - \frac{\partial^2 S_{13}^A}{\partial x_3 \partial x_3} = 0.$$

We will get the required estimate from the last terms in equations (2.32) and (2.33) and from the equation

$$\frac{\partial^2 v_3}{\partial x_3^2} = -\frac{\partial^2 v_1}{\partial x_1 \partial x_3} - \frac{\partial^2 v_2}{\partial x_2 \partial x_3}, \tag{2.34}$$

which follows from the "divergence-free" constraint, taking the derivative with respect to x_3. Let us denote

$$G_1 \equiv \frac{\partial}{\partial x_3} S_{13}^A(\mathbf{D}(\mathbf{v}), \mathbf{E}) \quad \text{and} \quad G_2 \equiv \frac{\partial}{\partial x_3} S_{23}^A(\mathbf{D}(\mathbf{v}), \mathbf{E}). \tag{2.35}$$

From the theorem on negative norms (cf. 5.1.8) and (2.32), (2.33) we have, for $r = 1, 2$,

$$\|G_r \xi_1\|_2 \leq \|G_r \xi_1\|_{-1,2} + \|G_r \nabla \xi_1\|_{-1,2} + \sum_{s=1}^{2} \sum_{i,j,k=1}^{3} \left\| \frac{\partial^2 S_{ij}^A}{\partial x_s \partial x_k} \xi_1 \right\|_{-1,2}$$

$$+ \left\| \operatorname{curl} \left(\frac{\partial \mathbf{v}}{\partial t} + [\nabla \mathbf{v}]\mathbf{v}_\varepsilon - \mathbf{f} + \chi^E [\nabla \mathbf{E}]\mathbf{E} \right) \xi_1 \right\|_{-1,2} \tag{2.36}$$

$$= I_6 + I_7 + I_8 + I_9.$$

One easily sees (cf. (1.32), (1.33), (2.35))

$$I_6 + I_7 \le c(\nabla \xi_1, A, \mathbf{E}_0) \|\nabla \mathbf{v}\|_2,$$

$$I_8 \le c(\mathbf{E}_0, A, \nabla^2 \xi_1) \Big(\sum_{s=1}^{2} \sum_{i=1}^{3} \Big\| \frac{\partial^2 \mathbf{v}}{\partial x_i \partial x_s} \xi_1 \Big\|_{2,\Omega_1} + \|\nabla \mathbf{v}\|_2 \Big), \qquad (2.37)$$

$$I_9 \le c(\nabla \xi_1, \varepsilon^{-1}) \Big(\Big\| \frac{\partial \mathbf{v}}{\partial t} \Big\|_2 + \|\nabla \mathbf{v}\|_2 \Big) + c(\mathbf{f}, \mathbf{E}_0, \nabla \xi_1).$$

Thus we deduce that for $r = 1, 2$

$$\int_0^T \|G_r \xi_1\|_2^2 \, dt \le c \Big(1 + \int_0^T \sum_{s=1}^{2} \sum_{i=1}^{3} \Big\| \frac{\partial^2 \mathbf{v}}{\partial x_i \partial x_s} \xi_1 \Big\|_{2,\Omega_1}^2 \, dt \Big), \qquad (2.38)$$

where we used (2.3) and (2.5). Applying the chain rule in (2.35) we obtain the system

$$A \zeta = \frac{1}{2} (\mathbf{G} - \mathbf{H} - \mathbf{F}), \qquad (2.39)$$

where

$$\mathbf{A} = \begin{pmatrix} \partial_{13} S_{13}^A & \partial_{23} S_{13}^A \\ \partial_{13} S_{23}^A & \partial_{23} S_{23}^A \end{pmatrix}, \qquad \zeta = \Big(\frac{\partial D_{13}}{\partial x_3}, \frac{\partial D_{23}}{\partial x_3} \Big),$$

$$\mathbf{G} = (G_1, G_2), \qquad \mathbf{H} = (H_1, H_2), \qquad \mathbf{F} = (F_1, F_2), \qquad (2.40)$$

with

$$H_r = \sum_{s,t=1}^{2} \partial_{st} S_{r3}^A \frac{\partial D_{st}}{\partial x_3} + \partial_{33} S_{r3}^A \frac{\partial^2 v_3}{\partial x_3^2}, \qquad F_r = \partial_n S_{r3}^A \frac{\partial E_n}{\partial x_3}, \qquad r = 1, 2.$$

In the last term of the definition of H_r, $r = 1, 2$ we apply (2.34) and conclude from (2.39) that (cf. (1.37), (1.47), (1.38), (2.5))

$$\sum_{r=1}^{2} \Big\| \frac{\partial D_{r3}}{\partial x_3} \xi_1 \Big\|_{2,\Omega_1} \le c(A, \mathbf{E}_0) \Big(1 + \sum_{r=1}^{2} \|G_r \xi_1\|_{2,\Omega_1} + \sum_{i,j=1}^{3} \sum_{r=1}^{2} \Big\| \frac{\partial^2 v_i}{\partial x_j \partial x_r} \xi_1 \Big\|_{2,\Omega_1} \Big), \qquad (2.41)$$

since the matrix $\mathbf{A}$ is positive definite thanks to (1.35). Integrating (2.41) with respect to time and using (2.38) we obtain

$$\int_0^T \|\zeta \xi_1\|_{2,\Omega_1}^2 \, dt \le c \Big(1 + \int_0^T \sum_{r=1}^{2} \Big\| \frac{\partial \nabla \mathbf{v}}{\partial x_r} \xi_1 \Big\|_{2,\Omega_1}^2 \, dt \Big). \qquad (2.42)$$

Moreover, the identity

$$\frac{\partial^2 v_r}{\partial x_3^2} = \frac{\partial D_{r3}(\mathbf{v})}{\partial x_3} - \frac{\partial^2 v_3}{\partial x_3 \partial x_r}, \qquad r = 1, 2,$$

together with (2.34) and the definition of ζ imply the pointwise inequality

$$\left|\frac{\partial^2 \mathbf{v}}{\partial x_3^2}\right| \leq |\zeta| + c \sum_{r=1}^{2} \left|\frac{\partial}{\partial x_r} \nabla \mathbf{v}\right|. \tag{2.43}$$

Thus we obtain

$$|\nabla^2 \mathbf{v}| \leq \left|\frac{\partial \mathbf{v}}{\partial x_3^2}\right| + \sum_{r=1}^{2} \left|\frac{\partial}{\partial x_r} \nabla \mathbf{v}\right| \leq |\zeta| + c \sum_{r=1}^{2} \left|\frac{\partial}{\partial x_r} \nabla \mathbf{v}\right|. \tag{2.44}$$

From this and (2.42) we derive

$$\int_0^T \|\nabla^2 \mathbf{v}\, \xi_1\|_{2,\Omega_1}^2 \, dt \leq c \left(1 + \int_0^T \sum_{r=1}^{2} \left\|\frac{\partial \nabla \mathbf{v}}{\partial x_r} \xi_1\right\|_{2,\Omega_1}^2 \, dt\right) \tag{2.45}$$

$$\leq c \left(1 + \int_0^T \sum_{r=1}^{2} \left\|\frac{\partial \nabla \mathbf{v}}{\partial \tau^r} \xi_1\right\|_{2,\Omega_1}^2 + \sup_{\Omega_1} \left(\sum_{s=1}^{2} \left|\frac{\partial a}{\partial x_s}\right|\right)^2 \|\nabla^2 \mathbf{v}\, \xi_1\|_{2,\Omega_1}^2 \, dt\right),$$

where we used in the last line the definition of $\frac{\partial}{\partial \tau_r}$, $r = 1, 2$ (cf. (5.1.29)), which yields

$$\sum_{r=1}^{2} \left|\frac{\partial}{\partial x_r} \nabla \mathbf{v}\right| \leq \left(1 + \sum_{r=1}^{2} \left|\frac{\partial a}{\partial x^r}\right|\right) \sum_{r=1}^{2} \left|\frac{\partial}{\partial \tau^r} \nabla \mathbf{v}\right| + \left(\sum_{r=1}^{2} \left|\frac{\partial a}{\partial x^r}\right|\right)^2 \left|\frac{\partial^2 \mathbf{v}}{\partial x_3^2}\right|. \tag{2.46}$$

As a consequence of (5.1.2) we can choose Ω_1 small enough such that

$$c \sup_{\Omega_1} \left(\sum_{s=1}^{2} \left|\frac{\partial a}{\partial x_s}\right|\right)^2 \leq \frac{1}{2} \tag{2.47}$$

and move the last term on the right-hand side of (2.45) to the left-hand side in (2.45). Finally, from this and (2.30) we arrive at

$$\int_0^T \|\nabla^2 \mathbf{v}\, \xi_1\|_{2,\Omega_1}^2 \, dt \leq c, \tag{2.48}$$

which together with (2.31) gives (2.7). The proof is complete. ∎

We will finish this section by proving a variant of the inequality (2.5), which will be useful in the following sections. For $t \in I$ we denote

$$F_A(t) \equiv 1 + \|U^A(|\mathbf{D}(\mathbf{v}^{\varepsilon,A}(t))|^2, |\mathbf{E}(t)|^2)\|_{1,\nu(t)}, \tag{2.49}$$

where $d\nu(t) = (1 + |\mathbf{E}(t)|^2)\, dx$.

Lemma 2.50. *Let $p_\infty \in [2,6)$ and set $\gamma_\mu(r) \equiv \frac{1}{1-\mu} r^{1-\mu}$, for $\mu \geq 0$, $\mu \neq 1$, and $\gamma_1(r) \equiv \ln(r)$. Then it holds for all $t \in I$, all $\mu \geq 0$ and $s > 1$*

$$\int_0^t \left\| \frac{\partial \mathbf{v}^{\varepsilon,A}}{\partial t} \right\|_2^2 \gamma_\mu'(F_A(\tau)) \, d\tau + \operatorname*{ess\,sup}_{\tau \in [0,t]} \gamma_\mu(F_A(\tau))$$

$$\leq c(\mathbf{f}, \mathbf{v}_0, \mathbf{E}_0) + \int_0^t \|[\nabla \mathbf{v}^{\varepsilon,A}] \mathbf{v}_\varepsilon^{\varepsilon,A}\|_2^2 \gamma_\mu'(F_A(\tau)) \, d\tau \tag{2.51}$$

$$+ c(\mathbf{E}_0) \int_0^t \gamma_\mu'(F_A(\tau)) \, \|U^A(|\mathbf{D}(\mathbf{v}^{\varepsilon,A})|^2, |\mathbf{E}|^2)\|_s^s \, d\tau \,.$$

PROOF : Let us denote $\mathbf{v} = \mathbf{v}^{\varepsilon,A}$ and $\gamma(r) = \gamma_\mu(r)$. Multiplying the first equation in $(1.2)_{\varepsilon,A}$ by $\frac{\partial \mathbf{v}}{\partial t} \gamma'(F_A(t))$ and integrating over $(0,t) \times \Omega$ yields

$$\int_0^t \left\| \frac{\partial \mathbf{v}}{\partial t} \right\|_2^2 \gamma'(F_A(\tau)) \, d\tau - \int_0^t \int_{\Omega_\tau} \frac{\partial}{\partial x_j} S_{ij}^A(\mathbf{D}(\mathbf{v}), \mathbf{E}) \frac{\partial v_i}{\partial t} \, dx \, \gamma'(F_A(\tau)) \, d\tau$$

$$\leq c(\mathbf{f}, \mathbf{E}_0) + \int_0^t \|[\nabla \mathbf{v}] \mathbf{v}_\varepsilon\|_2^2 \, \gamma'(F_A(\tau)) \, d\tau \,, \tag{2.52}$$

where we used that $\gamma'(F_A(t)) \leq 1$. The assertion of the lemma would follow by straight-forward manipulations of the second term on the left-hand side of (2.52), which we denote by J, if these manipulations were allowed. However the regularity of $\mathbf{v}$ established in Proposition 2.1, namely[5] $\mathbf{v} \in C(\bar{I}, V_2) \cap L^2(I, W^{2,2}(\Omega))$ and $\frac{\partial \mathbf{v}}{\partial t} \in L^2(Q_T)$, is not sufficient for some of the computations. Therefore we approximate $\mathbf{v}$ by a sequence $\{\mathbf{v}^k\} \subseteq C^\infty(\bar{I}, \mathcal{V})$ such that

$$\begin{aligned} \mathbf{v}^k &\to \mathbf{v} && \text{strongly in } L^2(I, W^{2,2}(\Omega))\,, \\ \mathbf{v}^k(t) &\to \mathbf{v}(t) && \text{strongly in } V_2 \text{ for all } t \in I\,, \\ \frac{\partial \mathbf{v}^k}{\partial t} &\to \frac{\partial \mathbf{v}}{\partial t} && \text{strongly in } L^2(I, L^2(\Omega))\,. \end{aligned} \tag{2.53}$$

We observe for all $t \in I$

$$\lim_{k \to \infty} \int_{\Omega_t} U^A(|\mathbf{D}(\mathbf{v}^k)|^2, |\mathbf{E}|^2) \, d\nu = \int_{\Omega_t} U^A(|\mathbf{D}(\mathbf{v})|^2, |\mathbf{E}|^2) \, d\nu \,. \tag{2.54}$$

This follows from the inequality (cf. (1.40))

$$|U^A(|\mathbf{B}|^2, |\mathbf{E}|^2) - U^A(|\mathbf{D}|^2, |\mathbf{E}|^2)| \leq c(A, p_0)(1 + |\mathbf{B}| + |\mathbf{D}|)|\mathbf{B} - \mathbf{D}|\,,$$

---

[5]It follows from the general theory of parabolic embeddings (cf. Gajewski, Gröger, Zacharias [35]) that a function $\mathbf{v}$ satisfying (2.5) and (2.7) belongs to the space $\mathbf{v} \in C(\bar{I}, V_2)$.

the regularity of $\mathbf{E}$ and $(2.53)_2$. In particular (2.54) yields

$$\lim_{k\to\infty} \|U^A(|\mathbf{D}(\mathbf{v}^k)|^2, |\mathbf{E}|^2)\|_{1,\Omega_t} = \|U^A(|\mathbf{D}(\mathbf{v})|^2, |\mathbf{E}|^2)\|_{1,\Omega_t}. \tag{2.55}$$

Observing that $F_A(\mathbf{v}^k(t)) \geq 1$ we see that $\|\gamma'(F_A(\mathbf{v}^k(t)))\|_\infty \leq 1$. Thus we obtain $\gamma'(F_A(\mathbf{v}^k(t))) \overset{*}{\rightharpoonup} \chi(t)$ in $L^\infty(I)$ at least for a subsequence. Moreover, from (2.54) follows that $\gamma'(F_A(\mathbf{v}^k(t))) \to \gamma'(F_A(\mathbf{v}(t)))$ for all $t \in I$ and it is easy to conclude that

$$\gamma'(F_A(\mathbf{v}^k(t))) \overset{*}{\rightharpoonup} \gamma'(F_A(\mathbf{v}(t))) \qquad \text{weakly}^* \text{ in } L^\infty(I). \tag{2.56}$$

From (2.53), (2.56) and the definition of $\mathbf{S}^A$ we deduce

$$J = -\lim_{k\to\infty} \int_0^t \gamma'(F_A(\mathbf{v}^k(\tau))) \int_\Omega \frac{\partial}{\partial x_j} S^A_{ij}(\mathbf{D}(\mathbf{v}^k), \mathbf{E}) \frac{\partial v^k_i}{\partial t} \, dx \, d\tau, \tag{2.57}$$

which leads to

$$J = \lim_{k\to\infty} \alpha_{31} \int_0^t \frac{d}{dt} \|U^A(|\mathbf{D}(\mathbf{v}^k)|^2, |\mathbf{E}|^2)\|_{1,\nu} \, \gamma'(F_A(\mathbf{v}^k(\tau)) \, d\tau$$

$$- \lim_{k\to\infty} \alpha_{31} \int_0^t \gamma'(F^A(\mathbf{v}^k(\tau))) \int_\Omega \frac{\partial(U^A(|\mathbf{D}(\mathbf{v}^k)|^2, |\mathbf{E}|^2)(1+|\mathbf{E}|^2))}{\partial E_n} \frac{\partial E_n}{\partial t} \, dx \, d\tau$$

$$= J_1 + J_2.$$

One easily sees that

$$\begin{aligned} J_1 &= \lim_{k\to\infty} \left(\gamma(F_A(\mathbf{v}^k(t))) - \gamma(F_A(\mathbf{v}^k(0))) \right) \\ &= \gamma(F_A(t)) - \gamma(F_A(0)), \end{aligned} \tag{2.58}$$

where we used in the last line (2.54). Applying (1.46) we obtain similarly to (2.54) that for all $t \in I$

$$\begin{aligned} &\lim_{k\to\infty} \int_{\Omega_t} \frac{\partial(U^A(|\mathbf{D}(\mathbf{v}^k)|^2, |\mathbf{E}|^2)(1+|\mathbf{E}|^2))}{\partial E_n} \frac{\partial E_n}{\partial t} \, dx \\ &= \int_{\Omega_t} \frac{\partial(U^A(|\mathbf{D}(\mathbf{v})|^2, |\mathbf{E}|^2)(1+|\mathbf{E}|^2))}{\partial E_n} \frac{\partial E_n}{\partial t} \, dx \end{aligned} \tag{2.59}$$

and therefore also strongly in $L^1(I)$, since from (1.48), (2.5), (2.7) and Lemma 5.1.17 we obtain that $\nabla \mathbf{v} \in L^{10/3}(Q_T)$, and thus we can use Vitali's theorem. The $L^1(I)$ convergence of (2.59) together with (2.56) justifies the limiting process in J_2. Finally, we obtain from (2.2) and (1.50) that the right-hand side of (2.59) is estimated by

$$c(\mathbf{E}_0) \left(1 + \|U^A(|\mathbf{D}(\mathbf{v})|^2, |\mathbf{E}|^2)\|_s^s \right). \tag{2.60}$$

The assertion follows from (2.52), (2.57)–(2.60). ∎

4.3 Limiting Process $A \to \infty$

Now, we will derive estimates independent of A, which enable the limiting process $A \to \infty$. Thus we will come to an approximation of the problem (1.1), (1.2) where only the convective term is mollified. In preparation for the limiting process $\varepsilon \to 0$ we will indicate the dependence of the estimates on ε. For this we denote

$$K_{\varepsilon,A}(t) \equiv \int_{\Omega_t} \left|[\nabla \mathbf{v}^{\varepsilon,A}]\mathbf{v}_\varepsilon^{\varepsilon,A}\right|^2 dx \,.$$

To shorten formulas let us introduce the following notation:

$$\theta_A = \theta_A(|\mathbf{D}|^2) = \begin{cases} (1+|\mathbf{D}|^2)^{\frac{1}{2}} & |\mathbf{D}| \le A \,, \\ (1+A^2)^{\frac{1}{2}} & |\mathbf{D}| \ge A \,, \end{cases} \tag{3.1}$$

and

$$I_{p,r,A}(\mathbf{u},\xi) = \int_\Omega (1+|\mathbf{E}|^2)^r |\mathbf{u}|^r \theta_A^{(p-2)r} \, \xi^r \, dx \,, \tag{3.2}$$

$$I_{p,A}(\mathbf{u},\xi) = \int_\Omega (1+|\mathbf{E}|^2)|\mathbf{u}|^2 \theta_A^{p-2} \, \xi^2 \, dx \,, \tag{3.3}$$

where $\mathbf{u}$ will usually be some second order derivative of $\mathbf{v}$, and ξ will be either identical 1 or one of the cut-off functions ξ_1 or ξ_0, defined above. Note, that it follows from Remark 1.43 that there are constants c_{12}, c_{13} such that

$$c_{12}\, I_{p,A}(\mathbf{D}(\nabla \mathbf{v}),\xi) \le \int_\Omega \partial_{kl} S_{ij}^A(\mathbf{D}(\mathbf{v}),\mathbf{E}) D_{ij}(\nabla \mathbf{v}) D_{kl}(\nabla \mathbf{v})\xi^2\, dx \tag{3.4}$$
$$\le c_{13}\, I_{p,A}(\mathbf{D}(\nabla \mathbf{v}),\xi) \,.$$

Recall that (cf. (2.49))

$$F_A(t) = 1 + \|U^A(|\mathbf{D}(\mathbf{v}^{\varepsilon,A}(t))|^2, |\mathbf{E}(t)|^2)\|_{1,\nu(t)} \,,$$

and that $\Omega_1^n = \Omega_0^n \cap \Omega$, where Ω_0^n, $n = 1, \cdots, N$, is a typical set from the appropriately chosen covering of the boundary $\partial\Omega$ and ξ_1^n are the corresponding cut–off function near the boundary (cf. Proof of Proposition 2.1).

Let us first give the definition of a solution to the problem (1.1), (1.2)$_\varepsilon$ and then derive the necessary estimates for the limiting process $A \to \infty$.

Definition 3.5. *The triple* $(\mathbf{E}, \mathbf{v}, \phi) = (\mathbf{E}, \mathbf{v}^\varepsilon, \phi^\varepsilon)$, *for* $\varepsilon > 0$ *given, is said to be a strong solution of the problem* (1.1) *and*

$$\frac{\partial \mathbf{v}}{\partial t} - \operatorname{div} \mathbf{S}(\mathbf{D}(\mathbf{v}),\mathbf{E}) + [\nabla \mathbf{v}]\mathbf{v}_\varepsilon + \nabla\phi = \mathbf{f} + \chi^E[\nabla \mathbf{E}]\mathbf{E}, \qquad \text{in } Q_T,$$
$$\operatorname{div} \mathbf{v} = 0$$
$$\mathbf{v} = 0 \qquad \text{on } I \times \partial\Omega, \tag{1.2}_\varepsilon$$
$$\mathbf{v}(0) = \mathbf{v}_0 \qquad \text{in } \Omega,$$

where $\mathbf{S}$ *is given by* (1.3)–(1.5), *if and only if*

$$\mathbf{E} \in C^1(\bar{Q}_T),$$

$$\mathbf{v} \in L^{2\frac{p_\infty-1}{p_0-1}}(I, V_{6\frac{p_\infty-1}{p_0-1}}) \cap L^{\frac{2}{p_0-1}}(I, W^{2,\frac{6}{p_0+1}}(\Omega)),$$

$$\mathbf{D}(\mathbf{v}) \in L^\infty(I, L^{p(|\mathbf{E}|^2)}(\Omega)) \cap L^{\frac{p_0+3}{2p_0}p(|\mathbf{E}|^2)}(Q_T),$$

$$\frac{\partial \mathbf{v}}{\partial t}, \phi \in L^2(Q_T),$$

$$(3.6)$$

the system $(1.1)_{1,2}$ *is satisfied almost everywhere in* Q_T, *for almost all* $t \in I$ *we have* $\mathbf{E}(t) - \mathbf{E}_0(t) \in \overset{\circ}{H}(\mathrm{div})$ *and the weak formulation of* $(1.2)_\varepsilon$

$$\int_{\Omega_t} \frac{\partial \mathbf{v}}{\partial t} \cdot \boldsymbol{\varphi} \, dx + \int_{\Omega_t} \mathbf{S}(\mathbf{D}(\mathbf{v}), \mathbf{E}) \cdot \mathbf{D}(\boldsymbol{\varphi}) \, dx + \int_{\Omega_t} [\nabla \mathbf{v}] \mathbf{v}_\varepsilon \cdot \boldsymbol{\varphi} \, dx - \int_{\Omega_t} \phi \, \mathrm{div} \, \boldsymbol{\varphi} \, dx$$
$$= \int_{\Omega_t} \mathbf{f} \cdot \boldsymbol{\varphi} \, dx - \chi^E \int_{\Omega_t} \mathbf{E} \otimes \mathbf{E} \cdot \mathbf{D}(\boldsymbol{\varphi}) \, dx$$

$$(3.7)$$

is satisfied for all $\boldsymbol{\varphi} \in \mathcal{D}(\Omega)$ *and almost all* $t \in I$.

Lemma 3.8. *Let* $2 \leq p_\infty \leq p_0 \leq 3$, $\varepsilon > 0$ *and* $A \geq A_0$ *be given. Then the solutions* $(\mathbf{E}, \mathbf{v}, \phi) = (\mathbf{E}, \mathbf{v}^{\varepsilon,A}, \phi^{\varepsilon,A})$ *of the problem* (1.1), $(1.2)_{\varepsilon,A}$ *satisfy the following estimates with constants* c *depending on* $\mathbf{f}, \mathbf{E}_0, \mathbf{v}_0, T$ *but not on* A *and* ε:

$$\|\mathbf{E}\|_{C^1(\bar{I}, W^{2,r}(\Omega))} \leq c,$$

$$(3.9)$$

$$\|\mathbf{v}\|_{L^\infty(I,H)}^2 + \int_0^T \|\nabla \mathbf{v}\|_{2,\nu}^2 + \|U^A(|\mathbf{D}(\mathbf{v})|^2, |\mathbf{E}|^2)\|_{1,\nu} \, dt \leq c,$$

$$(3.10)$$

$$\int_0^T \|\mathbf{S}^A(\mathbf{D}(\mathbf{v}), \mathbf{E})\|_{p_0'}^{p_0'} \, dt \leq c,$$

$$(3.11)$$

$$\int_0^T \|\phi\|_2^2 \, dt \leq c \left(1 + \int_0^T \|\mathbf{S}^A\|_{2s}^{2s} + K_{\varepsilon,A} \, dt\right),$$

$$(3.12)$$

and for almost all $t \in I$

$$\operatorname*{ess\,sup}_{\tau \in (0,t)} \gamma_\mu(F_A(\tau)) + \int_0^t \left\| \frac{\partial \mathbf{v}}{\partial t} \right\|_2^2 \gamma_\mu'(F_A(\tau)) \, d\tau$$
$$\leq c \left(1 + \int_0^t \left(\|\mathbf{S}^A\|_{2s,\Omega_\tau}^{2s} + K_{\varepsilon,A}(\tau) \right) \gamma_\mu'(F_A(\tau)) \, d\tau \right),$$

$$(3.13)$$

$$\int\limits_0^t \left(\left\| \nabla^2 \mathbf{v} \right\|_{r,\Omega_\tau}^u + \left\| \nabla \mathbf{S}^A \right\|_{r,\Omega_\tau}^u + Y(\mathbf{v}(\tau)) + Z(\mathbf{v}(\tau)) \right) \gamma'_\mu(F_A(\tau)) \, d\tau$$
$$\tag{3.14}$$
$$\leq c \left(1 + \int\limits_0^t \left(\left\| \mathbf{S}^A \right\|_{2s,\Omega_\tau}^{2s} + K_{\varepsilon,A}(\tau) \right) \gamma'_\mu(F_A(\tau)) \, d\tau \right),$$

where $r \in [1, \frac{6}{p_0+1}]$, $u \in [1, \frac{2}{p_0-1}]$, $\mu \geq 0$ and

$$Y(\mathbf{v}(t)) = I_{p,A}\big(\mathbf{D}(\nabla \mathbf{v}(t)), \xi_0\big) + \sum_{s=1}^{2} \sum_{n=1}^{N} I_{p,A}\big(\mathbf{D}(\frac{\partial \mathbf{v}(t)}{\partial \tau^s}), \xi_1^n\big),$$
$$\tag{3.15}$$
$$Z(\mathbf{v}(t)) = \left\| \nabla^2 \mathbf{v} \, \xi_0 \right\|_{2,\Omega_t}^2 + \sum_{s=1}^{2} \sum_{n=1}^{N} \left\| \nabla \frac{\partial \mathbf{v}}{\partial \tau^s} \, \xi_1^n \right\|_{2,\Omega_t}^2.$$

PROOF : We will drop the indices ε, A and write $\mathbf{v} = \mathbf{v}^{\varepsilon,A}$ and $\phi = \phi^{\varepsilon,A}$. The estimates (3.9) and (3.10) have already been proved in Proposition 2.1. Moreover, we derive from (1.38) and (1.34) that, for $s > 1$ and almost all $t \in I$

$$\left\| U^A \right\|_{\frac{2s}{1+s}}^{\frac{2s}{1+s}} \leq \left\| \mathbf{S}^A \right\|_{2s}^{2s} + c \left\| \nabla \mathbf{v} \right\|_2^2. \tag{3.16}$$

From (3.16) and (2.6) we immediately obtain (3.12). Moreover, using (3.16), (3.10) and $\gamma'_\mu(F_A) \leq 1$ we derive (3.13) from (2.51) with an appropriately chosen s. From (1.39) we obtain

$$\left\| \mathbf{S}^A \right\|_{p_0'}^{p_0'} \leq c \left(1 + \left\| U^A \right\|_1^1 \right), \tag{3.17}$$

which together with (3.10) gives (3.11) after integration with respect to time.

It remains to prove the crucial estimate (3.14), which will be accomplished now. Using (1.38) and Poincaré's inequality we get

$$\int\limits_\Omega \frac{\partial \mathbf{v}}{\partial t} \cdot \mathbf{v} \, dx \leq \delta \left\| \mathbf{v} \right\|_2^2 + c \left\| \frac{\partial \mathbf{v}}{\partial t} \right\|_2^2$$
$$\tag{3.18}$$
$$\leq \frac{1}{2} \int\limits_\Omega \mathbf{S}^A(\mathbf{D}, \mathbf{E}) \cdot \mathbf{D} \, dx + c \left\| \frac{\partial \mathbf{v}}{\partial t} \right\|_2^2.$$

Thus, choosing in (2.13) $\boldsymbol{\varphi} = \mathbf{v}$ and moving the term with the time derivative of $\mathbf{v}$ to the right-hand side we obtain

$$\int\limits_\Omega \mathbf{S}^A(\mathbf{D}, \mathbf{E}) \cdot \mathbf{D} \, dx \leq c(\mathbf{E}) \left(1 + \left\| \frac{\partial \mathbf{v}}{\partial t} \right\|_2^2 + \left\| \mathbf{f} \right\|_2 \right). \tag{3.19}$$

From Proposition 2.1 follows that $-\Delta \mathbf{v} \, \xi_0^2$ is an admissible test function in the weak formulation (1.53). Recall that $\xi_0 \in \mathcal{D}(\Omega)$ is a cut-off function for the interior regularity. Let us denote by $I_0, \dots, I_5$ the integrals in (1.53) for this test function. We observe that

$$|I_0| \le \frac{c_{15}}{16} \|\nabla^2 \mathbf{v} \xi_0\|_2^2 + c \left\| \frac{\partial \mathbf{v}}{\partial t} \right\|_2^2, \tag{3.20}$$

where c_{15} is the constant appearing in (3.5.18). For the terms $I_1, \dots, I_5$ we proceed exactly[6] as in the proof of Proposition 3.5.7 just setting $q = 2$ and $\alpha = 1$. Therefore we can deduce from (3.5.16)–(3.5.19) and (3.5.21)–(3.5.25) (recall the definition (3.2) of $I_{p,A}$) for almost all $t \in I$

$$I_{p,A}(\mathbf{D}(\nabla \mathbf{v}(t)), \xi_0) + \|\nabla^2 \mathbf{v}\, \xi_0\|_{2,\Omega_t}^2 \tag{3.21}$$

$$\le c(\mathbf{E}_0, \nabla \xi_0) \left(1 + \|\mathbf{f}\|_{2,\Omega_t}^2 + \|\mathbf{S}^A\|_{2s,\Omega_t}^{2s} + K_{\varepsilon,A} + \|\phi\|_{2,\Omega_t}^2 + \left\| \frac{\partial \mathbf{v}}{\partial t} \right\|_{2,\Omega_t}^2 + \|\nabla \mathbf{v}\|_{2,\Omega_t}^2 \right)$$

$$\le c(\mathbf{E}_0, \nabla \xi_0) \left(1 + \|\mathbf{f}\|_{2,\Omega_t}^2 + \|\mathbf{S}^A\|_{2s,\Omega_t}^{2s} + K_{\varepsilon,A} + \left\| \frac{\partial \mathbf{v}}{\partial t} \right\|_{2,\Omega_t}^2 + \|\nabla \mathbf{v}\|_{2,\Omega_t}^2 \right),$$

where $s > 1$ is chosen appropriately, and where we used (2.16). The same procedure works for the tangential derivatives near the boundary. Since we derived the estimates for the tangential derivatives in the previous section in detail, we will not repeat more or less the same steps here again. We obtain, for $r = 1, 2$, and almost all $t \in I$,

$$I_{p,A}\left(\mathbf{D}\left(\frac{\partial \mathbf{v}}{\partial \tau^r} \right), \xi_1 \right) + \left\| \nabla \frac{\partial \mathbf{v}}{\partial \tau^r} \, \xi_1 \right\|_{2,\Omega_t}^2 \tag{3.22}$$

$$\le c(\mathbf{E}_0, \nabla \xi_1) \left(1 + \|\mathbf{f}\|_{2,\Omega_t}^2 + \|\mathbf{S}^A\|_{2s,\Omega_t}^{2s} + K_{\varepsilon,A} + \left\| \frac{\partial \mathbf{v}}{\partial t} \right\|_{2,\Omega_t}^2 + \|\nabla \mathbf{v}\|_{2,\Omega_t}^2 \right).$$

Due to the terms $\|\mathbf{S}^A\|_{2s,\Omega_t}^{2s}$ appearing on the right-hand sides of (3.21) and (3.22), for which we have no apriori estimates, we have to establish global estimates for $\nabla \mathbf{S}^A$. For this we proceed as in the proof of Proposition 2.1, however we must be more careful, since the estimates have to be independent of A. We will use again equations (2.32)–(2.34). Similarly to (2.36) we notice, for $r \in [1, 2)$ and $s = 1, 2$, (cf. (2.35) for the definition of G_s)

$$\|G_s \xi_1\|_r \le \|G_s \xi_1\|_{-1,r} + \|G_s \nabla \xi_1\|_{-1,r} + \sum_{s=1}^{2} \sum_{i,j,k=1}^{3} \left\| \frac{\partial^2 S_{ij}^A}{\partial x_s \partial x_k} \xi_1 \right\|_{-1,r}$$

$$+ \left\| \text{curl} \left(\frac{\partial \mathbf{v}}{\partial t} + [\nabla \mathbf{v}] \mathbf{v}_\varepsilon - \mathbf{f} + \chi^E [\nabla \mathbf{E}] \mathbf{E} \right) \xi_1 \right\|_{-1,r} \tag{3.23}$$

$$= I_6 + I_7 + I_8 + I_9.$$

Since $2 < r'$, one easily checks

$$I_6 + I_7 \le c(\nabla \xi_1) \|\mathbf{S}^A\|_2, \tag{3.24}$$

$$I_9 \le c(\mathbf{E}_0, \nabla \xi_1) \left(1 + \|\mathbf{f}\|_2 + \left\| \frac{\partial \mathbf{v}}{\partial t} \right\|_2 + \|[\nabla \mathbf{v}] \mathbf{v}_\varepsilon\|_2 \right).$$

---

[6] The only difference is the treatment of the last term in (3.5.19), which due to Remark 1.43 can be estimated by $\|U^A(|\mathbf{D}|^2, |\mathbf{E}|^2)\|_s^s$ and then we use (3.16).

Furthermore, we obtain

$$I_8 \leq \sup_{\substack{\phi \in W_0^{1,r'}(\Omega) \\ \|\phi\|_{1,r'} \leq 1}} \sum_{s=1}^{2} \int_{\Omega} \left| \frac{\partial \mathbf{S}^A}{\partial x_s} \right| |\nabla \phi| \, \xi_1 + |\mathbf{S}^A| (|\nabla^2 \xi_1| |\phi| + |\nabla \xi_1| |\nabla \phi|) \, dx \tag{3.25}$$

$$= J_{8,1} + J_{8,2},$$

where we can estimate the first term by

$$J_{8,2} \leq c(\nabla^2 \xi_1) \|\mathbf{S}^A\|_2. \tag{3.26}$$

Concerning the term $J_{8,1}$ we get

$$J_{8,1} \leq \sup_{\substack{\phi \in W_0^{1,r'}(\Omega) \\ \|\phi\|_{1,r'} \leq 1}} \sum_{s=1}^{2} \int_{\Omega} \left(|\partial_{kl} \mathbf{S}^A D_{kl} (\frac{\partial \mathbf{v}}{\partial x_s})| + |\partial_n \mathbf{S}^A \frac{\partial E_n}{\partial x_s}| \right) |\nabla \phi| \, \xi_1 \, dx \tag{3.27}$$

$$\leq c \sum_{s=1}^{2} \left(I_{p,r,A} (\mathbf{D}(\frac{\partial \mathbf{v}}{\partial x_s}), \xi_1) \right)^{\frac{1}{r}} + c(\mathbf{E}_0)(1 + \|\partial_n \mathbf{S}^A \xi_1\|_r),$$

where we used (5.2.4) and the definition (3.3) of $I_{p,r,A}$. From (3.23)–(3.27) we conclude that

$$\sum_{s=1}^{2} \|G_s \xi_1\|_r \leq c(\mathbf{E}_0, \nabla \xi_1, \nabla^2 \xi_1) \left(1 + \|\mathbf{S}^A\|_2 + K_{\varepsilon,A}^{\frac{1}{2}} + \left\| \frac{\partial \mathbf{v}}{\partial t} \right\|_2 \right. \tag{3.28}$$

$$\left. + \|\mathbf{f}\|_2 + \sum_{s=1}^{2} \left(I_{p,r,A} (\mathbf{D}(\frac{\partial \mathbf{v}}{\partial x_s}), \xi_1) \right)^{\frac{1}{r}} + \|\partial_n \mathbf{S}^A \xi_1\|_r \right).$$

Similarly as in the proof of Proposition 2.1 (cf. (2.39)–(2.48)) we want to estimate all second derivatives near the boundary. From (1.35) follows that

$$\mathbf{A}\zeta \cdot \zeta \geq c_4 |\zeta|^2 \, \theta_A^{p-2}. \tag{3.29}$$

Moreover, due to the growth properties of $\partial_{ij} \mathbf{S}^A$ and the regularity of $\mathbf{E}$ we can estimate $\mathbf{H}$ and $\mathbf{F}$ as follows

$$|\mathbf{H}| \leq c(\mathbf{E}) \, \theta_A^{p-2} \sum_{s=1}^{2} |\nabla \frac{\partial \mathbf{v}}{\partial x_s}|, \tag{3.30}$$

$$|\mathbf{F}| \leq c(\mathbf{E}) |\partial_n \mathbf{S}^A|,$$

where we used in the first estimate (2.34).[7] Thus, (2.39) and (3.29), (3.30) imply

$$|\zeta| \theta_A^{p-2} \leq c(\mathbf{E}) \left(|\mathbf{G}| + \theta_A^{p-2} \sum_{s=1}^{2} |\nabla \frac{\partial \mathbf{v}}{\partial x_s}| + |\partial_n \mathbf{S}^A| \right).$$

---

[7]It is exactly this place where it is impossible to obtain an estimate involving only terms of the form $\sum_{s=1}^{2} \mathbf{D}(\frac{\partial \mathbf{v}}{\partial x_s})$. Thus we obtain slightly worse results compared to the case when $\mathbf{S}^A$ depends on $|\nabla \mathbf{v}|$ instead of $|\mathbf{D}(\mathbf{v})|$ (cf. Málek, Nečas, Růžička [72], [73], Růžička [113]).

Multiplying this by ξ_1, taking the L^r-norm and using (3.28) delivers (cf. (3.2) for the definition of $I_{p,r,A}$)

$$\left(I_{p,r,A}(\zeta,\xi_1)\right)^{\frac{1}{r}} \le c(\mathbf{E}_0,\nabla^2\xi_1)\left(1 + \|\mathbf{S}^A\|_2 + K_{\varepsilon,A}^{\frac{1}{2}} + \left\|\frac{\partial\mathbf{v}}{\partial t}\right\|_2 \right. \tag{3.31}$$
$$\left. + \|\mathbf{f}\|_2 + \sum_{s=1}^{2}\left(I_{p,r,A}(\nabla\frac{\partial\mathbf{v}}{\partial x_s},\xi_1)\right)^{\frac{1}{r}} + \|\partial_n\mathbf{S}^A\xi_1\|_r\right).$$

From this and (2.44) we obtain

$$\left(I_{p,r,A}(\nabla^2\mathbf{v},\xi_1)\right)^{\frac{1}{r}} \le c(\mathbf{E}_0,\nabla^2\xi_1)\left(1 + \|\mathbf{S}^A\|_2 + K_{\varepsilon,A}^{\frac{1}{2}} + \left\|\frac{\partial\mathbf{v}}{\partial t}\right\|_2 \right. \tag{3.32}$$
$$\left. + \|\mathbf{f}\|_2 + \sum_{s=1}^{2}\left(I_{p,r,A}(\nabla\frac{\partial\mathbf{v}}{\partial x_s},\xi_1)\right)^{\frac{1}{r}} + \|\partial_n\mathbf{S}^A\xi_1\|_r\right).$$

Due to (2.46) and (2.47) we conclude, for Ω_1 chosen small enough and almost all $t \in I$,

$$\left(I_{p,r,A}(\nabla^2\mathbf{v}(t),\xi_1)\right)^{\frac{1}{r}} \le c(\mathbf{E}_0,\nabla^2\xi_1)\left(1 + \|\mathbf{S}^A\|_{2,\Omega_t} + K_{\varepsilon,A}^{\frac{1}{2}}(t) + \left\|\frac{\partial\mathbf{v}}{\partial t}\right\|_{2,\Omega_t} \right. \tag{3.33}$$
$$\left. + \|\mathbf{f}\|_{2,\Omega_t} + \sum_{s=1}^{2}\left(I_{p,r,A}(\nabla\frac{\partial\mathbf{v}(t)}{\partial \tau^s},\xi_1)\right)^{\frac{1}{r}} + \|\partial_n\mathbf{S}^A\xi_1\|_{r,\Omega_t}\right).$$

We observe that

$$\|\nabla\mathbf{S}^A\|_r^r \le \int_\Omega |\partial_{kl}\mathbf{S}^A D_{kl}(\nabla\mathbf{v})|^r + |\partial_n\mathbf{S}^A\nabla E_n|^r\, dx$$
$$\le c(\mathbf{E})\left(I_{p,r,A}(\nabla^2\mathbf{v},1) + \|\partial_n\mathbf{S}^A\|_r^r\right),$$

where we also used (1.37) and (3.2). This and (5.2.28) thus imply

$$\|\nabla^2\mathbf{v}\|_r^r + \|\nabla\mathbf{S}^A\|_r^r \le c(\mathbf{E})\left(I_{p,r,A}(\nabla^2\mathbf{v},1) + \|\partial_n\mathbf{S}^A\|_r^r\right). \tag{3.34}$$

Extending our covering of $\partial\Omega$ (cf. proof of Proposition 2.1) to a covering of Ω and using a corresponding subordinated partition of unity we obtain for all $r \in [1,2)$

$$\|\nabla^2\mathbf{v}\|_r + \|\nabla\mathbf{S}^A\|_r \le \left(I_{p,r,A}(\nabla^2\mathbf{v},\xi_0)\right)^{\frac{1}{r}} + \sum_{n=1}^{N}\left(I_{p,r,A}(\nabla^2\mathbf{v},\xi_1^n)\right)^{\frac{1}{r}} + \|\partial_n\mathbf{S}^A\|_r$$
$$\le c\left(\left(I_{p,r,A}(\nabla^2\mathbf{v},\xi_0)\right)^{\frac{1}{r}} + \sum_{s=1}^{2}\sum_{n=1}^{N}\left(I_{p,r,A}(\nabla\frac{\partial\mathbf{v}}{\partial \tau^s},\xi_1^n)\right)^{\frac{1}{r}} + \|\partial_n\mathbf{S}^A\|_r\right.$$
$$\left. + 1 + \|\mathbf{S}^A\|_2 + K_{\varepsilon,A}^{\frac{1}{2}} + \left\|\frac{\partial\mathbf{v}}{\partial t}\right\|_2 + \|\mathbf{f}\|_2\right) \tag{3.35}$$
$$\le c\left(\left(\|\nabla^2\mathbf{v}\,\xi_0\|_2 + \sum_{s=1}^{2}\sum_{n=1}^{N}\|\nabla\frac{\partial\mathbf{v}}{\partial \tau^s}\xi_1^n\|_2\right)\|\theta_A^{p-2}\|_{\frac{2r}{2-r}}\right.$$
$$\left. + \|\partial_n\mathbf{S}^A\|_r + 1 + \|\mathbf{S}^A\|_2 + K_{\varepsilon,A}^{\frac{1}{2}} + \left\|\frac{\partial\mathbf{v}}{\partial t}\right\|_2 + \|\mathbf{f}\|_2\right),$$

where we used (3.33) and Hölder's inequality. Since for the first factor we have the estimates (3.21) and (3.22) it remains to bound $\|\partial_n \mathbf{S}^A\|_r$ and $\|\theta_A^{p-2}\|_{\frac{2r}{2-r}}$. From (1.41) we deduce

$$\|\theta_A^{p-2}\|_{\frac{2r}{2-r}} \le c\left(1 + \||\mathbf{S}^A|^{\frac{p-2}{p-1}}\|_{\frac{2r}{2-r}}\right) \le c\left(1 + \|\mathbf{S}^A\|_{\frac{2r}{2-r}\frac{p_0-2}{p_0-1}}^{\frac{p_0-2}{p_0-1}}\right)$$

$$\le c\left(1 + \|\mathbf{S}^A\|_1 + \|\nabla \mathbf{S}^A\|_r\right)^{\frac{p_0-2}{p_0-1}}, \tag{3.36}$$

whenever $r \in [1, \frac{6}{p_0+1}]$. Here we used the embedding $W^{1,r}(\Omega) \hookrightarrow L^{\frac{2r}{2-r}\frac{p_0-2}{p_0-1}}(\Omega)$, which holds for $r \in [1, \frac{6}{p_0+1}]$. From (3.3.47) and (1.47) we deduce for $1 < s < 3/2$

$$\|\partial_n \mathbf{S}^A\|_r \le \left(\int_\Omega \theta_A^{(p-2)sr} |\mathbf{D}|^r \, dx\right)^{\frac{1}{r}}$$

$$\le c\left(\int_\Omega \theta_A^{p-2}|\mathbf{D}|^2 \, dx\right)^{\frac{1}{2}}\left(\int_\Omega \theta_A^{(p-2)\frac{2r}{2-r}(s-\frac{1}{2})} \, dx\right)^{\frac{2-r}{2r}}$$

$$\le c\left(\int_\Omega \mathbf{S}^A \cdot \mathbf{D} \, dx\right)^{\frac{1}{2}}\left(1 + \|\mathbf{S}^A\|_{\frac{p_0-2}{p_0-1}\frac{2r}{2-r}(s-\frac{1}{2})}^{\frac{p_0-2}{p_0-1}(s-\frac{1}{2})}\right) \tag{3.37}$$

$$\le c\left(\int_\Omega \mathbf{S}^A \cdot \mathbf{D} \, dx\right)^{\frac{1}{2}}\left(1 + \|\mathbf{S}^A\|_1 + \|\nabla \mathbf{S}^A\|_r\right)^{\frac{p_0-2}{p_0-1}},$$

where we employed Hölder's inequality, (1.41), $s < 3/2$ and the above embedding. Using now in (3.35) the estimates (3.21), (3.22), (3.36), (3.37) and (3.19) we derive for $r \in [1, \frac{6}{p_0+1}]$

$$\|\nabla^2 \mathbf{v}\|_r + \|\nabla \mathbf{S}^A\|_r \le c\left(1 + \|\mathbf{S}^A\|_{2s}^s + K_{\varepsilon,A}^{\frac{1}{2}} + \left\|\frac{\partial \mathbf{v}}{\partial t}\right\|_2 + \|\mathbf{f}\|_2 + \|\nabla \mathbf{v}\|_2\right) \times$$

$$\times \left(1 + \|\mathbf{S}^A\|_1 + \|\nabla \mathbf{S}^A\|_r\right)^{\frac{p_0-2}{p_0-1}}$$

$$+ c\left(\int_\Omega \mathbf{S}^A(\mathbf{D}) \cdot \mathbf{D} \, dx\right)^{\frac{1}{2}}\left(1 + \|\mathbf{S}^A\|_1 + \|\nabla \mathbf{S}^A\|_r\right)^{\frac{p_0-2}{p_0-1}}$$

$$+ c\left(1 + \|\mathbf{S}^A\|_{2s}^s + K_{\varepsilon,A}^{\frac{1}{2}} + \left\|\frac{\partial \mathbf{v}}{\partial t}\right\|_2 + \|\mathbf{f}\|_2\right)$$

$$\le c\left(1 + \|\mathbf{S}^A\|_{2s}^s + K_{\varepsilon,A}^{\frac{1}{2}} + \left\|\frac{\partial \mathbf{v}}{\partial t}\right\|_2 + \|\mathbf{f}\|_2 + \|\nabla \mathbf{v}\|_2\right) \times$$

$$\times \left(1 + \|\mathbf{S}^A\|_1 + \|\nabla \mathbf{S}^A\|_r\right)^{\frac{p_0-2}{p_0-1}}$$

$$+ c\left(1 + \|\mathbf{S}^A\|_{2s}^s + K_{\varepsilon,A}^{\frac{1}{2}} + \left\|\frac{\partial \mathbf{v}}{\partial t}\right\|_2 + \|\mathbf{f}\|_2\right)$$

$$\le c\left(1 + \|\mathbf{S}^A\|_{2s}^{2s} + K_{\varepsilon,A} + \left\|\frac{\partial \mathbf{v}}{\partial t}\right\|_2^2 + \|\mathbf{f}\|_2^2 + \|\nabla \mathbf{v}\|_2^2\right)^{\frac{p_0-1}{2}} + \frac{1}{2}\|\nabla \mathbf{S}^A\|_r,$$

where we also used Young's inequality with $\delta = \frac{p_0-1}{p_0-2}$ and $p_0 \ge 2$. The last term can be absorbed into the left-hand side. The resulting inequality raised to the power u, leads, for $u \in [1, \frac{2}{p_0-1}]$, $r \in [1, \frac{6}{p_0+1}]$ and $1 < s < 3/2$ chosen appropriately, to

$$\|\nabla^2 \mathbf{v}\|_r^u + \|\nabla \mathbf{S}^A\|_r^u \le c\left(1 + \|\mathbf{S}^A\|_{2s}^{2s} + K_{\varepsilon,A} + \left\|\frac{\partial \mathbf{v}}{\partial t}\right\|_2^2 + \|\mathbf{f}\|_2^2 + \|\nabla \mathbf{v}\|_2^2\right). \tag{3.38}$$

Note that we can choose $u \geq 1$ for $p_0 \leq 3$, which gives the upper bound appearing in the lemma, and that the right-hand sides of (3.38) and (3.21), (3.22) coincide. Thus, the sum of inequalities (3.38) and (3.21), (3.22) yields, for almost all $t \in I$ and $u \in [1, \frac{2}{p_0-1}]$, $r \in [1, \frac{6}{p_0+1}]$ and $1 < s < 3/2$ chosen appropriately, (cf. (3.15) for the definition of $Y(\mathbf{v}(t))$, $Z(\mathbf{v}(t))$)

$$\left\|\nabla^2 \mathbf{v}\right\|_{r,\Omega_t}^u + \left\|\nabla \mathbf{S}^A\right\|_{r,\Omega_t}^u + Y(\mathbf{v}(t)) + Z(\mathbf{v}(t))$$
$$\leq c\left(1 + \left\|\mathbf{S}^A\right\|_{2s,\Omega_t}^{2s} + K_{\varepsilon,A}(t) + \left\|\frac{\partial \mathbf{v}}{\partial t}\right\|_{2,\Omega_t}^2 + \|\mathbf{f}\|_{2,\Omega_t}^2 + \|\nabla \mathbf{v}\|_{2,\Omega_t}^2\right). \tag{3.39}$$

Multiplying this inequality by $\gamma'_\mu(F_A(t))$, $\mu \geq 0$ (cf. (2.49), Lemma 2.50) and integrating over $(0, t)$ gives

$$\int_0^t \left(\left\|\nabla^2 \mathbf{v}\right\|_{r,\Omega_\tau}^u + \left\|\nabla \mathbf{S}^A\right\|_{r,\Omega_\tau}^u + Y(\mathbf{v}(\tau)) + Z(\mathbf{v}(\tau))\right)\gamma'_\mu(F_A(\tau))\, d\tau$$
$$\leq c(\mathbf{f}, \mathbf{v}_0, \mathbf{E})\left(1 + \int_0^t \left(\left\|\mathbf{S}^A\right\|_{2s,\Omega_\tau}^{2s} + K_{\varepsilon,A}(\tau)\right)\gamma'_\mu(F_A(\tau))\, d\tau\right), \tag{3.40}$$

where we also used (3.13) to bound the term with $\frac{\partial \mathbf{v}}{\partial t}$ and that $\gamma'_\mu(F_A) \leq 1$, $\mathbf{f} \in L^2(Q_T)$ and (3.10). This proves (3.14). ∎

On the right-hand sides of (3.12)–(3.14) appear terms with $\|\mathbf{S}^A\|_{2s}^{2s}$ for which we do not have any information in the moment. Thus we are now going to prove estimates without this term on the right-hand sides at the expense of further restricting the upper bound for p_0.

Lemma 3.41. Let $2 \leq p_\infty \leq p_0 < 8/3$, $\varepsilon > 0$ and $A \geq A_0$ be given. Then the solutions $(\mathbf{E}, \mathbf{v}, \phi) = (\mathbf{E}, \mathbf{v}^{\varepsilon,A}, \phi^{\varepsilon,A})$ of the problem (1.1), (1.2)$_{\varepsilon,A}$ satisfy the following estimates with constants c depending on $\mathbf{f}, \mathbf{v}_0, \mathbf{E}_0, T$ but independent of A and ε:

$$\int_0^T \left\|\mathbf{S}^A(\mathbf{D}(\mathbf{v}), \mathbf{E})\right\|_{\frac{2(p_0+3)}{3(p_0-1)}}^{\frac{2(p_0+3)}{3(p_0-1)}}\, dt \leq c\left(1 + \int_0^T K_{\varepsilon,A}\, dt\right)^{\frac{5}{3}}. \tag{3.42}$$

$$\int_0^T \left\|U^A(|\mathbf{D}(\mathbf{v})|^2, |\mathbf{E}|^2)\right\|_{\frac{p_0+3}{2p_0}}^{\frac{p_0+3}{2p_0}} + \|\mathbf{D}(\mathbf{v})\|_{\frac{p_0+3}{p_0}}^{\frac{p_0+3}{p_0}}\, dt \leq c\left(1 + \int_0^T K_{\varepsilon,A}\, dt\right)^{\frac{5}{3}}, \tag{3.43}$$

$$\int_0^T \|\phi\|_2^2\, dt \leq c\left(1 + \int_0^T K_{\varepsilon,A}\, dt\right), \tag{3.44}$$

and for almost all $t \in I$

$$\operatorname*{ess\,sup}_{\tau \in (0,t)} \left\| U^A (|\mathbf{D}(\mathbf{v}(\tau))|^2, |\mathbf{E}(\tau)|^2) \right\|_{1,\nu(\tau)} + \int_0^t \left\| \frac{\partial \mathbf{v}}{\partial t} \right\|_2^2 d\tau \le c \left(1 + \int_0^t K_{\varepsilon,A} \, d\tau \right), \quad (3.45)$$

$$\int_0^t \left\| \nabla^2 \mathbf{v} \right\|_{\frac{6}{p_0+1}}^{\frac{2}{p_0-1}} + \left\| \nabla \mathbf{S}^A (\mathbf{D}(\mathbf{v}), \mathbf{E}) \right\|_{\frac{6}{p_0+1}}^{\frac{2}{p_0-1}} d\tau \le c \left(1 + \int_0^t K_{\varepsilon,A} \, d\tau \right), \quad (3.46)$$

$$\int_0^t Y(\mathbf{v}(\tau)) + Z(\mathbf{v}(\tau)) \, d\tau \le c \left(1 + \int_0^t K_{\varepsilon,A} \, d\tau \right). \quad (3.47)$$

PROOF : We can rewrite (3.17) as

$$\left\| \mathbf{S}^A \right\|_{p_0',\Omega_t}^{p_0'} \le c \, F_A(t), \quad (3.48)$$

which together with (3.13) ($\mu = 0$) implies, for $1 < s$ appropriately chosen, and almost all $t \in I$

$$\operatorname*{ess\,sup}_{\tau \in (0,t)} \left\| \mathbf{S}^A \right\|_{p_0',\Omega_\tau}^{p_0'} \le c \left(1 + \int_0^t K_{\varepsilon,A}(\tau) + \left\| \mathbf{S}^A \right\|_{2s,\Omega_\tau}^{2s} d\tau \right). \quad (3.49)$$

We have the interpolation inequality

$$\begin{aligned} \left\| \mathbf{S}^A \right\|_q^q &\le \left\| \mathbf{S}^A \right\|_{p_0'}^{q(1-\lambda)} \left\| \mathbf{S}^A \right\|_{\frac{6}{p_0-1}}^{q\lambda} \\ &\le \left\| \mathbf{S}^A \right\|_{p_0'}^{q(1-\lambda)} \left\| \nabla \mathbf{S}^A \right\|_{\frac{6}{p_0+1}}^{q\lambda} + \left\| \mathbf{S}^A \right\|_{p_0'}^q, \end{aligned} \quad (3.50)$$

with $\lambda = \frac{6(p_0 q - p_0 - q)}{q(p_0-1)(6-p_0)}$, where we also used the embedding $W^{1,\frac{6}{p_0+1}}(\Omega) \hookrightarrow L^{\frac{6}{p_0-1}}(\Omega)$. Choosing now $q = \frac{2(3+p_0)}{3(p_0-1)}$ we see that $q\lambda = \frac{2}{p_0-1}$. For this q we thus obtain from (3.50), (3.49) and (3.14) with $\mu = 0$, $r = \frac{6}{p_0+1}$, $u = \frac{2}{p_0-1}$ that for almost all $t \in I$

$$\begin{aligned} \int_0^t \left\| \mathbf{S}^A \right\|_q^q d\tau &\le \left(\operatorname*{ess\,sup}_{(0,t)} \left\| \mathbf{S}^A \right\|_{p_0'}^{p_0'} \right)^{\frac{2}{3}} \int_0^t \left\| \nabla \mathbf{S}^A \right\|_r^u d\tau + \left(\operatorname*{ess\,sup}_{(0,t)} \left\| \mathbf{S}^A \right\|_{p_0'}^{p_0'} \right)^{\frac{6-p_0}{3p_0}} \int_0^t \left\| \mathbf{S}^A \right\|_{p_0'}^{p_0'} d\tau \\ &\le c \left(1 + \int_0^t K_{\varepsilon,A} \, d\tau \right)^{\frac{5}{3}} + c \left(\int_0^t \left\| \mathbf{S}^A \right\|_{2s}^{2s} d\tau \right)^{\frac{5}{3}}, \end{aligned} \quad (3.51)$$

where we also used (3.11) and $\frac{6-p_0}{2p_0} \le \frac{5}{3}$. Moreover, we dispose of the interpolation inequality

$$\left\| \mathbf{S}^A \right\|_{2s,Q_t} \le \left\| \mathbf{S}^A \right\|_{p_0',Q_t}^{1-\lambda} \left\| \mathbf{S}^A \right\|_{q,Q_t}^{\lambda} \quad (3.52)$$

with $\lambda = \frac{(3+p_0)(2s(p_0-1)-p_0)}{(6-p_0)(p_0-1)s}$. From this and (3.51), (3.11) we deduce for almost all $t \in I$

$$\|\mathbf{S}^A\|_{q,Q_t}^q \leq c\Big(1 + \int_0^t K_{\varepsilon,A}\, d\tau\Big)^{\frac{5}{3}} + c\,\|\mathbf{S}^A\|_{q,Q_t}^{\frac{10}{3}s\lambda}$$

$$\leq c\Big(1 + \int_0^t K_{\varepsilon,A}\, d\tau\Big)^{\frac{5}{3}}, \tag{3.53}$$

whenever

$$q > \frac{10}{3}s\lambda \qquad \Longleftrightarrow \qquad p_0 < \frac{8}{3}, \tag{3.54}$$

for $s > 1$ chosen appropriately. This proves (3.42). With the help of (3.52), (3.11) and (3.53) we arrive for almost all $t \in I$ at

$$\|\mathbf{S}^A\|_{2s,Q_t}^{2s} \leq c(\mathbf{f}, \mathbf{v}_0, \mathbf{E}_0)\|\mathbf{S}^A\|_{q,Q_t}^{2s\lambda}$$

$$\leq c(\mathbf{f}, \mathbf{v}_0, \mathbf{E}_0)\Big(1 + \int_0^t K_{\varepsilon,A}\, d\tau\Big), \tag{3.55}$$

whenever (3.54) holds, since $\lambda\frac{10s}{3q} < 1$. Due to (3.55) we immediately obtain (3.44)–(3.47) from (3.12)–(3.14) with $\mu = 0$, $r = \frac{6}{p_0+1}$, $u = \frac{2}{p_0-1}$. Furthermore, choosing in (3.16) $2s = \frac{2(p_0+3)}{3(p_0-1)}$ and using (3.42) and (3.10) we derive the estimate (3.43) for the first term on the left-hand side. Using in this estimate the lower bound (1.38) for U^A we get (3.43) also for the second term on the left-hand side. The proof is complete. ∎

Remark 3.56. Note, that the right-hand sides of (3.42)–(3.47) are bounded by constants $c(\mathbf{f}, \mathbf{v}_0, \mathbf{E}_0, \varepsilon^{-1})$, due to (3.10) and the properties of the mollifier. In particular (3.47) (cf. (3.15)) implies that $\nabla^2 \mathbf{v}^{\varepsilon,A}$ is bounded uniformly with respect to A in the space $L^2(I, L^2_{\text{loc}}(\Omega))$ and that $\nabla \frac{\partial \mathbf{v}^{\varepsilon,A}}{\partial \tau^s}$, $s = 1, 2$, are bounded uniformly with respect to A in the space $L^2(I, L^2(\Omega))$.

The estimates proved in Lemma 3.8 and Lemma 3.41 together with Remark 3.56 enable the limiting process $A \to \infty$.

Proposition 3.57. *Let* $2 \leq p_\infty \leq p_0 < 8/3$, $\varepsilon > 0$ *and assume that* Ω, T, $\mathbf{E}_0$, $\mathbf{f}$ *and* $\mathbf{v}_0$ *satisfy the assumptions of Proposition 2.1. Then there exist strong solutions* $(\mathbf{E}, \mathbf{v}^\varepsilon, \phi^\varepsilon)$ *of the problem* (1.1), (1.2)$_\varepsilon$, *which satisfy the following estimates with constants* c *depending on* $\mathbf{f}, \mathbf{E}_0, \mathbf{v}_0, T$ *but independent of* ε

$$\|\mathbf{E}\|_{C^1(I, W^{2,r}(\Omega))} \leq c, \tag{3.58}$$

$$\|\mathbf{v}^\varepsilon\|_{L^\infty(I,H)}^2 + \int_0^T \|U(|\mathbf{D}(\mathbf{v}^\varepsilon)|^2, |\mathbf{E}|^2)\|_{1,\nu}\, dt \leq c, \tag{3.59}$$

$$\int\limits_0^T \left\|\mathbf{S}(\mathbf{D}(\mathbf{v}^\varepsilon), \mathbf{E})\right\|_{p_0'}^{p_0'} dt \le c, \tag{3.60}$$

$$\int\limits_0^T \|\phi\|_2^2 \, dt \le c \left(1 + \int\limits_0^T K_\varepsilon \, dt\right), \tag{3.61}$$

$$\int\limits_0^T \left\|U(|\mathbf{D}(\mathbf{v}^\varepsilon)|^2, |\mathbf{E}|^2)\right\|_{\frac{p_0+3}{2p_0}}^{\frac{p_0+3}{2p_0}} + \left\|\mathbf{S}(\mathbf{D}(\mathbf{v}^\varepsilon), \mathbf{E})\right\|_{\frac{2(p_0+3)}{3(p_0-1)}}^{\frac{2(p_0+3)}{3(p_0-1)}} dt \le c \left(1 + \int\limits_0^T K_\varepsilon \, dt\right)^{\frac{5}{3}}, \tag{3.62}$$

$$\int\limits_0^T \left\|\mathbf{D}(\mathbf{v}^\varepsilon)\right\|_{\frac{p_0+3}{p_0}}^{\frac{p_0+3}{p_0}} dt \le c \left(1 + \int\limits_0^T K_\varepsilon \, dt\right)^{\frac{5}{3}}, \tag{3.63}$$

and for almost all $t \in I$

$$\operatorname*{ess\,sup}_{\tau \in (0,t)} \left\|U(|\mathbf{D}(\mathbf{v}^\varepsilon)|^2, |\mathbf{E}|^2)\right\|_{1,\nu(\tau)} + \int\limits_0^t \left\|\frac{\partial \mathbf{v}^\varepsilon}{\partial t}\right\|_2^2 d\tau \le c \left(1 + \int\limits_0^t K_\varepsilon \, d\tau\right), \tag{3.64}$$

$$\int\limits_0^t \left\|\nabla^2 \mathbf{v}^\varepsilon\right\|_r^u + \left\|\nabla \mathbf{S}(\mathbf{D}(\mathbf{v}^\varepsilon), \mathbf{E})\right\|_r^u d\tau \le c \left(1 + \int\limits_0^t K_\varepsilon \, d\tau\right), \tag{3.65}$$

$$\int\limits_0^t Y(\mathbf{v}^\varepsilon(\tau)) + Z(\mathbf{v}^\varepsilon(\tau)) \, d\tau \le c \left(1 + \int\limits_0^t K_\varepsilon \, d\tau\right), \tag{3.66}$$

$$\int\limits_0^t \left(I_{p,r}(\mathbf{D}(\nabla \mathbf{v}^\varepsilon), 1)\right)^{\frac{u}{r}} d\tau \le c \left(1 + \int\limits_0^t K_\varepsilon \, d\tau\right), \tag{3.67}$$

with $u = \frac{2}{p_0 - 1}$ and $r = \frac{6}{p_0 + 1}$.

PROOF : The right-hand sides of (3.9)–(3.11) are bounded by constants $c(\mathbf{f}, \mathbf{v}_0, \mathbf{E}_0)$ and the right-hand sides of (3.42)–(3.47) are bounded by constants $c(\mathbf{f}, \mathbf{v}_0, \mathbf{E}_0, \varepsilon^{-1})$, due to (3.10) and the properties of the mollifier. The inequalities (3.42)–(3.47) together with Remark 3.56 and the Aubin-Lions lemma imply the existence of $\mathbf{v}^\varepsilon, \phi^\varepsilon$ such that

$$
\begin{aligned}
\nabla \mathbf{v}^{\varepsilon,A} &\to \nabla \mathbf{v}^\varepsilon && \text{strongly in } L^u(I, L^m(\Omega)), \quad m < \tfrac{6}{p_0 - 1}, \\
\frac{\partial \mathbf{v}^{\varepsilon,A}}{\partial t} &\rightharpoonup \frac{\partial \mathbf{v}^\varepsilon}{\partial t} && \text{weakly in } L^2(Q_T), \\
\mathbf{v}^{\varepsilon,A} &\rightharpoonup \mathbf{v}^\varepsilon && \text{weakly in } L^q(I, V_q) \cap L^u(I, W^{2,r}(\Omega)), \\
\mathbf{v}^{\varepsilon,A} &\rightharpoonup^* \mathbf{v}^\varepsilon && \text{weakly* in } L^\infty(I, V_2), &&(3.68) \\
\phi^{\varepsilon,A} &\rightharpoonup \phi^\varepsilon && \text{weakly in } L^2(Q_T), \\
\nabla^2 \mathbf{v}^{\varepsilon,A} &\rightharpoonup \nabla^2 \mathbf{v}^\varepsilon && \text{weakly in } L^2(I, L^2_{\text{loc}}(\Omega)), \\
\nabla \frac{\partial \mathbf{v}^{\varepsilon,A}}{\partial \tau^s} &\rightharpoonup \nabla \frac{\partial \mathbf{v}^\varepsilon}{\partial \tau^s} && \text{weakly in } L^2(Q_T), \quad s = 1, 2,
\end{aligned}
$$

as $A \to \infty$ at least for a subsequence, where $u = \frac{2}{p_0-1}$, $r = \frac{6}{p_0+1}$ and $q = \frac{3+p_0}{p_0}$. From $(3.68)_1$ follows

$$\nabla \mathbf{v}^{\varepsilon,A} \to \nabla \mathbf{v}^{\varepsilon} \qquad \text{a.e. in } Q_T \tag{3.69}$$

and thus using Vitali's theorem and $(3.68)_3$ we obtain

$$\nabla \mathbf{v}^{\varepsilon,A} \to \nabla \mathbf{v}^{\varepsilon} \qquad \text{strongly in } L^m(I, L^m(\Omega)), \quad m < \frac{3+p_0}{p_0}. \tag{3.70}$$

Moreover, since $\mathbf{S}^A$ and U^A, respectively, converge locally uniformly to, respectively, $\mathbf{S}$ and U we also conclude from (3.69) that

$$\mathbf{S}^A(\mathbf{D}(\mathbf{v}^{\varepsilon,A}), \mathbf{E}) \to \mathbf{S}(\mathbf{D}(\mathbf{v}^{\varepsilon}), \mathbf{E}) \qquad \text{a.e. in } Q_T,$$
$$U^A(|\mathbf{D}(\mathbf{v}^{\varepsilon,A})|^2, |\mathbf{E}|^2) \to U(|\mathbf{D}(\mathbf{v}^{\varepsilon})|^2, |\mathbf{E}|^2) \qquad \text{a.e. in } Q_T. \tag{3.71}$$

Therefore (3.46), (3.42) and (3.45), (3.48) and show that

$$\mathbf{S}^A(\mathbf{D}(\mathbf{v}^{\varepsilon,A}), \mathbf{E}) \rightharpoonup \mathbf{S}(\mathbf{D}(\mathbf{v}^{\varepsilon}), \mathbf{E}) \qquad \text{weakly in } L^u(I, W^{1,r}(\Omega)) \cap L^n(Q_T),$$
$$\mathbf{S}^A(\mathbf{D}(\mathbf{v}^{\varepsilon,A}), \mathbf{E}) \rightharpoonup^* \mathbf{S}(\mathbf{D}(\mathbf{v}^{\varepsilon}), \mathbf{E}) \qquad \text{weakly* in } L^\infty(I, L^{p_0'}(\Omega)), \tag{3.72}$$

where $u = \frac{2}{p_0-1}$, $r = \frac{6}{p_0+1}$ and $n = \frac{2(3+p_0)}{3(p_0-1)}$, since weak and almost everywhere limits coincide. Moreover from (3.42) and Vitali's theorem we derive, for $m < \frac{2(3+p_0)}{3(p_0-1)}$,

$$\mathbf{S}^A(\mathbf{D}(\mathbf{v}^{\varepsilon,A}), \mathbf{E}) \to \mathbf{S}(\mathbf{D}(\mathbf{v}^{\varepsilon}), \mathbf{E}) \qquad \text{strongly in } L^m(Q_T). \tag{3.73}$$

Similarly, from $(3.71)_2$, (3.43) and Vitali's theorem we obtain, for $m < \frac{3+p_0}{2p_0}$,

$$U^A(|\mathbf{D}(\mathbf{v}^{\varepsilon,A})|^2, |\mathbf{E}|^2) \to U(|\mathbf{D}(\mathbf{v}^{\varepsilon})|^2, |\mathbf{E}|^2) \qquad \text{strongly in } L^m(Q_T) \tag{3.74}$$

and in particular for almost all $t \in I$

$$U^A(|\mathbf{D}(\mathbf{v}^{\varepsilon,A})|^2, |\mathbf{E}|^2) \to U(|\mathbf{D}(\mathbf{v}^{\varepsilon})|^2, |\mathbf{E}|^2) \qquad \text{strongly in } L^m(\Omega_t, \nu(t)), \tag{3.75}$$

where we used the regularity of $\mathbf{E}$. We also observe that, for almost all $t \in I$

$$\lim_{A\to\infty} \int_0^t K_{\varepsilon,A} \, d\tau = \int_0^t K_\varepsilon \, d\tau \equiv \int_0^t \|[\nabla \mathbf{v}^{\varepsilon}]\mathbf{v}^{\varepsilon}_\varepsilon\|_2^2 \, d\tau, \tag{3.76}$$

due to (3.70), since $2 < \frac{3+p_0}{p_0}$ and the properties of the mollifier. Because of (3.68)–(3.76) and the lower semicontinuity of norms the estimates (3.10), (3.11) and (3.42)–(3.47), except the term with $Y(\mathbf{v}^{\varepsilon,A})$, remain valid if we replace $\mathbf{v}^{\varepsilon,A}, \phi^{\varepsilon,A}, U^A, \mathbf{S}^A$ and $K_{\varepsilon,A}$ by $\mathbf{v}^{\varepsilon}, \phi^{\varepsilon}, U, \mathbf{S}$ and K_ε. This proves (3.58)–(3.66), except the term with $\int_0^t Y(\mathbf{v}^{\varepsilon}) \, d\tau$.

In order to show that also in the term $\int_0^t Y(\mathbf{v}^{\varepsilon,A}) \, d\tau$ we can replace $\mathbf{v}^{\varepsilon,A}$ by $\mathbf{v}^{\varepsilon}$ we have to proceed differently. From (3.70), (3.58) and Egerov's theorem we conclude that for all $\delta > 0$ there is a set Q_δ with $|Q \setminus Q_\delta| \leq \delta$, and a subsequence A_δ such that

$$\theta_{A_\delta}^{p(|\mathbf{E}|^2)-2}(\mathbf{D}(\mathbf{v}^{\varepsilon,A}))(1 + |\mathbf{E}|^2) \rightrightarrows (1 + |\mathbf{D}(\mathbf{v}^{\varepsilon})|^2)^{\frac{p(|\mathbf{E}|^2)-2}{2}}(1 + |\mathbf{E}|^2), \qquad \text{uniformly in } Q_\delta.$$

From this and the estimate (3.47) for $Z(\mathbf{v}^{\varepsilon,A})$ we obtain for almost all $t \in I$

$$\liminf_{A_\delta \to \infty} \int\limits_{Q_\delta \cap Q_t} (1 + |\mathbf{E}|^2)|\mathbf{D}(\nabla \mathbf{v}^{\varepsilon,A_\delta})|^2 \, \xi_0^2 \left(\theta_{A_\delta}^{p-2}(\mathbf{D}(\mathbf{v}^{\varepsilon,A_\delta})) - (1 + |\mathbf{D}(\mathbf{v}^\varepsilon)|^2)^{\frac{p-2}{2}}\right) dx \, dt = 0 \,,$$

which implies

$$\liminf_{A_\delta \to \infty} \int\limits_{Q_\delta \cap Q_t} (1 + |\mathbf{E}|^2)|\mathbf{D}(\nabla \mathbf{v}^{\varepsilon,A_\delta})|^2 \, \xi_0^2 \, \theta_{A_\delta}^{p-2}\left(\mathbf{D}(\mathbf{v}^{\varepsilon,A_\delta})\right) dx \, dt$$

$$= \liminf_{A_\delta \to \infty} \int\limits_{Q_\delta \cap Q_t} (1 + |\mathbf{E}|^2)|\mathbf{D}(\nabla \mathbf{v}^{\varepsilon,A_\delta})|^2 \, \xi_0^2 \, (1 + |\mathbf{D}(\mathbf{v}^\varepsilon)|^2)^{\frac{p-2}{2}}) \, dx \, dt \,.$$

Since, for $p_0 < 3$ we have due to $(3.68)_3$ that

$$1 \leq (1 + |\mathbf{D}(\mathbf{v}^\varepsilon)|^2)^{\frac{p(|\mathbf{E}|^2)-2}{2}} (1 + |\mathbf{E}|^2) \in L^2(Q_T) \tag{3.77}$$

and thus we can define a new measure $d\mu$ by

$$d\mu \equiv (1 + |\mathbf{D}(\mathbf{v}^\varepsilon)|^2)^{\frac{p(|\mathbf{E}|^2)-2}{2}} (1 + |\mathbf{E}|^2) \, dx \, dt \,.$$

For this measure we have $L^2(Q_T; d\mu) \hookrightarrow L^2(Q_T; dx \, dt)$. Due to $(3.68)_{6,7}$ we can identify the limits of $\mathbf{D}(\nabla \mathbf{v}^{\varepsilon,A}) \, \xi_0$ and $\sum_{s=1}^{2} \sum_{n=1}^{N} \mathbf{D}(\frac{\partial \mathbf{v}^{\varepsilon,A}}{\partial \tau^s}) \, \xi_1^n$, respectively, as $\mathbf{D}(\nabla \mathbf{v}^\varepsilon) \, \xi_0$ and $\sum_{s=1}^{2} \sum_{n=1}^{N} \mathbf{D}(\frac{\partial \mathbf{v}^\varepsilon}{\partial \tau^s}) \, \xi_1^n$, respectively, in the space $L^2(Q_T)$. Moreover we have

$$\lim_{A \to \infty} \int\limits_{Q_T} \mathbf{D}(\nabla \mathbf{v}^{\varepsilon,A}) \, \xi_0 \, \varphi \, d\mu = \lim_{A \to \infty} \int\limits_{Q_T} \mathbf{D}(\nabla \mathbf{v}^{\varepsilon,A}) \, \xi_0 \, \varphi \, (1 + |\mathbf{D}(\mathbf{v}^\varepsilon)|)^{\frac{p(|\mathbf{E}|^2)-2}{2}} (1 + |\mathbf{E}|^2) \, dx \, dt$$

$$= \int\limits_{Q_T} \mathbf{D}(\nabla \mathbf{v}^\varepsilon) \, \xi_0 \, \varphi \, (1 + |\mathbf{D}(\mathbf{v}^\varepsilon)|)^{\frac{p(|\mathbf{E}|^2)-2}{2}} (1 + |\mathbf{E}|^2) \, dx \, dt$$

$$= \int\limits_{Q_T} \mathbf{D}(\nabla \mathbf{v}^\varepsilon) \, \xi_0 \, \varphi \, d\mu \,, \tag{3.78}$$

for $\varphi \in L^\infty(Q_T; dx \, dt) \cap L^2(Q_T; d\mu)$, where we used $(3.68)_6$ and (3.77). Since one easily checks that functions from $L^\infty(Q_T; dx \, dt) \cap L^2(Q_T; d\mu)$ are dense in $L^2(Q_T; d\mu)$ we have shown that

$$\mathbf{D}(\nabla \mathbf{v}^{\varepsilon,A}) \, \xi_0 \rightharpoonup \mathbf{D}(\nabla \mathbf{v}^\varepsilon) \, \xi_0 \,, \qquad \text{weakly in } L^2(Q_T; d\mu) \,,$$

and thus

$$\liminf_{A_\delta \to \infty} \int\limits_{Q_\delta \cap Q_t} |\mathbf{D}(\nabla \mathbf{v}^{\varepsilon,A_\delta})|^2 \xi_0^2 \, d\mu \geq \int\limits_{Q_\delta \cap Q_t} |\mathbf{D}(\nabla \mathbf{v}^\varepsilon)|^2 \xi_0^2 \, d\mu \,.$$

The other term $\sum_{s=1}^{2} \sum_{n=1}^{N} \mathbf{D}(\frac{\partial \mathbf{v}^{\varepsilon,A}}{\partial \tau^s}) \, \xi_1^n$ is treated similarly and therefore we have shown that for almost all $t \in I$

$$\liminf_{A_\delta \to \infty} \int\limits_{Q_\delta \cap Q_t} Y(\mathbf{v}^{\varepsilon,A_\delta}) \, dt \geq \int\limits_{Q_\delta \cap Q_t} Y(\mathbf{v}^\varepsilon) \, dt \,.$$

Since $\delta > 0$ was arbitrary and the right-hand side of (3.47) is bounded independently of δ we proved that also in the term $\int_0^t Y(\mathbf{v}^{\varepsilon,A}) \, d\tau$ we can replace $\mathbf{v}^{\varepsilon,A}$ by $\mathbf{v}^{\varepsilon}$. It remains to show (3.67). Due to (5.2.35) we have

$$\int\limits_0^t \left(I_{p,r}(\mathbf{D}(\nabla\mathbf{v}^{\varepsilon}), 1) \right)^{\frac{u}{r}} d\tau \leq c \int\limits_0^t \left(\|\mathbf{S}\|_1^u + \|\nabla\mathbf{S}\|_r^u + \int\limits_\Omega \mathbf{S} \cdot \mathbf{D} \, dx \right) d\tau$$

$$\leq c \left(1 + \int\limits_0^t K_\varepsilon \, d\tau \right), \tag{3.79}$$

where we also used (3.60), $u = \frac{2}{p_0 - 1} \leq p_0'$, (3.65), the equivalence of the quantities $\mathbf{S}(\mathbf{D}, \mathbf{E}) \cdot \mathbf{D}$ and $U(|\mathbf{D}|^2, |\mathbf{E}|^2)$ (cf. (1.43)) and (3.59).

Finally, the limiting process $A \to \infty$ in the weak formulation (1.53) is clear in all terms except the term with $\mathbf{S}^A$. For this term we obtain for almost all $t \in I$ and all $\varphi \in \mathcal{D}(\Omega)$

$$\lim_{A \to \infty} \int\limits_{\Omega_t} \mathbf{S}^A(\mathbf{D}(\mathbf{v}^{\varepsilon,A}), \mathbf{E}) \cdot \mathbf{D}(\varphi) \, dx = \int\limits_{\Omega_t} \mathbf{S}(\mathbf{D}(\mathbf{v}^{\varepsilon}), \mathbf{E}) \cdot \mathbf{D}(\varphi) \, dx$$

due to (3.73). The proof is complete. ∎

Remark 3.80. It follows from (1.26), (1.42) (cf. (3.2.14)) and (3.59), (3.64) that

$$\int\limits_0^T \|\mathbf{D}(\mathbf{v}^{\varepsilon})\|_{p(|\mathbf{E}(t)|^2), \nu(t)} + \|\nabla\mathbf{v}^{\varepsilon}\|_{p_\infty, \Omega_t}^{p_\infty} \, dt \leq c(\mathbf{f}, \mathbf{v}_0, \mathbf{E}_0),$$

$$\operatorname*{ess\,sup}_{\tau \in (0,t)} \left(\|\mathbf{D}(\mathbf{v}^{\varepsilon})\|_{p(|\mathbf{E}(\tau)|^2)} + \|\nabla\mathbf{v}^{\varepsilon}\|_{p_\infty, \Omega_\tau}^{p_\infty} + \|\nabla\mathbf{v}^{\varepsilon}\|_{2, \Omega_\tau}^2 \right) \leq c \left(1 + \int\limits_0^t K_\varepsilon(\tau) \, d\tau \right). \tag{3.81}$$

We want to finish this section by two lemmata, which are consequences of (3.13) and (3.14) if we pass to the limit $A \to \infty$.

Lemma 3.82. *Let $2 \leq p_\infty \leq p_0 < 8/3$ and assume that γ_μ, $\mu \geq 0$, are defined as in Lemma 2.50. Then we have for almost all $t \in I$ and $1 < s < 1 + \delta$, δ sufficiently small,*

$$\operatorname*{ess\,sup}_{\tau \in (0,t)} \gamma_\mu(F(t)) \leq c(\mathbf{f}, \mathbf{v}_0, \mathbf{E}_0) \left(1 + \int\limits_0^t K_\varepsilon(\tau) \, \gamma_\mu'(F(\tau)) \, d\tau \right.$$

$$\left. + \int\limits_0^t \|\mathbf{S}(\mathbf{D}(\mathbf{v}^{\varepsilon}(\tau)), \mathbf{E}(\tau))\|_{2s}^{2s} \, \gamma_\mu'(F(\tau)) \, d\tau \right) \tag{3.83}$$

where $F(t) = 1 + \|U(|\mathbf{D}(\mathbf{v}^{\varepsilon}(t))|^2, |\mathbf{E}(t)|^2)\|_{1, \nu(t)}$.

PROOF :   The definition of F_A and F together with (3.75) yields for almost all $t \in I$

$$F_A(t) \to F(t),\qquad(3.84)$$

which in turn implies

$$\begin{aligned}\gamma_\mu(F_A(t)) &\to \gamma_\mu(F(t)) &&\text{for a.e. } t \in I\,,\\\gamma'_\mu(F_A) &\rightharpoonup^* \gamma'_\mu(F) &&\text{weakly}^* \text{ in } L^\infty(I)\,.\end{aligned}\qquad(3.85)$$

From (3.75) we also obtain for an appropriately chosen $s > 1$ and almost all $t \in I$

$$\left\|U^A(|\mathbf{D}(\mathbf{v}^{\varepsilon,A}(t))|^2, |\mathbf{E}(t)|^2)\right\|_s \to \left\|U(|\mathbf{D}(\mathbf{v}^\varepsilon(t))|^2, |\mathbf{E}(t)|^2)\right\|_s,\qquad(3.86)$$

moreover we can re-phrase (3.76) as

$$K_{\varepsilon,A} \to K_\varepsilon \qquad \text{strongly in } L^1(I)\,.\qquad(3.87)$$

Since the second term on the left-hand side of (3.13) is non-negative and since we can pass to the limit as $A \to \infty$ in the other terms, due to (3.85)–(3.87), the assertion follows, if we use (3.16) and (3.59).   ∎

Lemma 3.88. *Let $2 \le p_\infty \le p_0 < 8/3$ and assume that γ_μ, $\mu \ge 0$, are defined as in Lemma 2.50. Then we have for almost all $t \in I$ and $1 < s < 1 + \delta$, δ sufficiently small,*

$$\int_0^t \left(\left(I_{p,r}(\mathbf{D}(\nabla \mathbf{v}^\varepsilon)), 1)\right)^{\frac{u}{r}} + \left\|\nabla^2 \mathbf{v}^\varepsilon(\tau)\right\|_r^u \right) \gamma'_\mu(F(\tau))\, d\tau$$

$$+ \int_0^t \left(\left\|\nabla \mathbf{S}(\mathbf{D}(\mathbf{v}^\varepsilon(\tau)), \mathbf{E}(\tau))\right\|_r^u + Y(\mathbf{v}^\varepsilon(\tau)) + Z(\mathbf{v}^\varepsilon(\tau)) \right) \gamma'_\mu(F(\tau))\, d\tau \qquad(3.89)$$

$$\le c(\mathbf{f}, \mathbf{v}_0, \mathbf{E}_0) \left(1 + \int_0^t \left(\left\|\mathbf{S}(\mathbf{D}(\mathbf{v}^\varepsilon(\tau)), \mathbf{E}(\tau))\right\|_{2s}^{2s} + K_\varepsilon(\tau) \right) \gamma'_\mu(F(\tau))\, d\tau \right),$$

where $r = \frac{6}{p_0+1}$, $u = \frac{2}{p_0-1}$ and $\mu \ge 0$.

PROOF :   We want to pass to the limit as $A \to \infty$ in (3.14), which gives (3.89) for the last four terms on the left-hand side. From (3.87), (3.73) and (3.85) we deduce that we can pass to the limit in the terms on the right-hand side of (3.14). Moreover, from (3.85) and (3.84) we obtain for all $m < \infty$

$$\gamma'_\mu(F_A) \to \gamma'_\mu(F) \qquad \text{strongly in } L^m(I)\,.$$

Now, we proceed similarly as in the proof of Proposition 3.57. Therefore, by Egerov's theorem we know that for all $\delta > 0$ there is a set I_δ with $|I \setminus I_\delta| \le \delta$ and a subsequence $A_\delta \to \infty$ such that

$$\gamma'_\mu(F_A) \rightrightarrows \gamma'_\mu(F) \qquad \text{uniformly on } I_\delta\,.$$

We conclude from this and (3.46) that for almost all $t \in I$

$$\liminf_{A_\delta \to \infty} \int_{I_\delta \cap (0,t)} \left\| \nabla \mathbf{S}^{A_\delta}(\mathbf{D}(\mathbf{v}^{\varepsilon, A_\delta}), \mathbf{E}) \right\|_r^u \left(\gamma_\mu'(F_{A_\delta}) - \gamma_\mu'(F) \right) d\tau = 0 \,.$$

Thus we have

$$\liminf_{A_\delta \to \infty} \int_{I_\delta \cap (0,t)} \left\| \nabla \mathbf{S}^{A_\delta}(\mathbf{D}(\mathbf{v}^{\varepsilon, A_\delta}), \mathbf{E}) \right\|_r^u \gamma_\mu'(F_{A_\delta}) \, d\tau$$

$$= \liminf_{A_\delta \to \infty} \int_{I_\delta \cap (0,t)} \left\| \nabla \mathbf{S}^{A_\delta}(\mathbf{D}(\mathbf{v}^{\varepsilon, A_\delta}), \mathbf{E}) \right\|_r^u \gamma_\mu'(F) \, d\tau \,.$$

Since

$$0 < \gamma'(F(t)) \le 1 \tag{3.90}$$

we can define a new measure $d\mu$, which is absolutely continuous with respect to dt, by

$$d\mu \equiv \gamma'(F(t)) \, dt \,,$$

and thus we have that $L^m(I, dt; X) \hookrightarrow L^m(I, d\mu; X)$ for any Banach space X and $1 < m < \infty$. In particular functions from $\mathcal{D}(I, X)$ are dense in both spaces. In (3.72) we have identified the weak limit of $\nabla \mathbf{S}^A$ as $\nabla \mathbf{S}$ in the space $L^u(I, L^r(\Omega))$. This together with (3.90), the boundedness of the right-hand side of (3.14) and the uniqueness of weak limits leads to (cf. (3.78))

$$\nabla \mathbf{S}^A(\mathbf{D}(\mathbf{v}^{\varepsilon, A}), \mathbf{E}) \rightharpoonup \nabla \mathbf{S}(\mathbf{D}(\mathbf{v}^\varepsilon), \mathbf{E}) \qquad \text{weakly in } L^u(I, d\mu; L^r(\Omega))$$

and thus we have for almost all $t \in I$

$$\liminf_{A_\delta \to \infty} \int_{I_\delta \cap (0,t)} \left\| \nabla \mathbf{S}^{A_\delta}(\mathbf{D}(\mathbf{v}^{\varepsilon, A_\delta}), \mathbf{E}) \right\|_r^u \gamma_\mu'(F) \, dt \ge \int_{I_\delta \cap (0,t)} \left\| \nabla \mathbf{S}(\nabla \mathbf{v}^\varepsilon, \mathbf{E}) \right\|_r^u \gamma_\mu'(F) \, dt \,.$$

Since $\delta > 0$ was arbitrary and the right-hand side of (3.14) is bounded independently on δ we get (3.89) for the last four terms on the left-hand side. The estimate for the first term we obtain from (5.2.35)

$$\int_0^t \left(I_{p,r}(\mathbf{D}(\nabla \mathbf{v}^\varepsilon(\tau)), 1) \right)^{\frac{u}{r}} \gamma_\mu'(\tau) \, d\tau \le c \int_0^t \left(\|\mathbf{S}\|_1^u + \|\nabla \mathbf{S}\|_r^u + \int_\Omega \mathbf{S} \cdot \mathbf{D} \, dx \right) \gamma_\mu'(\tau) \, d\tau$$

$$\le c \left(1 + \int_0^t \left(\|\mathbf{S}\|_{2s}^{2s} + K_\varepsilon \right) \gamma_\mu'(\tau) \, d\tau \right), \tag{3.91}$$

where we used (3.60), (3.90), the estimate (3.89) for $\nabla \mathbf{S}$, the equivalence of the quantities $\mathbf{S}(\mathbf{D}, \mathbf{E}) \cdot \mathbf{D}$ and $U(|\mathbf{D}|^2, |\mathbf{E}|^2)$, (3.59) and (3.90). ∎

4.4   Limiting Process $\varepsilon \to 0$

In this section it remains to find conditions on p_∞ and p_0 such that the estimates proved in Proposition 3.57, Lemma 3.82 and Lemma 3.88 are independent of ε. This is already true for the following estimates (cf. (3.59), (3.81)$_1$, (3.60)):

$$\operatorname{ess\,sup}_{t \in I} \|\mathbf{v}^\varepsilon(t)\|_2^2 + \int_0^T \|U(|\mathbf{D}(\mathbf{v}^\varepsilon(t))|^2, |\mathbf{E}(t)|^2)\|_{1,\nu(t)}\, dt \le c(\mathbf{f}, \mathbf{v}_0, \mathbf{E}_0)\,, \qquad (4.1)$$

$$\int_0^T \|\mathbf{D}(\mathbf{v}^\varepsilon(t))\|_{p(|\mathbf{E}(t)|^2),\nu(t)} + \|\nabla\mathbf{v}^\varepsilon(t)\|_{p_\infty}^{p_\infty}\, dt \le c(\mathbf{f}, \mathbf{v}_0, \mathbf{E}_0)\,, \qquad (4.2)$$

$$\int_0^T \|\mathbf{S}(\mathbf{D}(\mathbf{v}^\varepsilon(t)), \mathbf{E}(t))\|_{p_0'}^{p_0'}\, dt \le c(\mathbf{f}, \mathbf{v}_0, \mathbf{E}_0)\,. \qquad (4.3)$$

On the right-hand side of all other estimates occurs the term $K_\varepsilon(t)$, which must be handled by the quantities on the left-hand sides. We distinguish two cases for which we establish estimates independent of ε. After that we will discuss how to use these estimates for the existence of weak and strong solutions, respectively.

(i) The Case $12/5 \le p_\infty \le p_0 < 8/3$

In this case we estimate K_ε as follows

$$
\begin{aligned}
K_\varepsilon &\le \|\nabla\mathbf{v}^\varepsilon\|_{p_\infty}^2 \|\mathbf{v}^\varepsilon\|_{\frac{2p_\infty}{p_\infty-2}}^2 \\
&\le c\,\|\mathbf{v}^\varepsilon\|_2^{2\frac{5p_\infty-12}{5p_\infty-6}} \|\nabla\mathbf{v}^\varepsilon\|_{p_\infty}^{p_\infty\frac{16-5p_\infty}{5p_\infty-6}} \|\nabla\mathbf{v}^\varepsilon\|_{p_\infty}^{p_\infty} =: g\|\nabla\mathbf{v}^\varepsilon\|_{p_\infty}^{p_\infty}\,,
\end{aligned}
\qquad (4.4)
$$

where we used the interpolation of $L^{\frac{2p_\infty}{p_\infty-2}}$ between L^2 and $L^{\frac{3p_\infty}{3-p_\infty}}$. Observe that for the p_∞ considered here we have $\frac{16-5p_\infty}{5p_\infty-6} \le 1$. Thus the function g in (4.4) belongs to $L^1(I)$, due to (4.2) and we obtain from (3.81)$_2$ and (4.4)

$$\|\nabla\mathbf{v}^\varepsilon(t)\|_{p_\infty}^{p_\infty} \le c + c\int_0^t g(\tau)\|\nabla\mathbf{v}^\varepsilon(\tau)\|_{p_\infty}^{p_\infty}\, d\tau\,,$$

which by Gronwall's lemma implies for almost all $t \in I$

$$\operatorname{ess\,sup}_{t \in I} \|\nabla\mathbf{v}^\varepsilon(t)\|_{p_\infty}^{p_\infty} \le c(\mathbf{f}, \mathbf{v}_0, \mathbf{E}_0)\,.$$

Therefore also the right-hand sides in (3.61), (3.62), (3.64)–(3.67) and (3.81)$_2$ are bounded independently of ε, which yields

$$\int_0^T \|\phi^\varepsilon\|_2^2 + \left\|\frac{\partial \mathbf{v}^\varepsilon}{\partial t}\right\|_2^2 dt \le c(\mathbf{f}, \mathbf{v}_0, \mathbf{E}_0),$$

$$\operatorname{ess\,sup}_{t \in I} \left\{ \|\nabla \mathbf{v}^\varepsilon(t)\|_{p_\infty}^{p_\infty} + \|\mathbf{D}(\mathbf{v}^\varepsilon(t))\|_{p(|\mathbf{E}(t)|^2), \nu(t)} \right\} \le c(\mathbf{f}, \mathbf{v}_0, \mathbf{E}_0),$$

$$\int_0^T \|\nabla^2 \mathbf{v}^\varepsilon\|_{\frac{6}{p_0+1}}^{\frac{2}{p_0-1}} + \|\nabla \mathbf{S}(\mathbf{D}(\mathbf{v}^\varepsilon), \mathbf{E})\|_{\frac{6}{p_0+1}}^{\frac{2}{p_0-1}} + Y(\mathbf{v}^\varepsilon) + Z(\mathbf{v}^\varepsilon) \, dt \le c(\mathbf{f}, \mathbf{v}_0, \mathbf{E}_0), \tag{4.5}$$

$$\int_0^T \|U(|\mathbf{D}(\mathbf{v}^\varepsilon)|^2, |\mathbf{E}|^2)\|_{\frac{p_0+3}{2p_0}}^{\frac{p_0+3}{2p_0}} + \|\mathbf{S}(\mathbf{D}(\mathbf{v}^\varepsilon), \mathbf{E})\|_{\frac{2(p_0+3)}{3(p_0-1)}}^{\frac{2(p_0+3)}{3(p_0-1)}} \, dt \le c(\mathbf{f}, \mathbf{v}_0, \mathbf{E}_0),$$

$$\int_0^T \left(I_{p,r}(\mathbf{D}(\nabla \mathbf{v}^\varepsilon), 1) \right)^{\frac{p_0+1}{3(p_0-1)}} dt \le c(\mathbf{f}, \mathbf{v}_0, \mathbf{E}_0),$$

From $(4.5)_{2,5}$ and $(5.2.41)$ we derive

$$\int_0^T \|\nabla \mathbf{v}^\varepsilon\|_{6\frac{p_\infty-1}{p_0-1}}^{2\frac{p_\infty-1}{p_0-1}} \, dt \le c(\mathbf{f}, \mathbf{v}_0, \mathbf{E}_0). \tag{4.6}$$

(ii) The Case $2 \le p_\infty \le p_0 < 8/3$, $p_\infty < 12/5$

Recall that $F(t) = 1 + \|U(|\mathbf{D}(\mathbf{v}^\varepsilon(t))|^2, |\mathbf{E}(t)|^2)\|_{1, \nu(t)}$ and $\gamma_\mu(s) = (1-\mu)^{-1} s^{1-\mu}$ if $\mu \ge 0$ and $\mu \ne 1$ and $\gamma(s) = \ln s$. Note, that (4.1) implies that

$$F \in L^1(I). \tag{4.7}$$

The inequalities (3.83), (3.89) and Lemma $5.2.40$ lead, for $1 < s < 1+\delta$, δ sufficiently small and almost all $t \in I$, to

$$\gamma_\mu(F(t)) + \int_0^t \left(\|\nabla \mathbf{v}^\varepsilon(\tau)\|_{6\frac{p_\infty-1}{p_0-1}}^{2\frac{p_\infty-1}{p_0-1}} + \|\nabla^2 \mathbf{v}^\varepsilon(\tau)\|_{\frac{6}{p_0+1}}^{\frac{2}{p_0-1}} \right) \gamma_\mu'(F(\tau)) \, d\tau \tag{4.8}$$

$$+ \int_0^t \left(\|\nabla \mathbf{S}(\mathbf{D}(\mathbf{v}^\varepsilon(\tau)), \mathbf{E}(\tau))\|_{\frac{6}{p_0+1}}^{\frac{2}{p_0-1}} + Y(\mathbf{v}^\varepsilon(\tau)) + Z(\mathbf{v}^\varepsilon(\tau)) \right) \gamma_\mu'(F(\tau)) \, d\tau$$

$$\le c(\mathbf{f}, \mathbf{v}_0, \mathbf{E}_0) \left(1 + \int_0^t K_\varepsilon(\tau) \, \gamma_\mu'(F(\tau)) \, d\tau + \int_0^t \|\mathbf{S}(\mathbf{D}(\mathbf{v}^\varepsilon(\tau)), \mathbf{E}(\tau))\|_{2s}^{2s} \gamma_\mu'(F(\tau)) \, d\tau \right)$$

$$+ c(\mathbf{f}, \mathbf{v}_0, \mathbf{E}_0) \left(\int_0^t \gamma_{\mu - \frac{2}{p_0-1} \frac{p_\infty-1}{p_\infty}}'(F(\tau)) \, d\tau \right).$$

Since we obtain an information from the first term on the left-hand side of (4.8) only for $\mu \leq 1$, and since we got this inequality as a sum of two separate inequalities, we will omit the first term if $\mu > 1$. Let us now discuss the terms on the right-hand side of (4.8). Since $2 \leq p_\infty \leq p_0$, we see that

$$\frac{2}{p_\infty} \frac{p_\infty - 1}{p_0 - 1} - \mu \leq 1 - \mu \leq 1,$$

which together with (4.2) implies that the last term in (4.8) is bounded by $c(\mathbf{f}, \mathbf{v}_0, \mathbf{E}_0)$. Let us therefore estimate the remaining two terms on the right-hand side of (4.8). We compute

$$K_\varepsilon \leq \|\nabla \mathbf{v}^\varepsilon\|_{p_\infty}^2 \|\mathbf{v}^\varepsilon\|_{\frac{6p_\infty}{5p_\infty - 2}}^2$$

$$\leq \|\nabla \mathbf{v}^\varepsilon\|_{p_\infty}^{\frac{2(2-\lambda)}{p_\infty}} \|\nabla \mathbf{v}^\varepsilon\|_{6\frac{p_\infty - 1}{p_0 - 1}}^{2\lambda} \tag{4.9}$$

$$\leq F^{\frac{2(2-\lambda)}{p_\infty}} \|\nabla \mathbf{v}^\varepsilon\|_{6\frac{p_\infty - 1}{p_0 - 1}}^{2\frac{p_\infty - 1}{p_0 - 1} \frac{\lambda(p_0 - 1)}{p_\infty - 1}},$$

where we used the interpolation inequality

$$\|\mathbf{f}\|_{\frac{6p_\infty}{5p_\infty - 6}} \leq \|\mathbf{f}\|_{p_\infty}^{1-\lambda} \|\mathbf{f}\|_{6\frac{p_\infty - 1}{p_0 - 1}}^{\lambda},$$

with $\lambda = \frac{(p_\infty - 1)(5p_\infty - 12)}{p_\infty p_0 - 7p_\infty + 6}$ and

$$\|\nabla \mathbf{v}^\varepsilon\|_{p_\infty}^{p_\infty} \leq F. \tag{4.10}$$

Furthermore, we observe, for $0 < \kappa < p_0'/(2s)$,

$$\|\mathbf{S}(\mathbf{D}(\mathbf{v}^\varepsilon), \mathbf{E})\|_{2s}^{2s} \leq \int_\Omega |\mathbf{S}(\mathbf{D}(\mathbf{v}^\varepsilon), \mathbf{E})|^{2s(1-\kappa)} \left(1 + U(|\mathbf{D}(\mathbf{v}^\varepsilon)|^2, |\mathbf{E}|^2)\right)^{\frac{2s\kappa}{p_0'}} dx$$

$$\leq \|\mathbf{S}(\mathbf{D}(\mathbf{v}^\varepsilon), \mathbf{E})\|_{\frac{2sp_0'(1-\kappa)}{p_0' - 2s\kappa}}^{2s(1-\kappa)} F^{\frac{2s\kappa}{p_0'}}, \tag{4.11}$$

where we employed the definition of F and

$$|\mathbf{S}(\mathbf{D}(\mathbf{v}), \mathbf{E})| \leq c\left(1 + U(|\mathbf{D}(\mathbf{v})|^2, |\mathbf{E}|^2)\right)^{\frac{p-1}{p}}, \tag{4.12}$$

which can be proved in the same way as (1.39). Due to the embedding $W^{1, \frac{6}{p_0 + 1}}(\Omega) \hookrightarrow L^{\frac{6}{p_0 - 1}}(\Omega)$ we require now $\frac{2sp_0'(1-\kappa)}{p_0' - 2s\kappa} = \frac{6}{p_0 - 1}$ or equivalently $\kappa = \frac{p_0(3 - s(p_0 - 1))}{s(p_0 - 1)(6 - p_0)}$ and thus we arrive at

$$\|\mathbf{S}(\mathbf{D}(\mathbf{v}^\varepsilon), \mathbf{E})\|_{2s}^{2s} \leq c F^{\frac{2s\kappa}{p_0'}} \left(\|\nabla \mathbf{S}(\mathbf{D}(\mathbf{v}^\varepsilon), \mathbf{E})\|_r + \|\mathbf{S}(\mathbf{D}(\mathbf{v}^\varepsilon), \mathbf{E})\|_1\right)^{2s(1-\kappa)}$$

$$\leq c F^{\frac{2s\kappa}{p_0'}} \|\nabla \mathbf{S}(\mathbf{D}(\mathbf{v}^\varepsilon), \mathbf{E})\|_r^{2s(1-\kappa)} + F^{\frac{2s}{p_0'}}, \tag{4.13}$$

where we have taken into account again (4.12) to bound the L^1-norm of $\mathbf{S}$ in terms of F. From (4.8), (4.9) and (4.13) we derive, for $1 < s < 1 + \delta$, δ sufficiently small and almost all $t \in I$,

$$\gamma_\mu(F(t)) + \int\limits_0^t \left(\|\nabla \mathbf{v}^\varepsilon(\tau)\|_{6\frac{p_\infty-1}{p_0-1}}^{2\frac{p_\infty-1}{p_0-1}} + \|\nabla^2 \mathbf{v}^\varepsilon(\tau)\|_{\frac{p_0-1}{p_0+1}}^{\frac{2}{p_0+1}} \right) \gamma_\mu'(F(\tau))\, d\tau \qquad (4.14)$$

$$+ \int\limits_0^t \left(\|\nabla \mathbf{S}(\mathbf{D}(\mathbf{v}^\varepsilon(\tau)), \mathbf{E}(\tau))\|_{\frac{p_0+1}{6}}^{\frac{2}{p_0+1}} + Y(\mathbf{v}^\varepsilon(\tau)) + Z(\mathbf{v}^\varepsilon(\tau)) \right) \gamma_\mu'(F(\tau))\, d\tau$$

$$\leq c \left[1 + \int\limits_0^t \left(\|\nabla \mathbf{v}^\varepsilon(\tau)\|_{6\frac{p_\infty-1}{p_0-1}}^{2\frac{p_\infty-1}{p_0-1}} \gamma_\mu'(F(\tau)) \right)^{\lambda\frac{p_0-1}{p_\infty-1}} F(\tau)^{\frac{2(2-\lambda)}{p_\infty} + \mu[\lambda\frac{p_0-1}{p_\infty-1}-1]}\, d\tau \right. $$

$$\left. + \int\limits_0^t \left(\|\nabla \mathbf{S}(\mathbf{D}(\mathbf{v}^\varepsilon(\tau)), \mathbf{E}(\tau))\|_{\frac{p_0+1}{6}}^{\frac{2}{p_0+1}} \gamma_\mu'(F(\tau)) \right)^{s(1-\kappa)(p_0-1)} F(\tau)^{\frac{2s\kappa}{p_0'} + \mu[s(1-\kappa)(p_0-1)-1]}\, d\tau \right].$$

In order to bound the terms on the right-hand side of (4.14) we will use Young's inequality, which enables us to absorb one term on the left-hand side of (4.14) and to bound the other one by (4.7). For that we require for the first term on the right-hand side

$$\lambda \frac{p_0-1}{p_\infty-1} \alpha = 1, \qquad \left\{ \frac{2(2-\lambda)}{p_\infty} + \mu\left(\lambda\frac{p_0-1}{p_\infty-1} - 1\right) \right\} \alpha' = 1, \qquad \frac{1}{\alpha} + \frac{1}{\alpha'} = 1$$

and thus μ_1 must satisfy [8]

$$\mu_1 = \frac{2(2p_\infty - 3 + 2p_0 - p_\infty p_0)}{3 + p_\infty - 2p_0(3 - p_\infty)}. \qquad (4.15)$$

The right-hand side of (4.15) is for p_0 and p_∞ considered here finite and positive as long as

$$p_0 < \frac{3 + p_\infty}{2(3 - p_\infty)}. \qquad (4.16)$$

For the second term we proceed similarly and require

$$s(\kappa - 1)(p_0 - 1)\beta = 1, \qquad \left\{ \frac{2s\kappa}{p_0'} + \mu\big(s(1 - \kappa)(p_0 - 1) - 1\big) \right\} \beta' \leq 1, \qquad \frac{1}{\beta} + \frac{1}{\beta'} = 1,$$

which implies

$$\mu_2 \geq \frac{2s(p_0 - 1) - p_0}{3 + p_0 - 3s(p_0 - 1)}.$$

For $s > 1$ chosen appropriately we deduce

$$\mu_2 > \frac{p_0 - 2}{6 - 2p_0}, \qquad (4.17)$$

---

[8]Since $F^{-\alpha} \leq 1$, we will compute the optimal values for μ, denoted by μ_1 and μ_2 respectively, for both terms separately and than choose the larger one.

which is finite and positive for p_∞ and p_0 considered here. Note, that for $p_0 < 8/3$ we can always select μ_2 satisfying (4.17) such that

$$\mu_2 < 1. \tag{4.18}$$

Moreover, we see that we can also choose $\mu_1 \leq 1$ if

$$\frac{9}{4} \leq p_\infty \leq p_0 \leq \frac{3(3 - p_\infty)}{2(5 - 2p_\infty)}. \tag{4.19}$$

This means, that in the case

$$p_\infty \in \left[\frac{9}{4}, \frac{12}{5}\right), \qquad p_\infty \leq p_0 \leq \min\left(\frac{3(3 - p_\infty)}{2(5 - 2p_\infty)}, \frac{8}{3}\right) \tag{4.20}$$

we can select $\mu \leq 1$ such that μ satisfies the restrictions (4.15) and (4.17). Using Young's inequality in the way described above we deduce from (4.14) and (4.7), for almost all $t \in I$,

$$\gamma_\mu(F(t)) + \int\limits_0^t \left(\left\|\nabla \mathbf{v}^\varepsilon(\tau)\right\|_{6\frac{p_\infty-1}{p_0-1}}^{2\frac{p_\infty-1}{p_0-1}} + \left\|\nabla^2 \mathbf{v}^\varepsilon(\tau)\right\|_{\frac{6}{p_0+1}}^{\frac{2}{p_0-1}}\right) \gamma_\mu'(F(\tau))\, d\tau \tag{4.21}$$

$$+ \int\limits_0^t \left(\left\|\nabla \mathbf{S}(\mathbf{D}(\mathbf{v}^\varepsilon(\tau)), \mathbf{E}(\tau))\right\|_{\frac{6}{p_0+1}}^{\frac{2}{p_0-1}} + Y(\mathbf{v}^\varepsilon(\tau)) + Z(\mathbf{v}^\varepsilon(\tau))\right) \gamma_\mu'(F(\tau))\, d\tau \leq c(\mathbf{f}, \mathbf{v}_0, \mathbf{E}_0).$$

Since $\mu \leq 1$ inequality (4.21) implies that $(4.5)_2$ is satisfied and this in turn implies that

$$\int\limits_0^t \left\|\nabla \mathbf{v}^\varepsilon(\tau)\right\|_{6\frac{p_\infty-1}{p_0-1}}^{2\frac{p_\infty-1}{p_0-1}} + \left\|\nabla^2 \mathbf{v}^\varepsilon(\tau)\right\|_{\frac{6}{p_0+1}}^{\frac{2}{p_0-1}} d\tau \leq c(\mathbf{f}, \mathbf{v}_0, \mathbf{E}_0),$$

$$\int\limits_0^t \left\|\nabla \mathbf{S}(\mathbf{D}(\mathbf{v}^\varepsilon(\tau)), \mathbf{E}(\tau))\right\|_{\frac{6}{p_0+1}}^{\frac{2}{p_0-1}} + Y(\mathbf{v}^\varepsilon(\tau)) + Z(\mathbf{v}^\varepsilon(\tau)) d\tau \leq c(\mathbf{f}, \mathbf{v}_0, \mathbf{E}_0). \tag{4.22}$$

Since $2\lambda \leq 2\frac{p_\infty-1}{p_0-1}$ for p_0 and p_∞ satisfying (4.20) we obtain from (4.9) and $(4.22)_1$ that

$$\int_0^t K_\varepsilon\, d\tau \leq c(\mathbf{f}, \mathbf{v}_0, \mathbf{E}_0),$$

and thus we deduce from (3.61), (3.62), (3.64)–(3.67) and $(3.81)_2$ that (4.5) and (4.6) hold.

In the other case, i.e. when either

$$p_\infty \in \left[2, \frac{9}{4}\right), \qquad p_0 < \min\left(\frac{3 + p_\infty}{2(3 - p_\infty)}, \frac{8}{3}\right), \tag{4.23}$$

or

$$\frac{9}{4} < p_\infty \leq p_0, \qquad \frac{3(3 - p_\infty)}{2(5 - 2p_\infty)} < p_0 < \frac{8}{3} \tag{4.24}$$

we have to choose $\mu = \mu_1 > 1$ satisfying (4.15). In these cases we obtain using Young's inequality from (4.14) and (4.7) that

$$
\int_0^T \left(\|\nabla \mathbf{S}(\mathbf{D}(\mathbf{v}^\varepsilon(t)), \mathbf{E}(t))\|_{\frac{6}{p_0+1}}^{\frac{2}{p_0-1}} + Y(\mathbf{v}^\varepsilon(t)) + Z(\mathbf{v}^\varepsilon(t)) \right) F(t)^{-\mu} \, dt
$$

$$
+ \int_0^T \left(\|\nabla \mathbf{v}^\varepsilon(t)\|_{6\frac{p_\infty-1}{p_0-1}}^{2\frac{p_\infty-1}{p_0-1}} + \|\nabla^2 \mathbf{v}^\varepsilon(t)\|_{\frac{6}{p_0+1}}^{\left(\frac{2}{p_0-1}\right)} \right) F(t)^{-\mu} \, dt \leq c(\mathbf{f}, \mathbf{v}_0, \mathbf{E}) \,.
\tag{4.25}
$$

This estimate is not sufficient to conclude that the right-hand side in (3.64) is bounded independently of ε. Thus we need a different estimate for $\frac{\partial \mathbf{v}^\varepsilon}{\partial t}$. We have, using the weak formulation (3.7) and (4.2), (4.3),

$$
\left\| \frac{\partial \mathbf{v}^\varepsilon}{\partial t} \right\|_{L^{q'}(I,(V_q)^*)} = \sup_{\substack{\varphi \in L^q(I,V_q) \\ \|\varphi\|_{L^q(I,V_q)} \leq 1}} \left| \int_0^T \int_\Omega \frac{\partial \mathbf{v}^\varepsilon}{\partial t} \cdot \varphi \, dx \, dt \right|
\tag{4.26}
$$

$$
\leq \sup_{\substack{\varphi \in L^q(I,V_q) \\ \|\varphi\|_{L^q(I,V_q)} \leq 1}} \left| \int_0^T \int_\Omega -\mathbf{S}(\mathbf{D}(\mathbf{v}^\varepsilon), \mathbf{E}) \cdot \mathbf{D}(\varphi) - [\nabla \mathbf{v}^\varepsilon] \mathbf{v}_\varepsilon^\varepsilon \cdot \varphi \right.
$$

$$
\left. + \mathbf{f} \cdot \varphi - \chi^E \mathbf{E} \otimes \mathbf{E} \cdot \mathbf{D}(\varphi) \, dx \, dt \right|
$$

$$
\leq \|\mathbf{S}(\mathbf{D}(\mathbf{v}^\varepsilon), \mathbf{E})\|_{p_0', Q_T} \|\nabla \varphi\|_{p_0, Q_T} + \|\mathbf{f}\|_{2, Q_T} \|\varphi\|_{p_0, Q_T}
$$

$$
+ c\|\mathbf{E}\|_{2p_0', Q_T}^2 \|\nabla \varphi\|_{p_0, Q_T} + \int_0^T \|\mathbf{v}^\varepsilon\|_s \|\nabla \mathbf{v}^\varepsilon\|_{p_\infty} \|\nabla \varphi\|_m \, dt
$$

$$
\leq c(\mathbf{f}, \mathbf{v}_0, \mathbf{E}_0) \|\nabla \varphi\|_{p_0, Q_T} + \|\mathbf{v}^\varepsilon\|_{L^\infty(I,H)}^{1-\lambda} \int_0^T \|\nabla \mathbf{v}^\varepsilon\|_{p_\infty}^{1+\lambda} \|\nabla \varphi\|_m \, dt \,,
$$

where $\frac{1}{s} = 1 - \frac{1}{p_\infty} - \frac{3-m}{3m}$ and where we employed div $\varphi = 0$ and the interpolation inequality

$$
\|\mathbf{v}\|_s \leq \|\mathbf{v}\|_2^{1-\lambda} \|\nabla \mathbf{v}\|_{p_\infty}^\lambda \,,
$$

with $\lambda = \frac{5p_\infty m - 6p_\infty - 6m}{m(6 - 5p_\infty)}$. Now, requiring for the last term in (4.26) that

$$
(1 + \lambda) m' = p_\infty
$$

leads to

$$
m = \frac{5p_\infty}{5p_\infty - 6} \,.
$$

Thus, if we choose

$$
q = \max \left(p_0, \frac{5p_\infty}{5p_\infty - 6} \right)
\tag{4.27}
$$

we obtain that

$$\left\| \frac{\partial \mathbf{v}^\varepsilon}{\partial t} \right\|_{L^{q'}(I,(V_q)^*)} \le c(\mathbf{f}, \mathbf{v}_0, \mathbf{E}_0) \,. \tag{4.28}$$

Now, we have at our disposal all estimates we need to show the existence of weak and strong solutions, respectively, to the problem (1.1), (1.2).

(iii) Strong Solutions

In the cases (i) and (4.20), i.e. when

$$\frac{9}{4} \le p_\infty \le p_0 < \min \left(\frac{8}{3}, \frac{3(3 - p_\infty)}{2(5 - 2p_\infty)} \right), \tag{4.29}$$

we have established the estimates (4.1)–(4.3), (4.5)$_{1-4}$ and (4.6) which are independent of ε. These estimates together with the Aubin-Lions lemma imply the existence of $\mathbf{v}, \phi$ such that (cf. Remark 3.56)

$$\begin{aligned}
\nabla \mathbf{v}^\varepsilon \to \nabla \mathbf{v} \qquad & \text{strongly in } L^{\frac{2}{p_0-1}}(I, L^m(\Omega)), \quad m < \frac{6}{p_0 - 1}, \\
\frac{\partial \mathbf{v}^\varepsilon}{\partial t} \rightharpoonup \frac{\partial \mathbf{v}}{\partial t} \qquad & \text{weakly in } L^2(Q_T), \\
\mathbf{D}(\mathbf{v}^\varepsilon) \rightharpoonup \mathbf{D}(\mathbf{v}) \qquad & \text{weakly in } L^{p(|\mathbf{E}|^2)}(Q_T) \cap L^{2\frac{p_\infty-1}{p_0-1}}(I, L^{6\frac{p_\infty-1}{p_0+1}}(\Omega)), \\
\mathbf{v}^\varepsilon \overset{*}{\rightharpoonup} \mathbf{v} \qquad & \text{weakly* in } L^\infty(I, E_{p(|\mathbf{E}(t)|^2)}) \cap L^\infty(I, V_{p_\infty}), \\
\phi^\varepsilon \rightharpoonup \phi \qquad & \text{weakly in } L^2(Q_T), \\
\nabla^2 \mathbf{v}^\varepsilon \rightharpoonup \nabla^2 \mathbf{v} \qquad & \text{weakly in } L^2(I, L^2_{\mathrm{loc}}(\Omega)) \cap L^{\frac{2}{p_0-1}}(I, L^{\frac{6}{p_0-1}}(\Omega)), \\
\nabla \frac{\partial \mathbf{v}^\varepsilon}{\partial \tau^s} \rightharpoonup \nabla \frac{\partial \mathbf{v}}{\partial \tau^s} \qquad & \text{weakly in } L^2(Q_T), \quad s = 1, 2,
\end{aligned} \tag{4.30}$$

as $\varepsilon \to 0$ at least for a subsequence. From (4.30)$_1$ we conclude

$$\nabla \mathbf{v}^\varepsilon \to \nabla \mathbf{v} \qquad \text{a.e. in } Q_T \,, \tag{4.31}$$

and thus

$$\begin{aligned}
\mathbf{S}(\mathbf{D}(\mathbf{v}^\varepsilon), \mathbf{E}) \to \mathbf{S}(\mathbf{D}(\mathbf{v}), \mathbf{E}) \qquad & \text{a.e. in } Q_T \,, \\
U(|\mathbf{D}(\mathbf{v}^\varepsilon)|^2, |\mathbf{E}|^2) \to U(|\mathbf{D}(\mathbf{v})|^2, |\mathbf{E}|^2) \qquad & \text{a.e. in } Q_T \,.
\end{aligned} \tag{4.32}$$

This in turn together with Vitali's theorem and (4.5)$_4$ implies for $m < \frac{2(p_0+3)}{3(p_0-1)}$

$$\mathbf{S}(\mathbf{D}(\mathbf{v}^\varepsilon), \mathbf{E}) \to \mathbf{S}(\mathbf{D}(\mathbf{v}), \mathbf{E}) \qquad \text{strongly in } L^m(Q_T). \tag{4.33}$$

Thus we can justify the limiting process $\varepsilon \to 0$ in the weak formulation (3.7) as in the proof of Proposition 3.57. Due to the lower semicontinuity of norms, the identicalness of weak and almost everywhere limits, Fatou's lemma and (4.30)–(4.33) the estimates (4.1)–(4.3), (4.5)$_{1-4}$ and (4.6) remain valid for the limiting elements $\mathbf{v}$,

ϕ, $\frac{\partial \mathbf{v}}{\partial t}$, $\mathbf{S}(\mathbf{D}(\mathbf{v}), \mathbf{E})$ and $U(|\mathbf{D}(\mathbf{v})|^2, |\mathbf{E}|^2)$. Therefore we have proved the existence part of Theorem 1.14 as far as strong solutions are concerned.

Let us prove their uniqueness. First of all it follows from Proposition 2.3.35 that a solution $\mathbf{E}$ of (1.1) which is orthogonal to $H_N(\Omega)$ is uniquely determined. Assume that $\mathbf{u}, \mathbf{v}$ are two strong solutions of (1.2) for the same data $\mathbf{f}$ and $\mathbf{v}_0$ and the uniquely determined $\mathbf{E}$. Let us denote their difference by $\mathbf{w}$. From the weak formulation (1.13) we obtain (cf. proof of Proposition 3.2.38)

$$\frac{1}{2}\frac{d}{dt}\|\mathbf{w}\|_2^2 + \int_{\Omega} (\mathbf{S}(\mathbf{D}(\mathbf{u}), \mathbf{E}) - \mathbf{S}(\mathbf{D}(\mathbf{v}), \mathbf{E})) \cdot \mathbf{D}(\mathbf{w})\, dx = \int_{\Omega} [\nabla \mathbf{u}]\mathbf{w} \cdot \mathbf{w}\, dx\,,$$

and (2.5) leads to

$$\int_{\Omega} (\mathbf{S}(\mathbf{D}(\mathbf{u}), \mathbf{E}) - \mathbf{S}(\mathbf{D}(\mathbf{v}), \mathbf{E})) \cdot \mathbf{D}(\mathbf{w})\, dx \geq \alpha_{31} \int_{\Omega} |\mathbf{D}(\mathbf{w})|^2\, dx\,,$$

while the convective term is estimated by

$$\|\nabla \mathbf{u}\|_{p_\infty} \|\mathbf{w}\|_{\frac{2p_\infty}{p_\infty-1}}^2 \leq c\|\nabla \mathbf{u}\|_{p_\infty} \|\mathbf{w}\|_2^{\frac{2p_\infty-3}{p_\infty-1}} \|\nabla \mathbf{w}\|_2^{\frac{3}{p_\infty}}$$

$$\leq \alpha_{31}\|\mathbf{D}(\mathbf{w})\|_2^2 + c\|\nabla \mathbf{u}\|_{p_\infty}^{\frac{2p_\infty}{2p_\infty-3}} \|\mathbf{w}\|_2^2\,,$$

where we also used Korn's inequality. Because of the last three relations we see that

$$\frac{d}{dt}\|\mathbf{w}\|_2^2 \leq c\|\nabla \mathbf{u}\|_{p_\infty}^{\frac{2p_\infty}{2p_\infty-3}}\|\mathbf{w}\|_2^2\,. \tag{4.34}$$

Since $\mathbf{w}(0) = 0$ and $\nabla \mathbf{u} \in L^\infty(I, L^{p_\infty}(\Omega))$, Gronwall's lemma yields

$$\mathbf{w}(t) = 0 \qquad \forall t \leq T\,,$$

which gives the uniqueness and finishes the proof of Theorem 1.14 in the case of strong solutions.

(iv) Weak Solutions

In the remaining cases, i.e. either (4.23) or (4.24) is satisfied, we have established the estimates (4.1)–(4.3), (4.25) and (4.28), which are independent of ε. For the limiting process $\varepsilon \to 0$ it is essential to prove (4.31). This will be accomplished as a consequence of the last term on the left-hand side of (4.25). Firstly, we will show that (4.25) implies that

$$\int_0^T \|\nabla^2 \mathbf{v}^\varepsilon\|_r^{\alpha u}\, dt \leq c(\mathbf{f}, \mathbf{v}_0, \mathbf{E}_0)\,, \tag{4.35}$$

where $u = \frac{2}{p_0-1}$, $r = \frac{6}{p_0+1}$ and where α satisfies

$$\alpha \leq \frac{3 + p_\infty - 2p_0(3 - p_\infty)}{5p_\infty - 3 - 2p_0}. \tag{4.36}$$

Indeed, we obtain from (4.25)

$$\int_0^T \|\nabla^2 \mathbf{v}^\epsilon\|_r^{\alpha u}\, dt \leq \int_0^T \left\{\|\nabla^2 \mathbf{v}^\epsilon\|_r^u\, F^{-\mu}\right\}^\alpha F^{\alpha\mu}\, dt$$

$$\leq c \left(\int_0^T \|\nabla^2 \mathbf{v}^\epsilon\|_r^u\, F^{-\mu}\, dt\right)^\alpha \left(\int_0^T F^{\mu\frac{\alpha}{1-\alpha}}\, dt\right)^{1-\alpha}$$

$$\leq c(\mathbf{f}, \mathbf{v}_0, E)$$

as long as

$$\mu \frac{\alpha}{1-\alpha} \leq 1.$$

This condition is equivalent to (4.36). It can be checked that the right-hand side in (4.36) is always strictly larger than zero, if (4.23) or (4.24) are satisfied. In fact, one can compute that α ranges in the interval $(0, 1/3)$. Having at our disposal (4.35) we use the embedding

$$W^{2,r}(\Omega) \hookrightarrow W^{1+s,p_\infty}(\Omega) \tag{4.37}$$

with[9] $s = \frac{6 - p_\infty(p_0-1)}{2p_\infty}$, and the interpolation inequality

$$\|\mathbf{v}\|_{1+\sigma,p_\infty} \leq \|\mathbf{v}\|_{1,p_\infty}^{1-\frac{\sigma}{s}} \|\mathbf{v}\|_{1+s,p_\infty}^{\frac{\sigma}{s}}, \tag{4.38}$$

which holds for $0 < \sigma < s$ in order to show (4.31). For that we choose $\beta \in (1, p_\infty)$ and determine $\sigma > 0$ such that

$$\int_0^T \|\mathbf{v}^\epsilon\|_{1+\sigma,p_\infty}^\beta\, dt \leq c(\mathbf{f}, \mathbf{v}_0, E). \tag{4.39}$$

Using (4.38), (4.37) and Hölder's inequality with $\delta = \frac{p_\infty s}{\beta(s-\sigma)}$ we compute

$$\int_0^T \|\mathbf{v}^\epsilon\|_{1+\sigma,p_\infty}^\beta\, dt \leq \int_0^T \|\mathbf{v}^\epsilon\|_{1,p_\infty}^{\beta\frac{s-\sigma}{s}} \|\mathbf{v}^\epsilon\|_{1+s,p_\infty}^{\beta\frac{\sigma}{s}}\, dt$$

$$\leq \left(\int_0^T \|\mathbf{v}^\epsilon\|_{1,p_\infty}^{p_\infty}\, dt\right)^{\frac{1}{\delta}} \left(\int_0^T \|\mathbf{v}^\epsilon\|_{2,r}^{\beta\frac{\sigma}{s}\delta'}\, dt\right)^{\frac{1}{\delta'}}. \tag{4.40}$$

---

[9]Note, that $s > 0$ for p_∞, p_0 and r considered here.

The last inequality implies (4.39) due to (4.2) and (4.35) provided that

$$\beta \frac{\sigma}{s} \delta' = \alpha u,$$

or equivalently

$$1 = \frac{1}{\delta} + \frac{1}{\delta'} = \frac{\beta(s - \sigma)}{s\,p_\infty} + \frac{\beta \sigma}{s\,u\,\alpha},$$

which implies that

$$\sigma = \frac{s\,\alpha\,u(p_\infty - \beta)}{\beta(p_\infty - \alpha\,u)}. \tag{4.41}$$

One easily checks that for p_∞ and p_0 considered here and β chosen as above we obtain that $\delta \in (1, \infty)$ and $\sigma \in (0, s)$. The estimate (4.39) means that

$$\mathbf{v}^\varepsilon \text{ is bounded in } L^\beta(I, W^{1+\sigma, p_\infty}(\Omega)) \cap L^{p_\infty}(I, V_{p_\infty}),$$

where $\beta > 1$ and $\sigma > 0$, which together with (4.28), the Aubin-Lions lemma and (4.1)–(4.3) implies that there exists $\mathbf{v}$ such that

$$
\begin{aligned}
\nabla \mathbf{v}^\varepsilon &\to \nabla \mathbf{v} &\quad &\text{a.e. in } Q_T, \\
\mathbf{v}^\varepsilon &\rightharpoonup \mathbf{v} &\quad &\text{weakly in } E_{p(|\mathbf{E}|^2)}(Q_T) \cap L^{p_\infty}(I, V_{p_\infty}), \\
\mathbf{v}^\varepsilon &\overset{*}{\rightharpoonup} \mathbf{v} &\quad &\text{weakly* in } L^\infty(I, H), \\
\frac{\partial \mathbf{v}^\varepsilon}{\partial t} &\rightharpoonup \frac{\partial \mathbf{v}}{\partial t} &\quad &\text{weakly in } L^{q'}(I, V_q^*),
\end{aligned}
\tag{4.42}
$$

where q satisfies (4.27). From $(4.42)_1$ we deduce that (4.32) holds. Moreover, the estimates (4.1)–(4.3) and (4.28) remain valid for the limiting elements $\mathbf{v}$, $\frac{\partial \mathbf{v}}{\partial t}$, $\mathbf{S}(\mathbf{D}(\mathbf{v}), \mathbf{E})$ and $U(|\mathbf{D}(\mathbf{v})|^2, |\mathbf{E}|^2)$. From the weak formulation (3.7), integrated over $(0, T)$, we obtain

$$
\int_0^T \left\langle \frac{\partial \mathbf{v}^\varepsilon}{\partial t}, \boldsymbol{\varphi} \right\rangle_{V_q} dt + \int_0^T \int_\Omega \mathbf{S}(\mathbf{D}(\mathbf{v}^\varepsilon), \mathbf{E}) \cdot \mathbf{D}(\boldsymbol{\varphi}) + [\nabla \mathbf{v}^\varepsilon] \mathbf{v}_\varepsilon \cdot \boldsymbol{\varphi} \, dx \, dt
$$

$$
= \int_0^T \int_\Omega \mathbf{f} \cdot \boldsymbol{\varphi} - \chi^E \mathbf{E} \otimes \mathbf{E} \cdot \mathbf{D}(\boldsymbol{\varphi}) \, dx \, dt,
\tag{4.43}
$$

which holds for all $\boldsymbol{\varphi} \in \mathcal{D}(-\infty, T, \mathcal{V})$ and where q is given in (4.27). The limiting process as $\varepsilon \to 0$ in the first term is easy; in the second term we employ Vitali's theorem, which is possible due to $(4.32)_1$, (4.3) and the growth properties of $\mathbf{S}$; in the convective term we use the properties of the mollifier and that

$$
\begin{aligned}
\mathbf{v}^\varepsilon &\to \mathbf{v} &\quad &\text{strongly in } L^m(Q_T), \quad m < \frac{5}{3} p_\infty \\
\nabla \mathbf{v}^\varepsilon &\to \nabla \mathbf{v} &\quad &\text{strongly in } L^r(Q_T), \quad r < p_\infty,
\end{aligned}
$$

which follows from (4.42), the parabolic embedding 5.1.17 and Vitali's theorem. Via a standard argument (cf. Málek, Nečas, Rokyta, Růžička [70], Remark 5.3.66) we deduce from (4.43) that also the weak formulation (1.10) holds for almost all $t \in I$. This completes the proof of Theorem 1.14. ∎

5 Appendix

5.1 General Auxiliary Results

In this section we collect general definitions and results used in the previous chapters.

Definition 1.1. (Description of the boundary) *We say that a bounded domain[1] $\Omega \subset \mathbb{R}^3$ belongs to the class $C^{l,\beta}$, $l \in \mathbb{N}$, $\beta \in [0,1]$, if and only if there exist: N Cartesian coordinate systems X_n ($N \in \mathbb{N}$, $n = 1, \ldots, N$)*

$$X_n = (x_{n1}, x_{n2}, x_{n3}) = (x'_n, x_{n3}),$$

a number $\alpha > 0$ and N functions

$$a_n \in C^{l,\beta}([-\alpha, \alpha]^2),$$

such that

$$\frac{\partial a_n(0')}{\partial x_{ns}} = 0, \qquad s = 1, 2, \tag{1.2}$$

and sets

$$\Lambda^n = \{x_n = (x'_n, x_{n3}); |x'_n| \leq \alpha, x_{n3} = a_n(x'_n)\}$$
$$V^n_+ = \{x_n = (x'_n, x_{n3}); |x'_n| \leq \alpha, a_n(x'_n) < x_{n3} < a_n(x'_n) + \alpha\}$$
$$V^n_- = \{x_n = (x'_n, x_{n3}); |x'_n| \leq \alpha, a_n(x'_n) - \alpha < x_{n3} < a_n(x'_n)\}$$
$$V^n = V^n_+ \cup \Lambda^n \cup V^n_-$$

with the properties

$$\Lambda^n \subset \partial\Omega, \quad V^n_+ \subset \Omega, \quad V^n_- \subset \mathbb{R}^3 \backslash \Omega, \quad \bigcup_{n=1}^{N} \Lambda^n = \partial\Omega.$$

Proposition 1.3. (Korn's inequality) *Let $1 < q < \infty$ and let $\Omega \subset \mathbb{R}^d$ be of class C^1. Then there exists a constant $K_q = K_q(\Omega)$ such that the inequality*

$$K_q \|\mathbf{v}\|_{1,q} \leq \|\mathbf{D}(\mathbf{v})\|_q \tag{1.4}$$

is fulfilled for all $\mathbf{v} \in W_0^{1,q}(\Omega)$.

PROOF : see e.g. Nečas [92]. ∎

---

[1] For simplicity we restrict ourselves to the three-dimensional case.

Lemma 1.5. *Let* $1 < q < \infty$ *and let* $\Omega \subset \mathbb{R}^d$ *be a domain of class* C^1. *Then there exists a constant* $c_q = c_q(\Omega)$ *such that for all* $\mathbf{v} \in W_0^{1,q}(\Omega) \cap W^{2,q}(\Omega)$

$$c_q \|\nabla^2 \mathbf{v}\|_q \leq \|\mathbf{D}(\nabla \mathbf{v})\|_q. \tag{1.6}$$

PROOF :   The assertion follows immediately from the algebraic identity

$$\frac{\partial^2 v_i}{\partial x_j \partial x_k} = D_{ik}\left(\frac{\partial \mathbf{v}}{\partial x_j}\right) + D_{ij}\left(\frac{\partial \mathbf{v}}{\partial x_k}\right) - D_{jk}\left(\frac{\partial \mathbf{v}}{\partial x_i}\right). \tag{1.7}$$

∎

Theorem 1.8. (On negative norms) *Let* $1 < q < \infty$ *and let* $\mathbf{v} \in L^q(\Omega)$. *Then there exists a constant such that*

$$c\|\mathbf{v}\|_q \leq \|\mathbf{v}\|_{-1,q} + \|\nabla \mathbf{v}\|_{-1,q}. \tag{1.9}$$

PROOF :   see e.g. Nečas [92].   ∎

Theorem 1.10. *Let* $\mathbf{F} \in (W_0^{1,q}(\Omega))^*$ *be such that*

$$\langle \mathbf{F}, \boldsymbol{\varphi}\rangle_{1,q} = 0, \qquad \forall \boldsymbol{\varphi} \in V_q.$$

Then there exists a unique $\phi \in L^{q'}(\Omega)$, *with* $\int_\Omega \phi \, dx = 0$ *such that*

$$\langle \mathbf{F}, \boldsymbol{\varphi}\rangle_{1,q} = -\int_\Omega \phi \operatorname{div} \boldsymbol{\varphi} \, dx \qquad \forall \boldsymbol{\varphi} \in W^{1,q}(\Omega). \tag{1.11}$$

PROOF :   see e.g. De Rham [109], Bogovskii [15] or Amrouche, Girault [5].   ∎

Theorem 1.12. (Vitali) *Let* Ω *be a bounded domain and let* $f_n : \Omega \to \mathbb{R}$ *be integrable functions such that:*

(i) $\lim_{n\to\infty} f_n(x)$ *exists and is finite for almost all* $x \in \Omega$;

(ii) for all $\varepsilon > 0$ *there exists* $\delta > 0$ *such that*

$$\sup_{n\in\mathbb{N}} \int_G |f_n(x)| \, dx < \varepsilon \qquad \forall G \subset \Omega, |G| < \delta.$$

Then we have

$$\lim_{n\to\infty} \int_\Omega f_n(x) \, dx = \int_\Omega \lim_{n\to\infty} f_n(x) \, dx.$$

PROOF :   see e.g. Kufner, John, Fučík [57].   ∎

Proposition 1.13. (Embedding) *Let* $\Omega \subset \mathbb{R}^d$ *be a bounded domain,* $\partial\Omega \in C^1$ *and let* $0 \leq j < k$, $1 \leq q < \infty$. *If* $\frac{1}{m} \equiv \frac{1}{q} - \frac{k-j}{d}$, *we have*

$$W^{k,q}(\Omega) \hookrightarrow W^{j,m}(\Omega). \tag{1.14}$$

PROOF :   see e.g. Kufner, John, Fučík [57].   ∎

Proposition 1.15. (Interpolation) *Let* $1 \leq q_2 \leq q \leq q_1 \leq \infty$ *and let* $\mathbf{v} \in L^{q_1}(\Omega) \cap L^{q_2}(\Omega)$. *Then we have*

$$\|\mathbf{v}\|_q \leq \|\mathbf{v}\|_{q_1}^{\alpha} \|\mathbf{v}\|_{q_2}^{1-\alpha}, \tag{1.16}$$

where $\frac{1}{q} = \frac{\alpha}{q_1} + \frac{1-\alpha}{q_2}$, $\alpha \in [0,1]$.

PROOF :   The assertion follows directly from Hölder's inequality.   ∎

Lemma 1.17. (Parabolic Embedding) *Let* $f \in L^{\infty}(I, L^2(\Omega))$, $\nabla f \in L^r(I, L^r(\Omega))$, $r > 1$. *Then we have*

$$\int_0^T \|f\|_{\alpha}^{\alpha} \, dt \leq \left(\operatorname*{ess\,sup}_I \|f\|_2^2 \right)^{\frac{2}{3}} \int_0^T \|\nabla f\|_r^r \, dt + T \operatorname*{ess\,sup}_I \|f\|_2^{\alpha}, \tag{1.18}$$

whenever

$$\alpha = \frac{5}{3} r. \tag{1.19}$$

PROOF :   This follows from the interpolation inequality

$$\|f\|_{\alpha} \leq \|f\|_2^{1-\lambda} \|f\|_{1,r}^{\lambda} \leq \|f\|_2 + \|f\|_2^{1-\lambda} \|\nabla f\|_r^{\lambda}$$

with $\lambda = \frac{3r(\alpha-2)}{\alpha(5r-6)}$. Requiring $\lambda\alpha = r$ yields (1.18).   ∎

Lemma 1.20. (Aubin-Lions) *Let* $1 < \alpha, \beta < \infty$ *and let* X *be a Banach space. Further, let* X_0, X_1 *be separable, reflexive Banach spaces. Provided* $X_0 \hookrightarrow\hookrightarrow X \hookrightarrow X_1$ *we have*

$$\left\{ v \in L^{\alpha}(I, X_0), \frac{dv}{dt} \in L^{\beta}(I, X_1) \right\} \hookrightarrow\hookrightarrow L^{\alpha}(I, X).$$

PROOF :   see e.g. Lions [68] Section 1.5.   ∎

Lemma 1.21. *Let* $f, g \in L^q(\Omega)$, $q < 2$. *Then we have*

$$\left(\int_{\Omega} |g|^q \, dx \right)^{\frac{2}{q}} \leq c \, \|1 + |f|^2\|_{\frac{q}{2}}^{\frac{2-q}{2}} \int_{\Omega} (1 + |f|^2)^{\frac{q-2}{2}} |g|^2 \, dx. \tag{1.22}$$

PROOF :   Inequality (1.22) follows from Hölder's inequality.   ∎

Lemma 1.23. *For all* $s > 1$ *there exist constants* $c(s), \tilde{c}(s)$ *such that for all* $x, y \in \mathbb{R}^n$

$$c\big(1 + |x|^2 + |y|^2\big)^{\frac{s}{2}} \leq \int_0^1 \big(1 + |(1-\lambda)x + \lambda y|^2\big)^{\frac{s}{2}} \, d\lambda \leq \tilde{c}\big(1 + |x|^2 + |y|^2\big)^{\frac{s}{2}} \tag{1.24}$$

PROOF :   see e.g. Giusti [39] Lemma 8.3.   ∎

Proposition 1.25. *Let $\partial\Omega \in C^{0,1}$ be a bounded domain and assume $q > 1$ and $\mathbf{f} \in L_0^q(\Omega)$ are given. Then there exists a solution $\mathbf{u} \in W_0^{1,q}(\Omega)$ of the problem*

$$\operatorname{div} \mathbf{u} = f \qquad in \ \Omega$$

which satisfies the estimate

$$\|\nabla \mathbf{u}\|_q \leq C \|f\|_q.$$

PROOF :   see e.g. Bogovskii [14], [15] or Galdi [36].   ∎

The Method of Difference Quotient

For the regularity in the interior and near the boundary we used the method of difference quotient. Here we fix the notation and state some relevant results. Using the definition of the boundary we finally choose sets Ω_0^n covering $\partial\Omega$ such that $\Omega_0^n \subset V^n = V_-^n \cup \Lambda^n \cup V_+^n$, $\operatorname{dist}(\partial\Omega_0^n, \partial V^n) \geq h_0 > 0$. Let us fix n and drop for simplicity the index n. Setting

$$\hat{e}^1 \equiv (1,0) \qquad and \qquad \hat{e}^2 \equiv (0,1),$$

we can define for $s = 1, 2$ and $h \in (0, h_0)$ the translation mapping $T = T_{s,h} : \Omega_0 \to V$ by

$$x \to (x' + h\hat{e}^s, x_3 + a(x' + h\hat{e}^s) - a(x')) \equiv y. \tag{1.26}$$

Then the inverse mapping T^{-1} is given by $(x = T^{-1}(y))$

$$y \to (y' - h\hat{e}^s, x_3 + a(y' - h\hat{e}^s) - a(y')).$$

Put

$$\Delta^{\pm} a(x') = a(x' \pm h\hat{e}^s) - a(x').$$

Then

$$\left(\frac{\partial T_i}{\partial x_j}(x)\right)_{i,j=1,2,3} = \begin{pmatrix} 1 & 0 & 0 \\ 0 & 1 & 0 \\ \frac{\partial \Delta^+ a}{\partial x_1}(x') & \frac{\partial \Delta^+ a}{\partial x_2}(x') & 1 \end{pmatrix} \tag{1.27}$$

and

$$\left(\frac{\partial T_i^{-1}}{\partial y_j}(y)\right)_{i,j=1,2,3} = \begin{pmatrix} 1 & 0 & 0 \\ 0 & 1 & 0 \\ \frac{\partial \Delta^- a}{\partial y_1}(y') & \frac{\partial \Delta^- a}{\partial y_2}(y') & 1 \end{pmatrix}. \tag{1.28}$$

For both matrices (1.27) and (1.28) the determinant is equal to 1. The s-th tangential derivative $(s = 1, 2)$ of any (scalar, vector or tensorial) function g, denoted $\frac{\partial g}{\partial \tau^s}$, is defined by

$$\frac{\partial g}{\partial \tau^s}(x) \equiv \lim_{h \to 0} \frac{g(Tx) - g(x)}{h},$$

and

$$\frac{\partial g}{\partial \tau^s}(x) = \frac{\partial g}{\partial x_s}(x) + \frac{\partial g}{\partial x_3}(x)\frac{\partial a}{\partial x_s}(x').$$ (1.29)

holds. For the readers convenience let us show that if $g \in W_0^{1,p}(\Omega)$, $p > 1$, then for all $h \in (0, h_0)$

$$\int_{\Omega_0} \left|\frac{g(Tx) - g(x)}{h}\right|^p dx \le c(a)\|\nabla g\|_p^p.$$ (1.30)

Indeed, setting

$$T_\lambda(x) = (x' + \lambda h\hat{e}^s, x_3 + a(x' + \lambda h\hat{e}^s) - a(x')),$$ (1.31)

we can write

$$\int_{\Omega_0} \left|\frac{g(Tx) - g(x)}{h}\right|^p dx = \int_{\Omega_0} \left|\int_0^1 \frac{\partial g(T_\lambda(x))}{\partial \tau^s} d\lambda\right|^p dx$$

$$\le \int_{\Omega_0} \int_0^1 \left|\frac{\partial g(T_\lambda(x))}{\partial \tau^s}\right|^p d\lambda \, dx$$

$$\le \int_0^1 \int_\Omega \left|\frac{\partial g(y)}{\partial \tau^s}\right|^p dy \, d\lambda \stackrel{(1.29)}{\le} c(a)\|\nabla g\|_p^p.$$

On the other hand, if $g \in L^p(\Omega)$ and if for all $h \in (0, h_0)$

$$\int_\Omega \left|\frac{g(Tx) - g(x)}{h}\right|^p dx \le c_0 < \infty,$$ (1.32)

then $\frac{\partial g}{\partial \tau^s}$ exists (for $s = 1, 2$) in the sense of distributions, and

$$\int_\Omega \left|\frac{\partial g}{\partial \tau^s}\right|^p dx \le c_0.$$ (1.33)

Let $V' \subset\subset \Omega_0 \subset\subset \Omega$, be such that $\text{dist}(\partial\Omega_0, \partial\Omega) = h_0 > 0$. Let $\hat{e}^r$, $r = 1, 2, 3$, be a basis of a coordinate system in $\mathbb{R}^3$. For $r = 1, 2, 3$ and $h \in (0, h_0)$ we define the translation mapping $T_{k,h} : \Omega_0 \to \Omega$ by

$$x \to x + h\hat{e}^k.$$ (1.34)

The gradient of T and its inverse T^{-1} have determinant equal to 1. The properties (1.30)–(1.33) hold correspondingly. Moreover, we have

$$\frac{g(T_k x) - g(x)}{h} \to \frac{\partial g}{\partial x_k}(x) \qquad \text{strongly in } L^p_{\text{loc}}(\Omega).$$ (1.35)

5.2   Auxiliary Results for the Approximations

In this section we prove the technical assertions used in the previous chapters for deriving estimates independent of A and ε. Note, that all results can be used for the approximations $\mathbf{S}^A$ of the extra stress tensor $\mathbf{S}$ both in the steady case (cf. (3.3.13)) and in the unsteady case (cf. (4.1.24)) .

Recall the notations (cf. (3.5.2), (4.3.2), (4.3.3))

$$I_{p,r,A}(\mathbf{u},\xi) = \int\limits_{\Omega} \left(1+|\mathbf{E}|^2\right)^r\left(1+|\mathbf{D}(\mathbf{v})|^2\right)^{\frac{q-2}{2}r}|\mathbf{u}|^r\xi^r \times \tag{2.1}$$

$$\times \left\{\left(1+|\mathbf{D}(\mathbf{v})|^2\right)^{\frac{p-q}{2}r}\chi_A + \left(1+A^2\right)^{\frac{p-q}{2}r}(1-\chi_A)\right\}dx\,,$$

and

$$I_{p,A}(\mathbf{u},\xi) = \int\limits_{\Omega} \left(1+|\mathbf{E}|^2\right)\left(1+|\mathbf{D}(\mathbf{v})|^2\right)^{\frac{q-2}{2}}|\mathbf{u}|^2\xi^2 \times \tag{2.2}$$

$$\times \left\{\left(1+|\mathbf{D}(\mathbf{v})|^2\right)^{\frac{p-q}{2}}\chi_A + \left(1+A^2\right)^{\frac{p-q}{2}}(1-\chi_A)\right\}dx\,,$$

where $\mathbf{u}$ is usually some second order derivative of $\mathbf{v}$ and ξ will be either identical 1 or one of the cut-off functions ξ_0 or ξ_1.

Lemma 2.3. *For $1 < r < \infty$, there exist constants $c, \tilde{c}$ independent of A such that*

$$c\,I_{p,r,A}(\mathbf{D}(\nabla\mathbf{v}),\xi_i) \leq \int\limits_{\Omega} |\partial_{kl}\mathbf{S}^A(\mathbf{D}(\mathbf{v}),\mathbf{E})D_{kl}(\nabla\mathbf{v})|^r\xi_i^r\,dx \tag{2.4}$$

$$\leq \tilde{c}\,I_{p,r,A}(\mathbf{D}(\nabla\mathbf{v}),\xi_i) \qquad i = 0,1\,.$$

PROOF :   Since $I_{p,r,A}(\mathbf{D}(\nabla\mathbf{v}),\xi_i)$, $i = 1,2$ is the sum of integrals over Ω_A and $\Omega\setminus\Omega_A$, we will show the above inequalities only in the case $\Omega\setminus\Omega_A$. For brevity we omit the index i at ξ_i. On the set $\Omega\setminus\Omega_A$ we have for the integrand of $I_{p,r,A}(\mathbf{D}(\nabla\mathbf{v}),\xi)$

$$\left((1+|\mathbf{E}|^2)(1+|\mathbf{D}(\mathbf{v})|^2)^{\frac{q-2}{2}}(1+A^2)^{\frac{p-q}{2}}|\mathbf{D}(\nabla\mathbf{v})|^2\right)^{\frac{r}{2}} \times$$

$$\times \left((1+|\mathbf{E}|^2)(1+|\mathbf{D}(\mathbf{v})|^2)^{\frac{q-2}{2}}(1+A^2)^{\frac{p-q}{2}}\right)^{\frac{r}{2}}\xi^r\,dx$$

$$\leq c\left(\partial_{kl}S_{ij}^A(\mathbf{D}(\mathbf{v}),\mathbf{E})D_{ij}(\nabla\mathbf{v})D_{kl}(\nabla\mathbf{v})\right)^{\frac{r}{2}} \times$$

$$\times \left((1+|\mathbf{E}|^2)(1+|\mathbf{D}(\mathbf{v})|^2)^{\frac{q-2}{2}}(1+A^2)^{\frac{p-q}{2}})\right)^{\frac{r}{2}}\xi^r\,,$$

where we used (3.3.25) and (3.3.56), respectively. Thus Young's inequality delivers

$$I_{p,r,A}(\mathbf{D}(\nabla\mathbf{v}),\xi(1-\chi_A)) \tag{2.5}$$

$$\leq c \int_{\Omega\setminus\Omega_A} |\partial_{kl}\mathbf{S}^A(\mathbf{D}(\mathbf{v}),\mathbf{E})D_{kl}(\nabla\mathbf{v})|^{\frac{r}{2}}\xi^{\frac{r}{2}} \times$$

$$\times \left((1+|\mathbf{E}|^2)(1+|\mathbf{D}(\mathbf{v})|^2)^{\frac{q-2}{2}}(1+A^2)^{\frac{p-q}{2}}|\mathbf{D}(\nabla\mathbf{v})|\xi\right)^{\frac{r}{2}} dx$$

$$\leq \frac{1}{2}I_{p,r,A}(\mathbf{D}(\nabla\mathbf{v}),\xi(1-\chi_A)) + \int_{\Omega\setminus\Omega_A} |\partial_{kl}\mathbf{S}^A(\mathbf{D}(\mathbf{v}),\mathbf{E})D_{kl}(\nabla\mathbf{v})|^r\xi^r dx.$$

This proves the first inequality in (2.4). The second one follows easily from (3.3.27) and (3.3.58), respectively. ■

Remark 2.6. From the proof of Lemma 2.3 it is clear that we could also allow $r = r(x) \in (1,\infty)$. Moreover, it is possible to take $\xi \equiv 1$ in Lemma 2.3.

Lemma 2.7. *There exists a constant c such that*

$$\|\nabla^2\mathbf{v}\,\xi_0\|_2^2 \leq cI_{p,A}(\mathbf{D}(\nabla\mathbf{v}),\xi_0) \tag{2.8}$$

if $q = 2$, and

$$\|\nabla\mathbf{v}\,\xi_0^{\frac{2}{q}}\|_{3q}^q \leq c\big(1 + I_{p,A}(\mathbf{D}(\nabla\mathbf{v}),\xi_0) + \|\nabla\mathbf{v}\|_q^q\big) \tag{2.9}$$

if $q < 2$.

PROOF : Inequality (2.8) follows immediately from the definition of $I_{p,A}$ and the algebraic identity (1.7). For $q < 2$ we notice

$$\frac{\partial}{\partial x_k}(1+|\mathbf{D}|^2)^{\frac{q}{4}} = \frac{q}{2}(1+|\mathbf{D}|^2)^{\frac{q-4}{4}}D_{ij}(\mathbf{v})D_{ij}\left(\frac{\partial\mathbf{v}}{\partial x_k}\right),$$

which immediately implies

$$\|\nabla(1+|\mathbf{D}|^2)^{\frac{q}{4}}\xi_0\|_2^2 \leq cI_{p,A}(\mathbf{D}(\nabla\mathbf{v}),\xi_0).$$

From the embedding $W^{1,2}(\Omega) \hookrightarrow L^6(\Omega)$ we obtain

$$\|\nabla(1+|\mathbf{D}|^2)^{\frac{q}{4}}\xi_0\|_2^2 \geq c\|\nabla\mathbf{v}\,\xi_0^{\frac{2}{q}}\|_{3q}^q - c(\nabla\xi_0)\big(1+\|\nabla\mathbf{v}\|_q^q\big),$$

where we also used Korn's inequality and some straightforward computations. Inequality (2.9) follows from the last two inequalities. ■

Lemma 2.10. *Let $r \leq 2$ if $p_0 < 2$ and $r \leq p_0'$ if $p_0 \geq 2$. Then there exists a constant $c = c(\mathbf{E}_0)$ independent of A such that for $i = 0,1$*

$$I_{p,r,A}(\mathbf{D}(\nabla\mathbf{v}),\xi_i) \tag{2.11}$$

$$\leq c(\mathbf{E}_0)\big(I_{p,A}(\mathbf{D}(\nabla\mathbf{v}),\xi_i)\big)^{\frac{r}{2}}\left(\int_\Omega 1 + \partial U^A(|\mathbf{D}(\mathbf{v})|^2,|\mathbf{E}|^2)\cdot\mathbf{D}(\mathbf{v})\,d\nu\right)^{\frac{2-r}{2}}.$$

PROOF :   Again we will only consider the case $|\mathbf{D}(\mathbf{v})| \geq A$. For $p \leq p_0 < 2$ we see

$$(1 + |\mathbf{D}|^2)^{\frac{q-2}{2}} (1 + A^2)^{\frac{p-q}{2}} \leq (1 + |\mathbf{D}|^2)^{\frac{p-2}{2}} \leq 1 \tag{2.12}$$

and in this case inequality (2.11) follows immediately from (2.5). If $p_0 \geq 2$ we get from (2.5) and Hölder's inequality

$$I_{p,r,A}(\mathbf{D}(\nabla\mathbf{v}), \xi(1 - \chi_A))$$

$$\leq c \Big(\int_{\Omega \backslash \Omega_A} \partial_{kl} S_{ij}^A(\mathbf{D}(\mathbf{v}), \mathbf{E}) D_{ij}(\nabla\mathbf{v}) D_{kl}(\nabla\mathbf{v}) \xi^2 \, dx \Big)^{\frac{r}{2}} \times$$

$$\times \Big(\int_{\Omega \backslash \Omega_A} \{ (1 + |\mathbf{E}|^2)(1 + A^2)^{\frac{p-q}{2}} (1 + |\mathbf{D}(\mathbf{v})|^2)^{\frac{q-2}{2}} \}^{\frac{r}{2-r}} \, dx \Big)^{\frac{2-r}{2}}. \tag{2.13}$$

The expression in the squiggly brackets can be written as

$$(1 + |\mathbf{E}|^2) \{ (1 + A^2)^{\frac{p-q}{2}} (1 + |\mathbf{D}|^2)^{\frac{q}{2}} \}^{\frac{p-2}{p}} \Big(\frac{1 + A^2}{1 + |\mathbf{D}|^2} \Big)^{\frac{p-q}{p}}$$

$$\leq c(1 + |\mathbf{E}|^2) \{ 1 + \partial U^A(|\mathbf{D}|^2, |\mathbf{E}|^2) \cdot \mathbf{D} \}^{\frac{p-2}{p}}, \tag{2.14}$$

where we used (3.3.29) and (3.3.60), respectively, $|\mathbf{D}| \geq A$ and $p \geq q$. On the set where $p(|\mathbf{E}|^2) \leq 2$, the last term is bounded by a constant, while on the set where $p(|\mathbf{E}|^2) > 2$ we require

$$\frac{p-2}{p} \frac{r}{2-r} \leq 1, \tag{2.15}$$

which is equivalent to $r \leq p_0'$. Therefore the second integral in (2.13) is estimated by

$$c(\mathbf{E}_0) \int_{\Omega \backslash \Omega_A} 1 + \partial U^A(|\mathbf{D}|^2, |\mathbf{E}|^2) \cdot \mathbf{D} \, d\nu,$$

which yields $(2.11)^2$.   ∎

Lemma 2.16. *There exists a constant $c = c(\mathbf{E}_0)$ such that, if $p_0 < 2$*

$$\int_\Omega |\partial_n \mathbf{S}^A \nabla E_n|^r \xi_i^r \, dx \leq c(\mathbf{E}_0) \Big(1 + \|\partial U^A \cdot \mathbf{D} \xi_i\|_1 \Big) \qquad i = 0, 1 \tag{2.17}$$

holds for $r \leq 2$, and if $p_0 \geq 2$,

$$\int_\Omega |\partial_n \mathbf{S}^A \nabla E_n|^r \xi_i^r \, dx \leq c(\mathbf{E}_0) \Big(1 + \|\nabla\mathbf{v}\|_q^s \|\mathbf{S}^A \xi_i\|_{\frac{qs}{q-s}}^s \Big) \qquad i = 0, 1 \tag{2.18}$$

holds for $r \leq p_0'$ and $s > 1$.

---

[2] Note, that in the case $|\mathbf{D}(\mathbf{v})| \leq A$, one must divide the set Ω_A into two sets $\{x, |\mathbf{D}(\mathbf{v})| \leq 1\}$ and $\{x, 1 \leq |\mathbf{D}(\mathbf{v})| \leq A\}$. The first one is trivial and the second one can be handled as in the case $|\mathbf{D}(\mathbf{v})| \leq A$.

PROOF : Using (3.3.44) and (3.3.68), respectively, and a similar argument as in (2.14) we obtain, for $s > 1$,

$$\int_\Omega |\partial_n \mathbf{S}^A \nabla E_n|^r \xi^r \, dx \leq c(\mathbf{E}_0) \int_\Omega \left(1 + \partial U^A \cdot \mathbf{D}\right)^{sr \frac{p_0-1}{p_0}} \xi^r \, dx, \tag{2.19}$$

where we used $\frac{p-1}{p} \leq \frac{p_0-1}{p_0}$. Now, for $p_0 < 2$ we choose s such that $s\frac{p_0-1}{p_0} \leq 1/2$ and (2.17) follows. For $p_0 \geq 2$ and $r \leq p_0'$ we have $r\frac{p_0-1}{p_0} \leq 1$. Moreover, due to (3.3.28), (3.3.24) and (3.3.59), (3.3.55), respectively, it holds

$$|\partial U^A(|\mathbf{D}(\mathbf{v})|^2, |\mathbf{E}|^2) \cdot \mathbf{D}(\mathbf{v})| \leq c(\mathbf{E}_0)|\mathbf{S}^A(\mathbf{D}(\mathbf{v}), \mathbf{E})| \, |\mathbf{D}(\mathbf{v})|. \tag{2.20}$$

Therefore the right-hand side of (2.19) is estimated by

$$c(\mathbf{E}_0)\left(1 + \|\nabla \mathbf{v}\|_q^s \|\mathbf{S}^A \xi\|_{\frac{qs}{q-s}}^s\right), \tag{2.21}$$

which gives (2.18). ∎

Remark 2.22. From the proof of (2.17) and (2.18) it is clear that also global versions of these inequalities hold, i.e. $\xi \equiv 1$ is admissible.

Proposition 2.23. *There exists a constant* $c = c(\mathbf{E}_0)$ *such that*

$$\|\nabla \mathbf{S}^A \xi_i\|_2^2 \leq c(\mathbf{E}_0)\left(1 + I_{p,A}(\mathbf{D}(\nabla \mathbf{v}), \xi_i) + \|\partial U^A \cdot \mathbf{D}\, \xi_i\|_1\right) \qquad i = 0, 1, \tag{2.24}$$

if $p_0 < 2$*, and for* $s > 1$

$$\|\nabla \mathbf{S}^A \xi_i\|_{p_0'}^2 \leq c(\mathbf{E}_0) I_{p,A}(\mathbf{D}(\nabla \mathbf{v}), \xi_i)\left(1 + \|\partial U^A \cdot \mathbf{D}\|_1\right)^{\frac{p_0-2}{p_0}} \tag{2.25}$$

$$+ c(\mathbf{E}_0)\left(1 + \|\mathbf{S}^A(\mathbf{D}(\mathbf{v}), \mathbf{E})\xi_i\|_{sq'}\|\nabla \mathbf{v}\|_q\right)^{\frac{2s}{p_0}} \qquad i = 0, 1,$$

if $p_0 \geq 2$*.*

PROOF : For $i = 0, 1$ we observe

$$\int_\Omega |\nabla \mathbf{S}^A|^r \xi_i^r \, dx \leq \int_\Omega |\partial_{kl} \mathbf{S}^A D_{kl}(\nabla \mathbf{v})|^r \xi_i^r + |\partial_n \mathbf{S}^A \nabla E_n|^r \xi_i^r \, dx \tag{2.26}$$

$$= I_1 + I_2.$$

For $p_0 < 2$ we use (2.4) and (2.11) for $r = 2$ to obtain

$$I_1 \leq c(\mathbf{E}_0) I_{p,A}(\mathbf{D}(\nabla \mathbf{v}), \xi_i)$$

and from (2.17) we conclude for $r = 2$

$$I_2 \leq c(\mathbf{E}_0)\left(1 + \|\partial U^A \cdot \mathbf{D}\, \xi_i\|_1\right)$$

which immediately proves (2.24). In the case $p_0 \geq 2$ we put $r = p_0'$ and use (2.4) and (2.11) to obtain

$$I_1 \leq c(\mathbf{E}_0)\big(I_{p,A}(\mathbf{D}(\nabla \mathbf{v}), \xi_i)\big)^{\frac{p_0'}{2}}\Big(1 + \|\partial U^A \cdot \mathbf{D}\|_1\Big)^{\frac{p_0-2}{2(p_0-1)}}$$

and from (2.18) we deduce

$$I_2 \leq c(\mathbf{E}_0)\big(1 + \|\mathbf{S}^A \xi_i\|_{q's}^s \|\nabla \mathbf{v}\|_q^s\big),$$

and inequality (2.25) follows.   ∎

Lemma 2.27. *There exists a constant independent of A such that*

$$\|\nabla^2 \mathbf{v}\|_r^r \leq c I_{p,r,A}(\mathbf{D}(\nabla \mathbf{v}), 1) \tag{2.28}$$

if $p_\infty \geq 2$ and

$$\|\nabla^2 \mathbf{v}\|_s^{r(p_\infty-1)} \leq c\big(1 + \|\nabla \mathbf{v}\|_s^{r(p_\infty-1)} + I_{p,r,A}(\mathbf{D}(\nabla \mathbf{v}), 1)\big) \tag{2.29}$$

if $p_\infty < 2$, where $s = \frac{3r(p_\infty-1)}{3-r(2-p_\infty)}$.

PROOF : Inequality (2.28) follows immediately from the definition of $I_{p,r,A}$, observing that in this case $q = 2$. Let us therefore assume $p_\infty = q < 2$. We have, again using the algebraic identity (1.7),

$$\int_\Omega |\nabla^2 \mathbf{v}|^s \, dx \leq \int_\Omega \Big((1 + |\mathbf{D}(\mathbf{v})|^2)^{\frac{q-2}{2}r}|\mathbf{D}(\nabla \mathbf{v})|^r\Big)^{\frac{s}{r}}(1 + |\mathbf{D}(\mathbf{v})|)^{(2-q)s} \, dx$$

$$\leq c\big(I_{p,r,A}(\mathbf{D}(\nabla \mathbf{v}), 1)\big)^{\frac{s}{r}}\Big(\int_\Omega (1 + |\mathbf{D}(\mathbf{v})|)^{(2-q)\frac{sr}{r-s}} \, dx\Big)^{\frac{r-s}{r}}.$$

Requiring now

$$(2-q)\frac{sr}{r-s} = \frac{3s}{3-s},$$

which is equivalent to

$$s = \frac{3r(q-1)}{3-r(2-q)},$$

and using

$$\|1 + |\mathbf{D}(\mathbf{v})|\|_{\frac{3s}{3-s}} \leq c\big(1 + \|\mathbf{D}(\mathbf{v})\|_s + \|\nabla^2 \mathbf{v}\|_s\big),$$

which follows from Sobolev's embedding theorem, we obtain

$$\|\nabla^2 \mathbf{v}\|_s^s \leq c\big(I_{p,r,A}(\mathbf{D}(\nabla \mathbf{v}), 1)\big)^{\frac{s}{r}}\big(1 + \|\mathbf{D}(\mathbf{v})\|_s + \|\nabla^2 \mathbf{v}\|_s\big)^{s(2-q)}$$

$$\leq c\big(I_{p,r,A}(\mathbf{D}(\nabla \mathbf{v}), 1)\big)^{\frac{s}{r(q-1)}} + \tfrac{1}{2}\big(1 + \|\mathbf{D}(\mathbf{v})\|_s^s + \|\nabla^2 \mathbf{v}\|_s^s\big),$$

where we used Young's inequality. The last inequality implies (2.29).   ∎

Corollary 2.30. *Let* $p_\infty < 2 \le p_0$, *and* $s = \frac{3p_0(p_\infty-1)}{p_0(1+p_\infty)-3}$. *Then*

$$\|\nabla^2\mathbf{v}\|_s^{p_0'(p_\infty-1)} \le c\big(1 + \|\nabla\mathbf{v}\|_{p_\infty}^{p_\infty} + I_{p,p_0',A}(\mathbf{D}(\nabla\mathbf{v}),1)\big).$$

PROOF :   We use inequality (2.29) and observe that $p_0'(p_\infty-1) \le p_\infty$. Now, if $s \le p_\infty$ the assertion follows immediately, and if $s > p_\infty$ we interpolate between p_∞ and $\frac{3s}{3-s}$ and use Young's inequality.   ∎

Analogously to (2.1) we introduce the notation

$$I_{p,r}(\mathbf{D}(\nabla\mathbf{v}),\xi) = \int_\Omega (1+|\mathbf{E}|^2)^r(1+|\mathbf{D}(\mathbf{v})|^2)^{\frac{p-2}{2}r}|\mathbf{D}(\nabla\mathbf{v})|^r\xi^r\,dx, \qquad (2.31)$$

where ξ is either identical 1 or $\xi = \xi_0 \in \mathcal{D}(\Omega)$. In the same way as in Lemma 2.3 we prove

Lemma 2.32. *There exists a constant such that*

$$I_{p,r}(\mathbf{D}(\nabla\mathbf{v}),\xi) \le c\int_\Omega |\partial_{kl}\mathbf{S}(\mathbf{D}(\mathbf{v}),\mathbf{E})D_{kl}(\nabla\mathbf{v})|^r\xi^r\,dx, \qquad (2.33)$$

where $\xi \equiv 1$ *or* $\xi \in \mathcal{D}(\Omega)$.

Lemma 2.34. *There exists a constant* $c = c(\mathbf{E}_0)$ *such that for* $p_\infty \ge 2$ *we have*

$$\big(I_{p,r}(\mathbf{D}(\nabla\mathbf{v}),1)\big)^{\frac{u}{r}} \le c\Big(\|\mathbf{S}\|_1^u + \|\nabla\mathbf{S}\|_r^u + \int_\Omega \mathbf{S}\cdot\mathbf{D}\,dx\Big). \qquad (2.35)$$

PROOF : Similarly as in (4.3.37) one can show

$$\|\partial_n\mathbf{S}\|_r \le c\Big(\int_\Omega \mathbf{S}^A\cdot\mathbf{D}\,dx\Big)^{\frac{1}{2}}(1+\|\mathbf{S}\|_1+\|\nabla\mathbf{S}\|_r)^{\frac{p_0-2}{p_0-1}}. \qquad (2.36)$$

The definition of $I_{p,r}$ and the chain rule applied to $\nabla\mathbf{S}$ implies

$$\big(I_{p,r}(\mathbf{D}(\nabla\mathbf{v}),1)\big)^{\frac{1}{r}} \le c\Big(\|\nabla\mathbf{S}^A\|_r + \|\partial_n\mathbf{S}\|_r\Big) \qquad (2.37)$$

$$\le c\Big(\|\nabla\mathbf{S}\|_r + \Big(\int_\Omega \mathbf{S}\cdot\mathbf{D}\,dx\Big)^{\frac{1}{2}}(1+\|\mathbf{S}\|_1+\|\nabla\mathbf{S}\|_r)^{\frac{p_0-2}{p_0-1}}\Big)$$

$$\le c\Big(1+\|\mathbf{S}\|_1+\|\nabla\mathbf{S}\|_r + \Big(\int_\Omega \mathbf{S}\cdot\mathbf{D}\,dx\Big)^{\frac{p_0-2}{2}}\Big).$$

This inequality raised to the power $u = \frac{2}{p_0-1}$ gives (2.35).   ∎

We also obtain the analogue of Lemma 2.16.

Lemma 2.38. *There exists a constant $c = c(\mathbf{E}_0)$ such that for $p_\infty \geq 2$ we have, for $s > 1$,*

$$\int_\Omega |\partial_n \mathbf{S}(\mathbf{D}(\mathbf{v}), \mathbf{E}) \nabla E_n|^{p_0'} \, dx \leq c \left(1 + \|\nabla \mathbf{v}\|_{p_\infty}^s \|\mathbf{S}(\mathbf{D}(\nabla \mathbf{v}), \mathbf{E})\|_{\frac{p_\infty s}{p_\infty - s}}^s \right). \tag{2.39}$$

Finally, we have the following lower bound for $I_{p,r}(\mathbf{D}(\nabla \mathbf{v}), 1)$.

Lemma 2.40. *There exists a constant such that for $1 < r < 3$ we have*

$$\|\nabla \mathbf{v}\|_{\frac{3r}{3-r}(p_\infty - 1)}^{r(p_\infty - 1)} \leq c \left(1 + \|\nabla \mathbf{v}\|_{p_\infty}^{r(p_\infty - 1)} + I_{p,r}(\mathbf{D}(\nabla \mathbf{v}), 1) \right). \tag{2.41}$$

PROOF : If $p_\infty < 2$ the inequality (2.41) follows from (2.29), Sobolev's embedding theorem and

$$\|\nabla \mathbf{v}\|_s \leq \|\nabla \mathbf{v}\|_{p_\infty} + \frac{1}{2} \|\nabla^2 \mathbf{v}\|_s \, .$$

If $p_\infty \geq 2$ we have

$$I_{p,r}(\mathbf{D}(\nabla \mathbf{v}), 1) \geq \int_\Omega \left| \nabla (1 + |\mathbf{D}(\mathbf{v})|) \right|^r (1 + |\mathbf{D}(\mathbf{v})|)^{(p_\infty - 2)r} \, dx$$

$$\geq c \int_\Omega \left| \nabla (1 + |\mathbf{D}|)^{p_\infty - 1} \right|^r \, dx$$

$$\geq c \|\nabla \mathbf{v}\|_{\frac{3r}{3-r}(p_\infty - 1)}^{r(p_\infty - 1)} - c \|\nabla \mathbf{v}\|_{p_\infty}^{r(p_\infty - 1)} - c \, ,$$

which gives (2.41).   ∎

References

[1] B. ABU-JDAYIL AND P.O. BRUNN, *Effects of Nonuniform Electric Field on Slit Flow of an Electrorheological Fluid*, J. Rheol. **39** (1995), 1327–1341.

[2] B. ABU-JDAYIL AND P.O. BRUNN, *Effects of Electrode Morphology on the Slit Flow of an Electrorheological Fluid*, J. Non-New. Fluid Mech. **63** (1996), 45–61.

[3] B. ABU-JDAYIL AND P.O. BRUNN, *Study of the Flow Behaviour of Electrorheological Fluids at Shear- and Flow- Mode*, Chem. Eng. and Proc. **36** (1997), 281–289.

[4] E. ACERBI AND N. FUSCO, *Partial Regularity under Anisotropic (p, q) Growth Conditions*, J. Differential Equations **107** (1994), 46–67.

[5] C. AMROUCHE AND V. GIRAULT, *Decomposition of Vector Spaces and Application to the Stokes Problem in Arbitrary Dimension*, Czechoslovak Math. J. **44** (1994), 109–140.

[6] R.J. ATKIN, X. SHI, AND W.A. BULLOGH, *Solutions of the Constitutive Equations for the Flow of an Electrorheological Fluid in Radial Configuratons*, J. Rheology **35** (1991), 1441–1461.

[7] P. BAILEY, D.G. GILLIES, D.M. HEYES, AND L.H. SUTCLIFFE, *Experimental and Simulation Studies of Electro-Rheology*, Mol. Sim. **4** (1989), 137–151.

[8] M. LE BELLAC AND J.M. LÉVY-LEBLOND, *Galilean Electromagnetism*, Il Nuovo Cimento **14** (1973), 217–232.

[9] H. BELLOUT, F. BLOOM, AND J. NEČAS, *Young Measure-Valued Solutions for Non-Newtonian Incompressible Fluids*, Comm. PDE **19** (1994), 1763–1803.

[10] T. BHATTACHARYA AND F. LEONETTI, *A new Poincaré Inequality and its Applications to the Regularity of Minimizers of Integral Functions with Nonstandard Growth*, Nonlinear Anal. Theory Methods Appl. **17** (1991), 833–839.

[11] T. BHATTACHARYA AND F. LEONETTI, $W^{2,2}$–*Regularity for Weak Solutions of Elliptic Systems with Nonstandard Growth*, J. Math. Anal. Applications **176** (1993), 224–234.

[12] L. BOCCARDO, T. GALLOUET, AND P. MARCELLINI, *Anisotropic Equations in L^1*, Diff. Int. Equ. **9** (1996), 209–212.

[13] L. BOCCARDO, P. MARCELLINI, AND C. SBORDONE, L^∞-Regularity for Variational Problems with Sharp Non Standard Growth Condition, Boll. Un. Mat. Ital. A **4** (1990), 219–225.

[14] M.E. BOGOVSKII, Solution of the First Boundary Value Problem for the Equation of Continuity of an Incompressible Medium, Dokl. Akad. Nauk SSSR **248** (1979), 1037–1040, English transl. in Soviet Math. Dokl. **20** (1979), 1094–1098.

[15] M.E. BOGOVSKII, Solution of Some Vector Analysis Problems Connected with Operators Div and Grad, Trudy Seminar S.L. Sobolev, Akademia Nauk SSSR **80** (1980), 5–40.

[16] R.T. BONNECAZE AND J.F. BRADY, Yield Stresses in Electrorheological Fluids, J. Rheol. **36** (1992), 73–115.

[17] A.P. CALDERON AND A. ZYGMUND, On the Existence of Certain Singular Integrals, Acta Math. **88** (1952), 85–139.

[18] H. J. CHOE, Interior Behaviour of Minimizers for Certain Functionals with Nonstandard Growth, Nonlinear Anal. Theory Methods Appl. **19** (1992), 933–945.

[19] B. D. COLEMAN AND W. NOLL, The thermodynamics of elastic materials with heat conduction and viscosity, Arch. Rat. Mech. Anal. **13** (1963), 167–178.

[20] A. COSCIA AND G. MINGIONE, Hölder Continuity of the Gradient of $p(x)$-harmonic Mappings, CRAS **328** (1999), 363–368.

[21] R.C. DIXON AND A.C. ERINGEN, A Dynamical Theory of Polar Elastic Dielectrics I, Int. J. Engng. Sci. **3** (1965), 359–377.

[22] R.C. DIXON AND A.C. ERINGEN, A Dynamical Theory of Polar Elastic Dielectrics II, Int. J. Engng. Sci. **3** (1965), 379–398.

[23] N. DUNFORD AND J.T. SCHWARTZ, Linear Operators I, General Theory, Interscience Publ. Inc., New York–London, 1958.

[24] W. ECKART, Theoretische Untersuchungen von elektrorheologischen Flüssigkeiten bei homogenen und inhomogenen elektrischen Feldern, Shaker Verlag, Aachen, 2000.

[25] D.E. EDMUNDS AND J. RÁKOSNÍK, Density of Smooth Functions in $W^{k,p(x)}(\Omega)$, Proc. Roy. Soc. Lond. A **437** (1992), 229–236.

[26] D.E. EDMUNDS AND J. RÁKOSNÍK, Sobolev Embeddings with Variable Exponent, preprint 1999.

[27] A.C. ERINGEN AND G. MAUGIN, Electrodynamics of Continua, vol. I and II, Springer, New York, 1989.

[28] R.P. FEYNMAN, R.B. LEIGHTON, AND M. SANDS, The Feynman Lectures on Physics, vol. 2, Addison Wesley, London, 1965.

[29] F. FILISKO, *privat communication*.

[30] J. FREHSE, J. MÁLEK, AND M. STEINHAUER, *An Existence Result for Fluids with Shear Dependent Viscosity - Steady Flows*, Non. Anal. Theory Meth. Appl. **30** (1997), 3041–3049.

[31] K.O. FRIEDRICHS, *Differential Forms on Riemannian Manifolds*, Comm. Pure Appl. Math. **8** (1955), 551–590.

[32] N. FUSCO AND C. SBORDONE, *Local Boundedness of Minimizers in a Limit Case*, Manuscripta Math. **69** (1990), 19–25.

[33] N. FUSCO AND C. SBORDONE, *Some Remarks on the Regularity of Minima of Anitropic Integrals*, Comm. Partial Differential Equations **18** (1993), 153–167.

[34] M. GAFFNEY, *The Harmonic Operator for Exterior Differential Forms*, Proc. National Science, U.S.A. **37** (1951), 48–50.

[35] H. GAJEWSKI, K. GRÖGER, AND K. ZACHARIAS, *Nichtlineare Operatorgleichungen und Operatordifferentialgleichungen*, Akademie-Verlag, Berlin, 1974.

[36] G.P. GALDI, *An Introduction to the Mathematical Theory of the Navier-Stokes Equations, Linearized Steady Problems*, Tracts in Natural Philosophy, vol. 38, Springer, New York, 1994.

[37] M. GIAQUINTA, *Growth Conditions and Regularity, a Counterexample*, Manuscripta Math. **59** (1987), 245–248.

[38] V. GIRAULT AND P.A. RAVIART, *Finite Element Methods for Navier-Stokes Equations*, Series in Computational Mathematics, vol. 5, Springer, Berlin, 1986.

[39] E. GIUSTI, *Metodi Diretti nel Calcolo delle Variazioni*, Unione Matematica Italiana, Bologna, 1994.

[40] S.R. DE GROOT AND L.G. SUTTORP, *Foundations of Electrodynamics*, North-Holland, Amsterdam, 1972.

[41] R.A. GROT, *Relativistic Continuum Physics: Electromagnetic Interactions*, Continuum Physics (A.C. Eringen, ed.), Academic Press, 1976, pp. 130–221.

[42] T.C. HALSEY, J.E. MARTIN, AND D. ADOLF, *Rheology of Electrorheological Fluids*, Phys. Rev. Letters **68** (1992), 1519–1522.

[43] T.C. HALSEY AND W. TOOR, *Structure of Electrorheological Fluids*, Phys. Rev. Lett. **65** (1990), 2820–2823.

[44] H. HUDZIK, *The Problems of Separability, Duality, Reflexivity and of Comparison for Generalized Orlicz-Sobolev Spaces $W_M^k(\Omega)$*, Comment. Math. Prace Mat. **21** (1979), 315–324.

[45] R.R. HUILGOL, *Continuum Mechanics of Viscoelastic Liquids*, Hindustan Publishing Corporation, Delhi, 1975.

[46] K. HUTTER, *On thermodynamics and thermostatics of viscous thermoelastic solids in the electromagnetic fields. A Lagrangian formulation*, Arch. Rat. Mech. Anal. **58** (1975), 339–368.

[47] K. HUTTER, *A Thermodynamic Theory of Fluids and Solids in the Electromagnetic Fields*, Arch. Rat. Mech. Anal. **64** (1977), 269–298.

[48] K. HUTTER AND A.A.F. VAN DE VEN, *Field Matter Interactions in Thermoelastic Solids*, Lecture Notes in Physics, vol. 88, Springer, Berlin, 1978.

[49] J.D. JACKSON, *Klassische Elektrodynamik*, Walter de Gruyter, Berlin, 1983.

[50] P. KAPLICKÝ, J. MÁLEK, AND J. STARÁ, *Full Regularity of Weak Solutions to a Class of Nonlinear Fluids in Two Dimensions - Stationary Periodic Problem*, CMUC **38** (1997), 681–695.

[51] P. KAPLICKÝ, J. MÁLEK, AND J. STARÁ, *$C^{1,\alpha}$-Regularity of Weak Solutions to a Class of Nonlinear Fluids in Two Dimensions - Stationary Dirichlet Problem*, Zap. Nauchn. Sem. Pt. Odel. Mat. Inst. **259** (1999), 89–121.

[52] P. KATSIKOPOULOS AND C. ZUKOSKI, *Effects of Electrode Morphology on the Electrorheological Response*, Proceedings of the 4-th Int. Conf. on Electrorheological Fluids Mechanics Today (Singapore) (R. Tao, ed.), World Scientific, 1995.

[53] B. KIRCHHEIM, 1997, Privat Communication.

[54] D.J. KLINGENBERG, F. VAN SWOL, AND C.F. ZUKOSKI, *The Small Shear Rate Response of Electrorheological Suspensions II. Extension beyond the Point Dipole Limit*, J. Chem. Phys. **94** (1991), 6170–6178.

[55] O. KOVÁČIK AND J. RÁKOSNÍK, *On Spaces $L^{p(x)}$ and $W^{k,p(x)}$*, Czechoslovak Math. J. **41** (1991), 592–618.

[56] R. KRESS, *Grundzüge einer Theorie der Verallgemeinerten Harmonischen Vektorfelder*, Meth. Verfahren Math. Phys. **2** (1969), 49–83.

[57] A. KUFNER, O. JOHN, AND S. FUČÍK, *Function Spaces*, Academia, Prague, 1977.

[58] O.A. LADYZHENSKAYA, *The Mathematical Theory of Viscous Incompressible Flow*, Gordon and Beach, New York, 1969, 2nd edition.

[59] O.A. LADYZHENSKAYA AND G.A. SEREGIN, *On Smoothness of Solutions to Systems Describing the Flow of Generalized Newtonian Fluids and on Evaluation of Dimensions for their Attractors*, Russian Dokl. Acad. Sci. **354** (1997), 590–592.

[60] R. LEIS, *Zur Theorie elektromagnetischer Schwingungen in anisotropen inhomogenen Medien*, Math. Z. **106** (1968), 213–224.

[61] R. Leis, *Initial Boundary Value Problems in Mathematical Physics*, Teubner, Stuttgart, 1986.

[62] F. Leonetti, *Two-dimensional Regularity of Minima of Variational Functionals without Standard Growth Conditions*, Ric. Mat. **38** (1989), 41–50.

[63] F. Leonetti, *Weak Differentiability for Solutions to Nonlinear Elliptic Systems with p, q-Growth Conditions*, Ann. Mat. Pura Appl. **162** (1992), 349–366.

[64] F. Leonetti, *Higher Differentiability for Weak Solutions of Elliptic Systems with Nonstandard Growth Conditions*, Ric. Mat. **42** (1993), 101–122.

[65] F. Leonetti, *Higher Integrability for Minimizers of Integral Functionals with Nonstandard Growth*, J. Diff. Equ. **112** (1994), 308–324.

[66] J.M. Lévy-Leblond, *On the Conceptual Nature of the Physical Constants*, Rivista del Nuovo Cimento **7** (1977), 187–214.

[67] Gary M. Lieberman, *Gradient Estimates for a New Class of Degenerate Elliptic and Parabolic Equations*, Ann. Scu. Norm. Sup. **21** (1994), 497–522.

[68] J.L. Lions, *Quelques Méthodes de Résolution des Problèmes aux Limites Non Linéaires*, Dunod, Paris, 1969.

[69] I.S. Liu and I. Müller, *On the Thermodynamics and Thermostatics of Fluids in Electromagnetic Fields*, Arch. Rat. Mech. Anal. **46** (1972), 149–176.

[70] J. Málek, J. Nečas, M. Rokyta, and M. Růžička, *Weak and Measure-valued Solutions to Evolutionary Partial Differential Equations*, Applied Mathematics and Mathematical Computation, vol. 13, Chapman and Hall, London, 1996.

[71] J. Málek, J. Nečas, and M. Růžička, *On the Non-Newtonian Incompressible Fluids*, M^3AS **3** (1993), 35–63.

[72] J. Málek, J. Nečas, and M. Růžička, *On Weak Solutions to a Class of Non-Newtonian Incompressible Fluids in Bounded Three-dimensional Domains. The Case $p \geq 2$*, Preprint SFB 256 no. 481 (1996).

[73] J. Málek, J. Nečas, and M. Růžička, *On Weak Solutions to a Class of Non-Newtonian Incompressible Fluids in Bounded Three-dimensional Domains. The Case $p \geq 2$*, Adv. Diff. Equ. (2000), accepted.

[74] J. Málek, K.R. Rajagopal, and M. Růžička, *Existence and Regularity of Solutions and the Stability of the Rest State for Fluids with Shear Dependent Viscosity*, M^3AS **5** (1995), 789–812.

[75] J. Málek, M. Růžička, and G. Thäter, *Fractal Dimension, Attractors, and the Boussinesq Approximation in Three Dimensions*, Acta Appl. Math. **37** (1994), 83–97.

[76] P. MARCELLINI, *Un Exemple de Solution Discontinue d'un Problème Variationnel dans le cas scalaire*, Istituto Matematico U. Dini, Universitá di Firenze, preprint **No. 11**, (1987).

[77] P. MARCELLINI, *Regularity of Minimizers of Integrals of the Calculus of Variations with Non Standard Growth Conditions*, Arch. Rational Mech. Anal. **105** (1989), 267–281.

[78] P. MARCELLINI, *Regularity and Existence of Solutions of Elliptic Equations with p, q-Growdth Conditions*, J. Differential Equations **90** (1991), 1–30.

[79] P. MARCELLINI, *Regularity for Elliptic Equations with General Growdth Conditions*, J. Differential Equations **105** (1993), 296–333.

[80] E. MARTENSEN, *Potentialtheorie*, Teubner, Stuttgart, 1968.

[81] E. MASCOLO AND G. PAPI, *Local Boundedness of Minimizers of Integrals of the Calculus of Variations*, Ann. Mat. Pura Appl. **167** (1994), 323–339.

[82] G.A. MAUGIN AND A.C. ERINGEN, *On the Equations of the Electrodynamics of Deformable Bodies of Finite Extent*, J. Mécanique **16** (1977), 100–147.

[83] A. MILANI AND R. PICARD, *Decomposition Theorems and their Application to Non-Linear Electro- and Magneto-Static Boundary Value Problems*, 317–340, Springer, 1988, pp. 317–340, LNM 1357.

[84] HONG MIN-CHUNG, *Some Remarks on the Minimizers of Variational Integrals with Non Standard Growth Conditions*, Boll. Un. Mat. Ital. **A 6** (1992), 91–101.

[85] H. MINKOWSKI, *Die Grundgleichungen für die elektromagnetischen Vorgänge in bewegten Körpern*, Göttinger Nach. (1908), 53–111.

[86] P. Z. MKRTYCHYAN, *Singular Quasilinear Parabolic Equation Arising in Nonstationary Filtration Theory*, Izv. Akad. Nau. Armyan. SSSR. Mat. **24** (1989), 103–116, English transl. in Soviet J. Contemp. Math. 24 (1989), 1–13.

[87] P. Z. MKRTYCHYAN, *An Estimate of the Solution Gradient and the Classical Solvability of the First Initial-Boundary Value Problem for a Class of Quasilinear Nonuniformly Parabolic Equations*, Izv. Akad. Nauk Armyan. SSR. Ser. Mat. **24** (1989), 293–299, English transl. in Soviet J. Contemp Math. 24 (1989), 85–91.

[88] C.B. MORREY, *A Variational Method in the Theory of Harmonic Integrals II*, Amer. Jour. Math. **78** (1956), 137–170.

[89] G. MOSCARIELLO AND L. NANIA, *Hölder Continuity of Minimizers of Functionals with non Standard Growth Conditions*, Ric. Mat. **40** (1991), 259–273.

[90] C. MÜLLER, *Grundprobleme der mathematischen Theorie elektromagnetischer Schwingungen*, Springer, Berlin, 1957.

[91] J. MUSIELAK, *Orlicz Spaces and Modular Spaces*, Springer, Berlin, 1983.

[92] J. Nečas, *Sur le normes équivalentes dans* $W_p^k(\Omega)$ *et sur la coercivité des formes formellement positives*, Séminaire Equations aux Dérivées Partielles, Montreal **317** (1966), 102–128.

[93] J. Nečas, 1991, Privat Communication.

[94] J. Nečas and M. Šilhavý, *Viscous Multipolar Fluids*, Quart. Appl. Math. **49** (1991), 247–265.

[95] W. Panofsky and M. Phillips, *Classical Electricity and Magnetism*, Addison-Wesley, London, 1962.

[96] Y.H. Pao, *Electromagnetic Forces in Deformable Continua*, Mechanics Today (S. Nemat-Nasser, ed.), vol. 4, Pergamon Press, 1978, pp. 209–306.

[97] M. Parthasarathy and D.J. Klingenberg, *Electrorheology: Mechanism and Models*, Material Sci. Engng. R **17** (1996), 57–103.

[98] P. Penfield and H. A. Haus, *Electrodynamics of Moving Media*, M.I.T. Press, Cambridge, 1967.

[99] V. Chiadò Piat and A. Coscia, *Hölder Continuity of Minimizers of Functionals with Variable Growth Exponent*, Manus. Math. **93** (1997), 283–299.

[100] R. Picard, *Randwertaufgaben in der verallgemeinerten Potentialtheorie*, Math. Meth. Appl. Sci. **3** (1981), 218–228.

[101] R. Picard, *On the Boundary Value Problems of Electro- and Magnetostatics*, Proc. Royal Soc. Edinburgh **92A** (1982), 165–174.

[102] R. Picard, *An Elementary Proof for a Compact Imbedding Result in Generalized Electromagnetic Theory*, Math. Zeitschrift **187** (1984), 151–164.

[103] L. Pick, *A Remark on Continuous Imbeddings between Banach Function Spaces*, Coll. Math. Soc. János Bolyai **58** (1990), 571–581.

[104] K.R. Rajagopal and M. Růžička, *On the Modeling of Electrorheological Materials*, Mech. Research Comm. **23** (1996), 401–407.

[105] K.R. Rajagopal and M. Růžička, *Mathematical Modeling of Electrorheological Materials*, Continuum Mechanics and Thermodynamics (2000), accepted.

[106] K.R. Rajagopal, M. Růžička, and A. Srinivasa, *On the Oberbeck-Boussinesq Approximation*, M³AS **6** (1996), 1157–1167.

[107] K.R. Rajagopal and A.S. Wineman, *Flow of Electrorheological Materials*, Acta Mechanica **91** (1992), 57–75.

[108] K.R. Rajagopal, R.C. Yalamanchili, and A.S. Wineman, *Modeling Electrorheological Materials using Mixture Theory*, Int. J. Engng. Sci. **32** (1994), 481–500.

[109] G. DE RHAM, *Variétés Différentiables*, Hermann, Paris, 1960.

[110] A. ROMANO, *Analisi adimensionale delle equazioni di Maxwell ed approssimazione quasi-statica*, Lecture Notes, University of Naples.

[111] M. RŮŽIČKA, *Mathematical and Physical Theory of Multipolar Viscoelasticity*, BMS **233** (1992).

[112] M. RŮŽIČKA, *A Note on Steady Flow of Fluids with Shear Dependent Viscosity*, Non. Anal. Theory Meth. Appl. **30** (1997), 3029–3039.

[113] M. RŮŽIČKA, *Electrorheological Fluids: Modeling and Mathematical Theory*, 1998, Habilitationsschrift University Bonn.

[114] M. RŮŽIČKA, *Flow of Shear Dependent Electrorheological Fluids: Unsteady Space Periodic Case*, Applied Nonlinear Analysis (A. Sequeira, ed.), Plenum Press, 1999, pp. 485–504.

[115] G. SCHWARZ, *Hodge Decomposition - A Method for Solving Boundary Value Problems*, Lecture Notes in Mathematics, vol. 1607, Springer, Berlin, 1995.

[116] BAYER SILICONE, *Provisional Product Information Rheobay*, 1997, report.

[117] A. SOMMERFELD, *Vorlesungen über theoretische Physik, Elektrodynamik*, vol. III, Akademische Verlagsgesellschaft, Leipzig, 1961.

[118] A.J.M. SPENCER, *Theory of Invariants*, Continuum Physics (A.C. Eringen, ed.), vol. 1, Academic Press, 1971.

[119] E.M. STEIN, *Singular Integrals and Differentiability Properties of Functions*, Princeton University Press, Princeton, N.J., 1970.

[120] R. TEMAM, *Navier-Stokes Equations*, North-Holland, Amsterdam, 1977.

[121] H.F. TIERSTEN, *A Developoment of the Equations of Electromagnetism in Material Continua*, Tracts in Natural Philosophy, vol. 36, Springer, New York, 1990.

[122] R.A. TOUPIN, *A Dynamical Theory of Elastic Dielectrics*, Int. J. Engng. Sci. **1** (1963), 101–126.

[123] C. TRUESDELL AND R.A. TOUPIN, *Classical Field Theories*, Handbuch der Physik, eds. S. Függe, vol. III/1, Springer, Berlin, 1960.

[124] C. TRUESDELL AND W. NOLL, *The Non-Linear Field Theories of Mechanics*, Handbuch der Physik, vol. III/3, Springer, New York, 1965.

[125] S. VEL, K.R. RAJAGOPAL, AND R.C. YALAMANCHILI, *Compressibility Effects in Electrorheological Fluids*, in preparation.

[126] C. WEBER, *A Local Compactness Theorem for Maxwell's Equations*, Math. Meth. Appl. Sci. **2** (1980), 12–25.

[127] N. WECK, *Maxwell's Boundary Value Problem on Riemannian Manifolds with Nonsmooth Boundaries*, J. Math. Anal. Appl. **46** (1974), 410–437.

[128] P. WERNER, *Über das Verhalten elektromagnetischer Felder für kleine Frequenzen in mehrfach zusammenhängenden Gebieten. I*, J. Reine Ang. Math **278/279** (1975), 365–397.

[129] A. WINEMAN, *privat communication*.

[130] A.S. WINEMAN AND K.R. RAJAGOPAL, *On Constitutive Equations for Electrorheological Materials*, Continuum Mechanics and Thermodynamics **7** (1995), 1–22.

[131] W.M. WINSLOW, *Induced Fibration of Suspensions*, J. Applied Physics **20** (1949), 1137–1140.

[132] K.J. WITSCH, *A Remark on a Compactness Result in Electromagnetic Theory*, Math. Meth. Appl. Sci. **16** (1993), 123–129.

[133] M. WITTLE, *Computer Simulations of an Electrorheological Fluid*, J. Non-New. Fluid Mech. **37** (1990), 233–263.

[134] T. WUNDERLICH AND P.O. BRUNN, *Pressure Drop Measurements Inside a Flat Channel - with Flush Mounted and Protruding Electrodes of Varable Length - Using an Electrorheological Fluid*, Experiments in Fluids (1999), accepted.

[135] V.V. ZHIKOV, *Averaging of Functionals of the Calculus of Variations and Elasticity Theory*, Math. USSR Izv. **29** (1987), 33–66.

Index

Springer
Berlin
Heidelberg
New York
Barcelona
Hong Kong
London
Milan
Paris
Singapore
Tokyo

Lecture Notes in Mathematics 1748

Editors:
A. Dold, Heidelberg
F. Takens, Groningen
B. Teissier, Paris